$75.00

1

AF768772

SYNCHRONIZATION IN SCIENCE AND TECHNOLOGY

I.I. Blekhman

USSR Research and Design Institute for the
Mechanical Treatment of Mineral Products, Leningrad

Translated from the Russian and edited by Eugene I. Rivin,
Wayne State University, Detroit, Michigan

ASME PRESS NEW YORK 1988

The Russian edition of this book was published by Nauka Publishers, Moscow, in 1981.

Synchronization is seen in such varied examples as weakly coupled pendulums, organ tubes, celestial bodies, electrical, electromagnetic and quantum generators, vibrations in process equipment, turbine blades, populations of cells, swarms of fireflies, flocks of birds, schools of fish, and applauding or marching people. It reflects the ordered behavior of certain nonlinear differential equations, as opposed to the usual tendency towards chaos.

The book describes examples and applications of synchronization, and gives the problem a working mathematical formulation. It presents the basic principles and results, as well as some of the open problems. It is intended for a wide range of specialists who may encounter nonlinear vibrations problems: mathematicians, engineers, physicists, chemists, and biologists, as well as students in these fields.

Copyright © 1988 The American Society of Mechanical Engineers
345 East 47th Street, New York, NY 10017

Library of Congress Cataloging-in-Publication Data

Blekhman, Il'ia Izrailevich.
Synchronization in science and technology.

Translation of: Sinkhronizatsiia v prirode i tekhnike.
1. Synchronization. I. Title.
Q172.5.S96B5413 1988 500 88-22322
ISBN 0-7918-0003-2

Foreword

The behavior of nonlinear dynamic systems can differ dramatically from that of neighboring linear systems. Some characteristically nonlinear phenomena such as amplitude dependent natural frequencies, jumps in steady state response amplitudes, and the existence of limit cycles have been known for a long time. Other nonlinear phenomena have only recently been discovered with the introduction of the concept of chaos and the realization that many nonlinear systems possess chaotic solutions.

Synchronization is a distinctive nonlinear phenomena which is not new (it was first described more than three hundred years ago by Huygens) and which is in a certain sense just the opposite of chaotic behavior. Synchronization is the process of accommodation of the responses of two or more coupled nonlinear oscillators until a periodic steady state is achieved in which all oscillators have locked in to a common period. Synchronization theory is well developed and widely applied in the electric power industry and in electronic frequency multiplying and dividing circuits. There has not, however, been a comparable development of synchronization theory by mechanical scientists and engineers in America and Western Europe. Except for attempts to model the lock-in of vortex shedding behind vibrating cylinders there has been a tendency for research on synchronization in the West to go out of style. The situation is quite different in the Soviet Union.

Forty years ago at the Institute Mekhanobr, a mechanical ore-handling research and design institute in Leningrad, it was accidently discovered that two dissimilar induction motors driving unbalanced rotors mounted at different locations on the same vibratory structure tended to operate synchronously. Later it was discovered that in some circumstances the tendency to synchronism was so strong that the power to one of the motors could be turned off without disturbing the synchronization. These remarkable facts triggered an intensive investigation of synchronization phenomena in the Soviet Union. The leader in this effort was Professor I.I. Blekhman who gave the first theoretical explanations of these synchronization phenomena. He also received patents for a number of vibratory ore processing machines which exploited these phenomena. In 1971 he published an extensive account of this research in a large monograph (896 pages with 497 references) entitled *Synchronization of Dynamic Systems*. The principal subjects of investigation were coupled rotating vibro-exciters and coupled pendulum clocks and the principal tools were the small parameter procedures of Poincare and Pontryagin. Ten years later he published the Russian edition of the present volume which contains a clear explanation of the principal results of the earlier book plus a broad ranging overview of synchronization phenomena in other fields of science and technology from astronomy to biology.

The ASME Press is to be congratulated for making this translation available. It provides ready access for western scientists and engineers to a fascinating field of research that has been relatively inactive outside of the Soviet Union.

Stephen H. Crandall
Ford Professor of Engineering
Massachusetts Institute of Technology
October, 1988

Contents

PART THREE

MATHEMATICAL APPARATUS AND PRINCIPAL RESULTS OF THE GENERAL THEORY OF SYNCHRONIZATION OF WEAKLY COUPLED DYNAMIC OBJECTS

*From the preface to the book Synchronization of Dynamic Systems**

This book attempts to provide a systematic description of facts and results related to a fast developing branch of science and technology, the synchronization of dynamic systems.

Roughly speaking, synchronization may be defined as a property of material objects of very different natures, to develop a uniform rhythm of coexistence in spite of differences in individual rhythms and often extremely weak mutual connections. This marvelous property is associated with the properties of certain nonlinear differential equations which allow stable periodic solutions. In our opinion, it deserves a philosophical discussion.

Synchronization evidently represents one of the forms of self-organized matter and therefore, may be considered to be somewhat related to cybernetics.

It is no wonder that two shafts connected by identical gears rotate with the same angular velocities. This is synchronization by "strong," rigid connections.

However, when two pendulum clocks which have been running differently, start to run identically (synchronously) when attached to the same movable beam instead of to the wall, it is looked upon as a paradox. Here synchronization is realized with a "weak" connection (coupling) by a hardly noticeable beam vibration. (It is rumored that this effect, which had been described by Huygens in the mid-seventeenth century, was exploited many a time by dishonest clockmakers for cheating unsuspecting customers.)

The tendency toward synchronization in the presence of weak connections is noted in many man-made objects, such as closely situated organ tubes, unbalanced machines located on a common foundation, and electrical generators feeding the common load. It is also typical for the vital functions of living organisms.

More than three hundred years have passed since Huygens' celebrated memoirs were first published. Nowadays the phenomenon of synchronization is widely used in technology, for example, in electrical devices, radio devices and in vibratory machines. It can also explain the specifics of behavior of various biological systems.

Recently, especially in the past few years, there have been many articles dedicated to different specific problems of synchronization theory. However, this book is probably the first attempt to describe different aspects of the synchronization problem from a uniform point of view.

The author tried, above all, to emphasize the generic character of synchronization phenomena for objects of very different natures. An understanding of this generality will, possibly, help to transfer achievements in one area to other areas.

This book does not touch on the extremely interesting and important question of synchronization of objects in the presence of random excitations. This deficiency is, to a certain extent, made up for in the monograph by A.N. Malakhov (181).

The author is pleased that the teachings of synchronization as described in the book are optimistic. Synchronization phenomena testify to the presence in our environment of strikingly expressed tendencies toward hierarchy and order, and not only the opposites whose existence is reflected, for example, in the second law of

*See reference (57)

thermodynamics. We hope that stylistic flaws do not prevent the reader from feeling the enthusiasm for the most amazing features of the synchronization phenomena the author felt while working on the book.

A few words on terminology. By *synchronization* (also *mutual* or *internal synchronization*) we understand the most general case where definite frequency correlations are established due to interactions between objects considered to be of equal rank. In an idealized case where one object is considered to be so powerful as to impose its motion rhythm (supposedly existing beforehand and unperturbed) on the other self-oscillating objects, we will speak of *catching (entrainment)* or *external synchronization phenomena*, with the first term to be applied in the case of one synchronized object. It should be noted that, in the literature, synchronization frequently means only catching phenomena. In celestial mechanics and partially in electrical engineering, a different terminology is used.

The author's interest in the field of vibration exciters was sparked by D.A. Pliss, the inventor, investigator and designer of vibratory machinery. G.A. Finkelstein supported these investigations at the stage when their application importance was far from clear. B.P. Lavrov has converted the author's many vague ideas in the field of vibration exciter synchronization into real designs. Later, he suggested and implemented many of his own fruitful ideas.

Of great importance to the author was the permanent possibility to communicate with his teachers A.I. Lurie and G.Y. Janelidze, and also the frequent conversations with Y.G. Panovko.

An important influence on the character and content of many of this book's sections was the continuous creative contact with leading designers and investigators of vibratory machines: E.A. Agranovsky, I.I. Bykhovsky, G.A. Denisov, B.N Dubrovin, A.I. Makarov, R.F. Nagaev, G.I. Nicolenko, V.A. Olevsky, A.D. Rudin, K.A. Rundkwist, A.K. Rundkwist, I.A. Savin and K.S. Khodjaev.

The management of the "Mekhanobr" Institute was always interested in studies on synchronization and supported the related projects.

Considerable help in the publication process was provided by N.A. Talitskikh.

A lot of valuable remarks on the book were made by its editor, M.Z. Kolovsky.

To all the abovementioned people, the author expresses his deepest gratitude.

I. I. Blekhman
Leningrad
November 4, 1969

Preface

The idea of writing this book was given to me by G.A. Volpert, who advised me to present the basic results of the book *Synchronization of Dynamic Systems* (57) in a more compact form more easily understood by specialists working in the fields of science and technology where synchronization phenomena might be encountered. However, almost ten years after (57), valuable specific results were obtained by a number of investigators, and the general understanding of synchronization as a scientific phenomenon was extended. The most important results were obtained in the following areas:

1) Considerable progress was made in the general theory of synchronization of dynamic objects. Particularly, it was further confirmed that the tendency toward synchronization is a general property of objects of very different natures. This tendency, as one of the forms of orderly behavior, is opposite to an also existing tendency in dynamic systems toward stochastic behavior.

On the one hand, a class of weakly coupled dynamic objects which have a tendency toward synchronization was determined. It was proved that there is a possibility for stable synchronization in systems with a great number of objects of the same type. On the other hand, it was shown that contrary to previous beliefs, stochastic behavior may develop in systems of a low order of magnitude (starting from the third) and in a comparatively wide range of parameters.

2) The mathematical methods of synchronization theory have been further developed.

3) A number of new important facts were discovered and new results obtained which are related to synchronization in different fields of technology and science; new applications of synchronization theory were established. The importance of these results is not at all diminished by our statement about the general character of synchronization phenomena. These phenomena, for each class of man-made and natural objects, have their own clearly distinctive features.

All the latest developments have found a definite reflection in this book. In particular, the book contains the following new elements:

1) An attempt is made to determine the synchronization tendency in dynamic systems.

2) It is shown that the synchronization tendency is characteristic for a very broad class of dynamic systems; namely systems with so-called periodical (or quasi-periodical) quasi-potential function having the property of *m*-roughness. Such systems include, with reasonably general assumptions, orbital systems whose specific cases are systems with mechanical vibration exciters, as well as more complex systems which can be encountered in celestial mechanics and physics.

3) The concept of dynamic self-balancing for systems with rotating masses is formulated and illustrated by a number of examples. This principle can be considered as a wide generalization of the known principle of self-balancing of a disc rotating on a flexible shaft (Laval's principle). This general concept is a natural consequence of the so-called integral stability criterion (extremal property) of synchronized motions.

4) It is shown that the general results of synchronization theory, such as the abovementioned extremal property of synchronized motions, can be used for explaining and describing the remarkable whole number ("resonance") ratios observed

between angular rotational velocities and the circulations of celestial bodies, and also for explaining and predicting the direction of the evolutions of their motions.

5) The presentation of the theory of synchronization of mechanical vibration exciters is considerably reworked and seemingly simplified and improved, especially the important problem of the stability of working regimes of devices utilizing the self-synchronization phenomenon.

6) Presentation of synchronization problems in biochemical and biological systems is considerably expanded.

This book consists of three parts.

The short *first part* gives a formulation of the general problem of synchronization in dynamic systems. The principal concepts and definitions are introduced and the meaning of the resulting correlations for weakly coupled objects is explained. These correlations are obtained by the analysis of differential equations.

After becoming familiar with the first part, the reader interested in the specific applications may address himself to the corresponding section of *part two* which describes the specifics of synchronization of various man-made and natural objects. An expert in vibration theory (a mathematician, physicist or mechanic) may choose to go from "generic to specific." This means he may jump from the first to the third part, and only after this to the second part of the book.

The *third part* of the book describes mathematical methods used in solving the problems of synchronization theory. The basic results of general synchronization theory for weakly coupled dynamic objects are presented. Synchronization of some basic classes of such objects is considered. The objects are chosen by the character of corresponding differential equations and the properties of their solutions. As in the book (57), Poincare and Lyapunov's perturbation methods are used. However, in contrast to (57), there are no proofs of mathematical theorems. On the other hand, many effective alternative methods for solving synchronization problems are presented.

In the second part of the book the most detailed consideration is given to synchronization theory for mechanical vibration exciters and some other similar objects. This is due not only to the author's interests and expertise, but also to the fact that vibration exciters are now practically the only objects of a mechanical nature whose synchronization has already achieved a stage of important practical applications, exemplified by a new class of virbatory machinery. In addition, development of self-synchronization theory for vibration exciters has given a considerable push to the development of the general synchronization theory.

A thorough numbering of chapters and sections is used in this book. The numbering of formulas shows only the number of the section (the first number) and the formula number (the second number). If a reference is given for a formula inside a chapter, then the chapter number is omitted; the chapter number is indicated only for the formula references from other chapters.

This book was being prepared for printing when Anatoly Isakovich Lurie, Corresponding Member of the U.S.S.R Academy of Science, passed away. He had always been interested in the problem of synchronization. The result of his interest was a remarkable presentation at the International Symposium on Non-linear Vibration (175), partially described in this book, and also a detailed review (176) of the book (57). The author will always harbor feelings of the deepest respect toward Anatoly Isakovich, an outstanding mechanic and human being, my teacher and the teacher of my teachers.

A brilliant engineer and a talented scientist, the late Y.L. Granat had often paid attention to the most interesting manifestations of synchronization in motion of celestial bodies. The conversations with V.V. Beletsky and his brilliant publications always promoted the author's interest in problems of celestial mechanics.

O.P. Barzukov made a number of comments on the book (57). On this book, comments were made by I.S. Landa and I.S. Chudinov. These comments are greatly appreciated.

I. I. Blekhman
Leningrad
November 4, 1979

To the memory of my publishers and teachers:
Nicolai Andrianovich Talitsky
and
Georgi Abramovich Volpert,
selfless people who dedicated themselves to the
formation of a national literature on mechanics.

—The Author

Introduction

§ 1) Synchronization phenomena in nature and technology. Brief review of development and modern state of the art.

The synchronization phenomenon is described as follows: several man-made or natural objects performing in the absence of interaction oscillations or rotational motions with different frequencies (angular velocities); they start moving with the same multiple or commensurable frequencies (angular velocities) in the presence of even very weak interactions. In such cases, definite phase correlations between oscillations and rotations are being established.*

It seems that the first observation and description of a specific case of synchronization—the mutual synchronization of pendulum clocks—was made by Christian Huygens, who, at the beginning of the second half of the seventeenth century, had discovered that a pair of clocks which ran with different rates became synchronized when attached to a light-weight beam instead of a wall. In one of Huygens' memoirs (110), the following observation was made while he was sailing.

"The pendulums of the clocks were 9 inches long and weighed half a pound. The mechanism was brought into motion by weights enclosed in a box together with the mechanism. The box was 4 feet long. Its bottom was loaded with at least 100 pounds of lead to better keep its vertical position on the boat. An extremely interesting observation was made using these clocks.

"Two of these clocks were suspended under the same double supported beam. Both pendulums were always moving in opposite directions and their oscillation exactly coincided with each other and never diverged even slightly. The ticking of both clocks was heard at the same time. If this synchronism was artificially disturbed, it reinstated itself within a short time. At first I was surprised by this strange phenomenon, but finally, after attentive investigation, I found that the cause is an imperceptible motion of the beam itself. The pendulum's oscillation imparts a certain motion to the clocks no matter how heavy they are. This motion is transferred to the beam and if the pendulums were not moving in opposite directions, now it is forced on them, and only when it happens, then and only then the motion of the beam will stop. However, this cause would not be effective enough if the runs of both clocks were not very uniform and coordinated to each other from the very beginning."

It is remarkable that this description, astonishing in its correctness and depth, was made in an era when the basic laws of mechanics had not yet been rigorously formulated. Naturally, however, Huygens could not perform a quantitative analysis of the phenomenon he had discovered.

Presented in Fig. 1 is a photo of clocks which were used to reproduce the Huygens experiment at the "Mekhanobr" Institute in Leningrad. As can be seen, not only the motion of clocks in opposite phases can be stable (Fig. 1b) but also the syn-phased motion (Fig. 1a). The theory of the mutual synchronization of pendu-

* In some cases, the objects in the absence of interaction cannot perform stable oscillations or rotations, but in the presence of interaction they would perform synchronized oscillations or rotations (e.g., see § 6 chap. 3, § 2 chap. 6 and § 2 chap. 8).

FIG. 1.

lums was developed relatively recently in publications by Minorsky (194), by the author (42) and by Y.I. Marchenko with the author (60) (see chap. 5).

Synchronization in acoustic and electroacoustic systems was discovered by Rayleigh at the end of the nineteenth century. By observing two organ tubes with openings located next to each other he found that with a sufficiently small amount of detuning, they sounded in unison, i.e., the mutual synchronization of both self-vibrating systems takes place. In some cases the tubes may force an almost total mutual sound cancellation.

A similar phenomenon was also discovered by Rayleigh for two tuning forks with electromagnetic excitation which were coupled either electrically or mechanically (by a flexible thread), or by using a resonator box (262).

At the turn of the century the *phenomena of synchronization in electrical circuits and electromagnetic systems* were discovered. The application importance of the self-synchronization of electrical generators is difficult to overestimate. In spite of the availability of sufficiently accurate regulators, the frequencies at which independent generators operate differ somewhat from one another. If the common load which is driven by the generators were not a synchronizing factor, then the frequency differential would inevitably lead to beat–long period voltage oscillations in the overall circuit. (Perhaps the reader has had the chance to observe such fluctuations in light bulbs' intensity when the bulbs are fed from several small generators when "malfunction," or falling out of synchronization, occurs.)

There is available now a voluminous literature on the theory and practice of synchronization of electric generators [see, for example, reviews in (136, 137, 253), and also in chap. 5], although the problem cannot be considered conclusively solved due to its extreme complexity. Even recently, cases have been described of generators falling out of synchronization, sometimes causing serious accidents.

Of great practical importance is also *synchronization of electric and electromagnetic oscillations in electronics and radiophysics*. The synchronization theory of these generators—both quasi-linear, which generate oscillations close to harmonic, and relaxation-type, which generate essentially nonharmonic oscillations—has been developed in the first decades of this century in classic publications by Van der Pol

(334–336), Appleton (305), A.A. Andronov and A.A. Vitt (6–10). These investigations played an important role in the development of nonlinear vibration theory.*

In the works mentioned, entrainment was considered the simplest degenerative case of synchronization. The first papers on mutual synchronization of quasi-linear generators were (179, 319) and (98), and on relaxation-type generators (73). Nowadays, the theory of synchronizing generators, especially quasi-linear ones, can be considered to be adequately developed. Reviews of numerous investigations in this field widely used in designing important devices and instruments may be found, for example, in such books as (171, 181, 247).

An especially important case in the practical sense is the synchronization of natural modes in distributed systems where waves with frequencies close to commensurable are self-synchronizing and self-adjusting in phases due to various weak interactions (27, 92, 169, 170, 274). The modal synchronization in lasers finds important application in generating extremely powerful light pulses (92, 170). However, sometimes such a synchronization should be abated, e.g., when its effects create disturbances for instrument operation (92, 274).

In the past, the principal technological applications of synchronization were connected with the synchronization of electrical generators and the generators of electromagnetic oscillations.

The situation changed after 1947–48 when the *self-synchronization phenomenon of mechanical vibro-exciters mounted on the same vibrating organ* was discovered by chance in the U.S.S.R. at the "Mekhanobr" Institute in Leningrad. In the years between 1950 and 1956, the first foreign publications [in the form of patent descriptions (327, 328)] appeared. It was found that such exciters, which in a simple case were unbalanced rotors driven by asynchronous motors, were running synchronously under certain conditions. This was occurring in spite of possible parameter differences and the absence of any kinematic or electrical connections between rotors. (See Fig. 2; Fig. 2a shows the usual method of synchronization using gears, while Fig. 2b uses self-synchronization.)

In many cases, the tendency of vibro-exciters toward synchronous rotation is so strong that this rotation is not disturbed even after switching off one or several motors. Such an event, caused by an accidental wire rupture, led to the discovery of the self-synchronization effect with vibro-exciters (see § 3, chap. 3). *This is the phenomenon of maintaining an unbalanced rotor rotation by means of a vibrating support structure,* which presents a limiting special case for synchronization of the vibro-exciter phenomenon.

It is noteworthy that it took about three hundred years for the synchronization effect discovered by Huygens for pendulum oscillations to be rediscovered (accidentally) for *rotations of unbalanced rotors*. This can be explained only by the fact that there was not yet developed a concept of the universality of the synchronization phenomena. Such concepts were developed in the following years under the considerable influence of the discovery of the synchronization of vibro-exciters and some other objects.

The first theoretical explanation and study of the self-synchronization effect in exciters (31) was reported in 1953. Currently, the theory of self-synchronization and the forced synchronization on vibro-exciters is being developed in detail and is a suf-

*It should be noted that the problems of synchronization theory, not to mention degenerative cases, are essentially nonlinear.

ficiently reliable basis for designing new vibro-machines. The importance of this theory is determined by the fact that the basic features of the synchronization effect for vibro-exciters are, in many cases, paradoxical (see § 14, chap. 3). Therefore, a possibility for the application of the self-synchronization effect in technological devices cannot be predicted, as a rule, on the basis of intuitive considerations, and a theoretical and/or computational analysis is required.

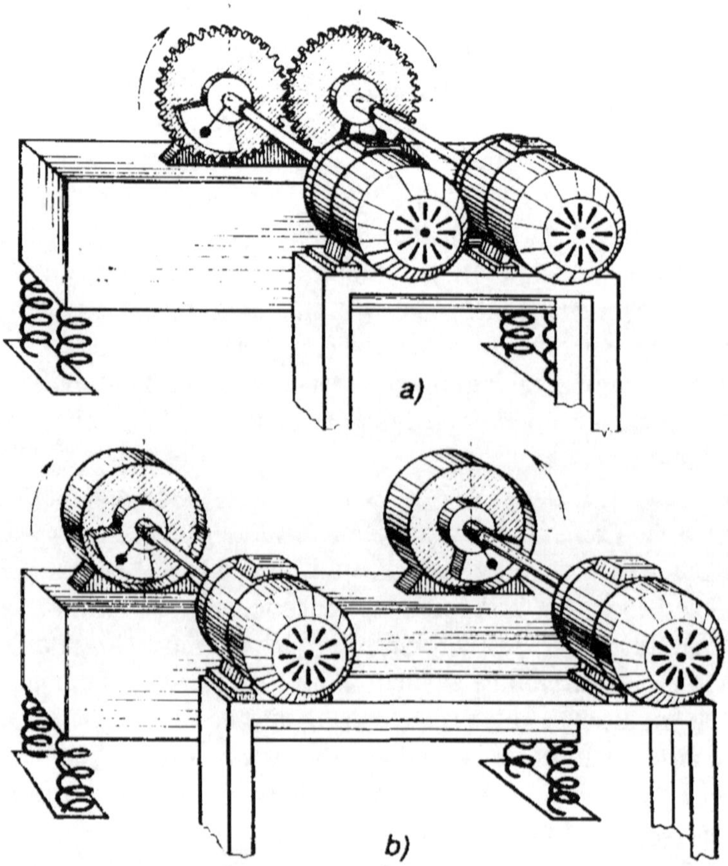

FIG. 2.

Such an analysis may be facilitated by the so-called *integral stability criterion (extremal property) of synchronized motions*, which has been formulated and proved in (40, 41). According to this criterion, the phases of rotation of rotor exciters in stable synchronized motions correspond to minimum points of a certain function of these phases called *potential function*. This function plays the same role as potential energy in equilibrium problems for mechanical systems. In the simplest but most frequently encountered cases, the potential function turned out to be equal to a mean value of the Lagrange function of the oscillatory part of the system, which is calculated with an adequate approximation. The importance of the integral criterion is that it allows us to prove, in a closed form, the presence of a tendency toward synchronization for a wide class of systems with vibro-exciters.

The theoretical and experimental studies, as well as practical utilization of the self-synchronization effect, started immediately after its discovery at the "Mekhanobr" Institute in 1947–48. Later, other investigators and groups have joined these efforts. Foreign publications appeared much later, in the early 1960's (307, 320, 323, 329, 330). These efforts have resulted in the development of a new class of vibro-machines and devices—sizing screens, conveyors, feeders, mills, breakers, flotation machines, shakers and many other devices widely used in various industries. There are more than a hundred patents in which the self-synchronization effect for vibro-exciters is used, and their number is growing continually.

Wide practical application of self-synchronization effects for vibro-exciters and vibrating support of rotation is determined, in many cases, by the fact that with real parameters a lot of power (on the order of or exceeding 1000 kW, see § 14, chap. 3) can be transmitted between the vibrators. This theoretically proven fact has been completely confirmed experimentally.

There is a case when the self-synchronization of vibro-exciters located on the same foundation may be harmful and even cause accidents. Such a case occurs where several nonbalanced machines are installed on a so-called group foundation. If the self-synchronization is characterized by synphaseous or quasi-synphaseous motion of the machines, then the individual unbalanced forces are summed up, causing undesirable consequences. However, with the proper design of the group foundation, it is possible to achieve a self-synchronization regime with mutual compensation of unbalanced forces and moments generated by individual machines. Thus, the vibration level of the foundation can be decreased considerably.

The feasibility and stability of such compensatory phasing of self-synchronization also explain the operation of a number of automatic balancing devices similar to systems using vibro-exciters (see chap. 4). For exciters, as well as for unbalanced machines on a group foundation and for autobalancers, a principle can be applied which may be considered as a generalization of the known Laval's principle of dynamic auto-balancing of a disc on a flexible shaft in the postcritical region. The correctness of this generalized principle easily follows from the aforementioned integral stability criterion.

Publications on the theory and applications of synchronization of vibro-exciters published up to 1970 were summarized in the book (57); an additional review considering later studies is given in part 7, § 8, chap. 3.

A striking example of the synchronization of natural objects is the remarkable *whole number ratios between average angular* velocities of rotations and circulations (orbital motions) or celestial bodies.*

The simplest regularity of this type had been known since time immemorial: the moon is always directed toward the earth by one side of its surface (see Fig. 3). It means the equality of average angular velocities of the moon in axial and orbital motions (resonance of the type 1:1, as they say in Celestial Mechanics). G.D. Cassini first formulated the laws of the moon's rotation in 1693. These laws not only reflect this regularity but also the synchronization (resonance) of the same type between the motion of the moon's axis of rotation and the disturbed precession of its orbit. Many similar, as well as more complicated correlations between the observed "average motions" of celestial bodies, are now known. Thus, for example, as a result of Mercury and Venus radiolocation data, it was found that Mercury's circulation

*In publications on celestial mechanics and in a number of mathematical studies, such cases are called *resonance correlations*, simply *resonances* or cases of a *strict commensurability of the average motions* (see § 1, chap. 1 and also chap. 7), while in the general vibration theory, mechanical engineering, electrical and radio engineering, the word *synchronization* is used, and by resonance one considers amplification of forced vibrations within certain ranges of frequency variation of acting force. The latter wording, assumed in this book, corresponds to international terminology recommendations. (See, e.g., recommendations SEV 5.301-75.) (SEV stands for Council of Mutual Economic Assistance—COMECON. E.R.)

period (≈ 88 days) relates to its rotation period as 3:2, and the angular circulation velocities of Earth and Venus, ω_e and ω_v, are related to Venus's angular velocity around its axis Ω as $\Omega = 4\omega_v - 5\omega_e$. The last relationship corresponds to the fact that in the moments when the centers of the Sun, Earth and Venus lie on the same line,* Venus is always facing toward the Earth with the same section of its surface.

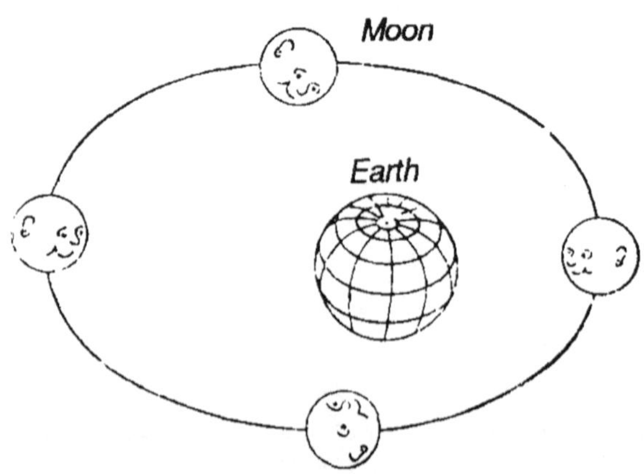

FIG. 3.

The theory of synchronization (resonances) at the motions of celestial bodies was initiated by Laplace and Lagrange, who studied, in particular, linearized equations of the Moon's motion, whose solution gives a motion approximating Cassini's laws. A further important development of the theory (now in a nonlinear approach) was achieved in a recent time. Reviews of corresponding studies together with presentation of a number of interesting original results may be found in the book (23), and also in (281), (100) and in chap. 7 of this book. These publications show, in particular, that Cassini's empirical laws correspond to the stable solutions of nonlinear differential equations that describe (in a reasonably good approximation) the motion of a Moon-like body. On the basis of these results a number of stabilization systems for artificial satellites, etc., were proposed. At the same time, the theory of synchronization of celestial bodies is far from complete. One important conclusion which follows from observations and theoretical studies of various special cases seems to be firmly established now, namely a statement about deterministic, not accidental, commensurability of the average motions of celestial bodies, i.e., the tendency of these bodies toward synchronization. Considering what has been said before, the theoretical establishment of such a tendency in a sufficiently general case would be a very important achievement. As will be shown in chap. 7, the aforementioned conclusion of selectivity, or of what is sometimes called "elitism" of synchronized motions, actually follows (with some restrictions) from the general theory of synchronization of dynamic systems which was developed in recent years. Namely, the tendency toward synchronization under some conditions follows from the aforementioned integral criterion of stability (extremal property†) of synchronized motions, according to which the stable synchronized motions may correspond to points coinciding or approaching minimum points of averaged Lagrange function for the system or parts of this function taken with proper signs, by initial phases of the motions of the bodies.

It is noteworthy that the "selection rule" of stable phases in synchronized motions which follows from the extremal property, is in good correlation with astronomical observations and calculation data. Thus, for example, the connections of Saturn's satellites Encelada and Dione always take place close to the pericenter of Encelada, as predicted by the extremal property.

* Such moments of opposition are repeated every year and a half.

† The term "extremal property" was proposed by V. V. Beletsky.

In connection with the synchronization problem of the motions of celestial bodies, A.M. Molchanov's hypothesis should be mentioned. It postulates synchronization (complete resonance) of averaged angular velocities of circulation of the large planets of the solar system (202). In light of the mentioned results, this hypothesis, which provoked heated polemics, has gained if not complete substantiation, at least an important argument in its favor.

Relatively recently, in the past few decades, attention has been paid to the fact that *synchronization phenomena* are also characteristic of *biological objects*, from ensembles of cells to ensembles of animals. Some examples are, on the one hand, synchronized oscillations of the nuclei and cells of malignant tumors; synchronization of the action of muscular tissues and of transfer of nervous excitation, etc. (85, 125, 197, 247). On the other hand, one may discover a coordinated light-twinkling generated by several beetles (fireflies); synchronized wing strokes that occur during the flight of bird flocks; or the fin motions of schools of fish.

Synchronization also takes place in the collective behavior of humans. Some examples are synchronized applause in a large audience and groups of people walking "in step."

The theory of synchronization phenomena in biological systems is only in its beginning stages. For its development the classic models (sometimes slightly improved) of quasi-linear and relaxation Van der Pol autogenerators are used, as is a modified "predator-prey" model of Lottky-Volterra (see chap. 8).

In connection with what was said before, it should be noted that discovery of the synchronization of a new class of objects is of essential engineering and scientific importance, since it allows the achievement of a new level in corresponding fields. Thus, the discovery of the synchronization of vibro-exciters, as was mentioned, led to the creation of a new class of machines. Not less, and possibly more important consequences would be brought about by the discovery and investigation of synchronization phenomena in biology and in the motions of celestial bodies.

Universal occurrence of synchronization naturally arouses the idea that *synchronization* is a fundamental property of oscillating and rotating objects, and therefore must play an essential role in the *organization of the microworld*. Some preliminary considerations in this regard are mentioned in § 2, chap. 9.

The discovery of the self-synchronization of vibro-exciters, of biological objects and also the evolved understanding of commonality of synchronization phenomena, caused the development of the *general theory of synchronization of dynamic systems* (see chap. 12). Publications on the synchronization theory of objects similar to weakly coupled autogenerators are proliferating (see, for example, 57, 171, 181, 247). The synchronization theory for systems with almost uniform rotational motions has been developed (41, 42, 57), as well as the theory of synchronization of quasi-conservative objects (15, 208–211, 213, 215) and of the objects approximating arbitrary nonlinear objects (207). The concept of orbital systems was developed, combining systems with mechanical vibro-exciters, celestial-mechanical systems, and also systems which are of interest in physics (63, 66).

The development of synchronization theory in turn required the additional development of the corresponding mathematical apparatus used in the theory of periodic solutions of nonlinear differential equations with a small parameter and, after all, the classical Poincare-Lyapunov theory (15, 33, 34, 50, 184, 216) (see chaps. 10

and 11). The already mentioned integral stability criterion (extremal property) of periodical and synchronized motions (57, 81, 175, 209–211, 285, 286) was generalized and expanded on many classes of objects. This criterion allows the reduction of the problem of studying the existence and stability of the aforementioned motions to the question of the minima of a certain function which is called a potential (or quasi-potential) function. By using this criterion, it was possible to prove, in general form, the existence of a tendency toward synchronization (see chap. 2) in a wide class of weakly coupled objects—vibro-exciters, quasi-conservative objects, objects with almost uniform rotations and also objects forming orbital systems (63, 66). Thus, it was theoretically proved that the tendency toward synchronization is a real general property of interacting dynamic objects of very different natures.

§ 2) General outlook on synchronization. What is the meaning of the tendency toward synchronization?

It follows from § 1 that a tendency toward synchronization is a peculiar feature of material objects of very different natures. Undoubtedly, this represents one of the manifestations of the tendency of material forms toward self-organization. The latter was often mentioned by the representatives of various sciences and especially by the distinguished scientist and thinker V.I. Vernadsky, whose ideas are gaining more and more appreciation (236). Many of V.I. Vernadsky's ideas were naturally incorporated into the basics of cybernetics.

It can also be said that we meet with synchronization wherever we deal with oscillations or rotations, i.e., practically everywhere.

At the same time, this view of synchronization is not yet finalized, and sometimes we hear very different opinions about it. It seems incorrect to extrapolate on the general case, as is sometimes done, from the statement by A. M. Molchanov that synchronizing (property of resonance) is characteristic for any evolutionarily mature dynamic system. Evidently, Molchanov himself had in mind only a certain class of orbital dynamic systems. From the author's point of view, which was expressed in the preface to the book (57), and which, it seems, has gained additional support in this book and is shared by a number of investigators, not the actual establishment of synchronized motions, but only *the tendency toward synchronization* is universal.*

In the presence of sufficiently strong desynchronizing factors, the tendency toward synchronization is suppressed and the system, however "evolutionarily mature," is not synchronized. In such cases, a trend opposite to the tendency toward synchronization prevails.

In other words, this case also represents one of the manifestations of a known general phenomenon, namely "a struggle" between order and disorder (see, for example, 141). It is described in more detail in § 1, chap. 9.

§ 3) Basic features of synchronization.

The features of the synchronization of objects of different natures are characterized by a certain specificity. Thus, there are distinctive differences between the fea-

*Naturally, in this case, it is important to define the tendency toward synchronization. In this book it is suggested that such a tendency exists if the system has at least one motion which is stable in relation to the corresponding coordinates or functions of coordinates (see § 1, chap. 1, for details).

tures of the synchronization of vibrating objects (e.g., pendulum clocks) and of objects with rotational motions (vibro-exciters, celestial bodies).

At the same time, it is possible to list a number of general features for dynamic objects of very different natures. Following are some of these basic features. More specific features characteristic of the synchronization of individual classes of dynamic objects are described in later chapters of this book (see, for example, § 14, chap. 3; § 3, chap. 5 and § 2, chap. 8).

1. *Absence of a synchronization threshold.* Synchronization can occur with coupling between objects having any degree of weakness, provided that the difference of corresponding object parameters is small enough. Similarly, catching is possible at an external action of any degree of weakness. This feature is manifested in a peculiar form in the presence of parameter fluctuations of both the objects and coupling systems.

2. *Correlation between synchronization and partial frequencies of the objects.* At the most significant degree, the possibility or impossibility of the mutual synchronization of self-oscillating objects depends on the values of their partial frequencies (angular velocities) ω_s, i.e., on the frequencies (velocities) in the absence of interaction. If, for example, all respective partial frequencies were sufficiently close to each other or equal, a simple mutual synchronization would be, as a rule, possible regardless of the values of the other parameters of the objects and coupling systems. It should be noted, however, that even at weak mutual couplings, the tendency of the objects toward synchronization may sometimes be so strong that the objects would synchronize with even quite different partial frequencies (angular velocities).

3. *Worsening of synchronization conditions with an increase in the order of a synchronized regime.* The regions of existence and stability, as well as "regions of attraction" for synchronous regimes in the phase space of a system, are narrowed with an increase in the "order" of the synchronous regime, i.e., a certain number characterizing values of whole-number coefficients in the appropriate correlations between frequencies of oscillation or rotations. The best conditions in this sense usually correspond to the case of a simple synchronization when frequencies (velocities) are equal. However, exceptions to this rule are possible.

4. *Establishment of certain correlations between initial phases of the objects' motions.* The synchronous regime is characterized by a definite set of initial phase values for objects' motions. Frequently, at fixed system parameters it is possible to have not one, but several (locally) stable synchronous motions which differ by specific values of the initial phases. There may exist other (nonsynchronous) locally stable motions. In such cases, the character of an actually developing motion is determined by the initial conditions. In some systems, however, there is only one globally stable synchronous regime developing at any initial conditions. In the first and most general case, it is possible, somewhat widening the common terminology, to speak of a *rigid excitation of a synchronous regime* or a *non-self-stabilized synchronous regime*. In the second case, one speaks of the *soft excitation* or *self-stabilizing regime*.

5. *Extremal property of synchronous motions and their tendency toward synchronization.* Stable synchronous motions are, in many cases, distinguished from other possible motions of a system of mutually intercoupled objects by the fact that they coincide with the minimum of a certain function D (potential function). This function often has a definite physical sense and represents the sum of lagrangians of

elements of the coupling system averaged for the period, and the so-called *potential of excessive forces* taken with definite signs. In systems complying with the integral stability criterion, as well as for systems approximating the latter in a certain sense, the tendency toward synchronization is manifested at sufficiently wide assumptions.

6. *Correlation of synchronous motions with the character of the coupling system.* The character and number of stable synchronous motions of a system may essentially depend on the number of degrees of freedom and the properties of the coupling system. The latter is characteristic for objects with rotational motions. However, there are known objects with oscillating motions (for example, pendulum clocks) whose synchronous motions only weakly depend on the character of the coupling system.

7. *Paradox of nonworking couplings.* For the mutual synchronization of identical objects (for example, mechanical vibro-exciters and pendulum clocks), there exist stable synchronous motions at which the displacement in the connection (coupling) system is absent. The connection system is "turned on" only by the occasional disturbance of the synchronous motion of the objects or when their parameters are changing.

8. *Effect of frequency averaging.* The synchronous frequency (velocity) ω at a simple synchronization (see § 1, chap. 1) of self-oscillatory objects, is often a nondecreasing function of a certain average (or weighted average) of partial frequencies (angular velocities) ω_s of individual objects. At some properties of the coupling system between objects, ω becomes merely equal to the mentioned average or weighted average. Then ω is no more than the largest and not less than the smallest from the partial frequencies (angular velocities) ω_s of individual objects:

$$\mathrm{Inf}\,[\,\omega_1, \ldots, \omega_k\,] \leq \omega \leq \mathrm{Sup}\,[\,\omega_1, \ldots, \omega_k].$$

In many cases, especially at close values of ω_s, the synchronous frequency ω is close to the weighted arithmetic average values of the partial frequencies, i.e., $\omega \approx b_1\omega_1 + \ldots + b_k\omega_k$ ($b_1 + \ldots + b_k = 1$), where $b_1 > 0, \ldots, b_k > 0$ are certain functions of system parameters.

It follows that the mean square deviation of ω at a random deviation of ω_s may be considerably less than the mean square deviation of ω_s itself.

PART ONE
GENERAL FORMULATION OF THE PROBLEMS AND CHARACTERISTICS OF THE BASIC MATHEMATICAL APPARATUS FOR THE SYNCHRONIZATION THEORY OF DYNAMIC SYSTEMS

Chapter 1
GENERAL FORMULATION OF SYNCHRONIZATION PROBLEMS AND THEIR BASIC FEATURES

§ 1) Formulation of synchronization problems: basic terms and definitions.

We will try to give a sufficiently general mathematical formulation of synchronization problems in order to cover as many particular cases as possible. At the same time we will further explain the terms used and give basic definitions.

First of all, we will limit ourselves to the study of the synchronization of the so-called dynamic systems, whose basic property is that their condition is characterized by a certain set of parameters (phase coordinates). Assigning these coordinates at some fixed moment would completely determine the coordinate values in any moment in time. More precise formulations of the dynamic system concept can be found, for example, in the monographs (218, 220).

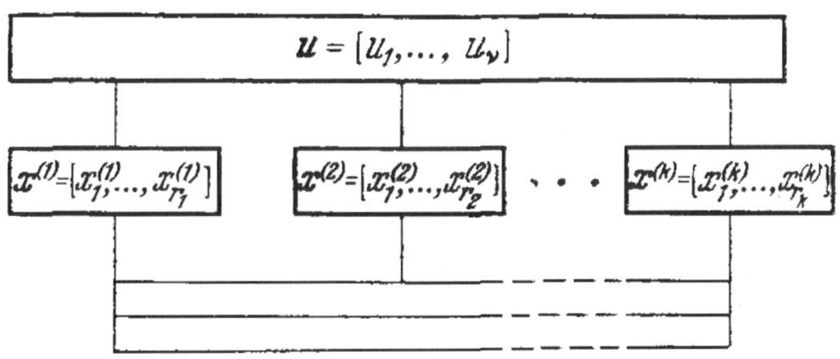

FIG. 4.

Consider a certain number k of dynamic objects connected to one another in a united system (Fig. 4). Let the condition of object s be determined by r_s-dimensional vector $x^{(s)} = [x_1^{(s)}, \ldots, x_{r_s}^{(s)}]$ $(s = 1, \ldots, k)$, whose components $x_j^{(s)}$ are coordinates of the object in the phase space of the system.

The condition of the system as a whole is determined by both the totality of $x^{(s)}$ vectors, and by ν–dimensional vector $u = [u_1, \ldots, u_\nu]$ characterizing the condition of the connection system between the objects. Thus, the phase space of the total system has $l = r_1 + \ldots + r_k + \nu$ dimensions.

We will say that a system performs a *synchronous motion* if its phase coordinates follow the law:

$$x_j^{(s)} = \tilde{n}_j^{(s)}\omega t + y_j^{(s)}\left(m_j^{(s)}\omega t\right) \qquad (j = 1, \ldots, r_s;\ s = 1, \ldots, k),$$
$$u_\rho = \tilde{n}_\rho \omega t + v_\rho (m_\rho \omega t) \quad (\rho = 1, \ldots, \nu), \tag{1.1}$$

where ω is a positive constant, $\tilde{n}_j^{(s)}$ and $\tilde{n}_\rho$ are whole numbers, $m_j^{(s)}$ and m_ρ are whole positive numbers, and $y_j^{(s)}$ and v_ρ are periodic functions with periods $2\pi/m_j^{(s)}$ and $2\pi/m_\rho$ by ωt, respectively: (i.e., with a common period 2π by ωt) the numbers $|\tilde{n}_j^{(s)}|$, $|\tilde{n}_\rho|$, $m_j^{(s)}$, and m_ρ may be considered mutually simple without violating the generality. In the case of any number $\tilde{n}_j^{(s)}$ or $\tilde{n}_\rho$ equal to zero, we will conditionally call the corresponding coordinate $x_j^{(s)}$ or u_ρ *oscillating* and in the case of $\tilde{n}_j^{(s)} \neq 0$ or $\tilde{n}_\rho \neq 0$, *rotational*.

The presence of rotational coordinates is characteristic for mechanical systems. In a real system, the rotational coordinate would not necessarily represent a body rotation, but would represent any uniformity in average motion (for example, rectilinear motion with a constant average velocity).

From (1.1) we have:

$$\langle \dot{x}_j^{(s)} \rangle = \tilde{n}_j^{(s)}\omega, \quad \langle \dot{u}_\rho \rangle = \tilde{n}_\rho \omega, \quad \text{where} \quad \langle \ldots \rangle = \frac{\omega}{2\pi}\int_0^{2\pi/\omega} \ldots\, d\tau \tag{1.2}$$

so that oscillating or "uniform in average" motions in each of the phase coordinates with equal or multiple frequencies (average angular or linear velocities) correspond to the synchronous motions of the system.

If all numbers $\tilde{n}_j^{(s)}$ are equal to 0, +1 or –1 and all numbers $m_j^{(s)}$ are equal to 1, then we will speak of *simple*, and in the opposite case, of *multiple-synchronous motions*. Accordingly, we will differentiate between the *problems of simple* and *multiple* synchronization of dynamic objects. The value $\omega > 0$ will be called *synchronous velocity (frequency)*.

If a system of connected objects permits, at certain conditions, at least one stable synchronous motion,† we will speak of objects under these conditions as *revealing a tendency toward synchronization*. If, under some conditions, the system motion at $t \longrightarrow \infty$ would unlimitedly approach a certain synchronous motion, then we will say that the objects at these conditions are *synchronized*.

In many cases, the behavior of dynamic objects and the connecting systems between them can be adequately described using differential equations which, according to the structural scheme (see Fig. 4), may be written in the following typical format:

$$\dot{\boldsymbol{x}}^{(s)} = \boldsymbol{X}^{(s)}\left(\boldsymbol{x}^{(s)}\right) + \boldsymbol{F}^{(s)}\left(\boldsymbol{x}^{(1)}, \ldots, \boldsymbol{x}^{(k)}, \boldsymbol{u}\right) \qquad (s = 1, \ldots, k),$$
$$\dot{\boldsymbol{u}} = \boldsymbol{U}\left(\boldsymbol{x}^{(1)}, \ldots, \boldsymbol{x}^{(k)}, \boldsymbol{u}\right), \tag{1.3}$$

where

$$\boldsymbol{X}^{(s)} = \left[X_1^{(s)}, \ldots, X_{r_s}^{(s)}\right], \quad \boldsymbol{F}^{(s)} = \left[F_1^{(s)}, \ldots, F_{r_s}^{(s)}\right], \quad \boldsymbol{U} = [U_1, \ldots, U_\nu]$$

are correspondingly r_s- and ν-dimensional vector functions which comply with very general conditions at which the system (1.3) is dynamic, and a certain special condition which will be shown below. Vector functions $\boldsymbol{F}$ and $\boldsymbol{U}$, which characterize the connections between the individual objects, we will call *connection (coupling) functions*.

† Here and later only stability according to A.M. Lyapunov or orbital stability are considered. In some cases, with special explanations, stability with respect to a fraction of phase coordinates by functions of the coordinates, or conditional stability, will be considered (178, 183).

From the structural diagram of the system and also from the equations in (1.3), we see that each of the objects may be connected with the rest either directly or through a *connection system* whose condition is characterized by the phase coordinates u_ρ. Also, it can be seen that coordinates $x_j^{(s)}$, determining the condition of the objects, and coordinates u_ρ of the connection system, have essentially identical status in equation (1.3). Specifics of each group of variables, typical for many synchronization problems, will be explained below.

As the *principal problem of synchronization theory*, we will define the problem of determining the conditions of existence and the stability of the solutions for equations (1.3) having the format of (1.1), i.e., solutions corresponding to synchronous motions.

Let us now specify the format of the right-hand sides of equations (1.3). We assume that the functions $X_j^{(s)}$, $F_j^{(s)}$ and U_ρ depend on their arguments in such a way, that after substitution, instead of $x_j^{(s)}$ and U_ρ,, their expressions according to formulas (1.1), these functions become periodic functions of "dimensionless time" $\tau = \omega t$ with period 2π.

The latter condition is not mandatory for the existence of synchronous motions in the system (1.3). However, it is completed within all basic specific problems of synchronization and significantly simplifies their solution. Therefore, compliance with this condition is further assumed. It should be mentioned that for the compliance to exist, it is enough for $X_j^{(s)}$ and U_ρ to be periodical functions of rotational coordinates with period 2π, and possibly also to be functions of the differences $x_j^{(s)}/\tilde{n}_j^{(s)} - x_\alpha^{(\beta)}/\tilde{n}_\alpha^{(\beta)}$, $u_j/\tilde{n}_j - u_\rho/\tilde{n}_\rho$, $x_j^{(s)}/\tilde{n}_j^{(s)} - u_\rho/\tilde{n}_\rho$ where $x_j^{(s)}$, $x_\alpha^{(\beta)}$, u_j *and* u_ρ are rotational coordinates.

Changing from variables x, u and t to variables y, v and $\tau = \omega t$ in (1.3) using the formulas in (1.1) we obtain the system in the following format:

$$\frac{dy^{(s)}}{d\tau} = Y^{(s)}\left(y^{(s)}, \tau\right) + \Phi^{(s)}\left(y^{(1)}, \ldots, y^{(k)}, v, \tau\right) \qquad (s = 1, \ldots, k)$$

$$\frac{dv}{d\tau} = V\left(y^{(1)}, \ldots, y^{(k)}, v, \tau\right), \tag{1.4}$$

where, in accordance with assumptions on the character of vector functions $X^{(s)}$ and U, the functions $Y^{(s)}$ and V are periodical with period 2π with respect to dimensionless time τ.

Thus, the principal problem of synchronization is reduced to determining the conditions of existence and stability of periodic solutions having period 2π in the system (1.4).

In addition to the principal problem of synchronization formulated above, solutions of the following problems are also frequently of interest:

1) *Actual calculation of synchronous velocity (frequency)* ω_1, *and also solutions of equations (1.2) corresponding to stable synchronous motions.* In some cases, we may confine ourselves to determination of average values of the functions: $y^{(s)}(\tau)$ and v (τ) during period 2π, i.e.:

$$\alpha^{(s)} = \frac{1}{2\pi}\int_0^{2\pi} y^{(s)}(\tau)\,d\tau, \quad \alpha = \frac{1}{2\pi}\int_0^{2\pi} v(\tau)\,d\tau, \tag{1.5}$$

and also maximal deviations $|y_j^{(s)}(\tau) - \alpha_j^{(s)}|_{max}$ and $|v_\rho(\tau) - \alpha_\rho|_{max}$ from these average values.

2) *Selection of the connection system at which the existence and stability of synchronous motion (1.1) of the given form are realized.* This problem, which may be called the *synthesis problem*, is, in some sense, opposite to the principal problem.

In some cases, the solution of an extremely difficult problem is also of interest—i.e., determining in the phase space of the system of such regions of initial values, the coordinates ("entrainment regions") for which, with time, the motion is asymptotically approaching a certain synchronous motion.

Naturally, for the possibility of using synchronization phenomena, it is necessary that the time of formation of a synchronous regime not be too long, and that the basic characteristics of synchronous motion have enough "stability" with respect to different kinds of errors of system manufacturing. Therefore, for some applications, it is important to make an estimate of the time of practical formation of a stable synchronous regime at given initial conditions; also to evaluate the "sensitivity" of these characteristics of a synchronous regime in the system with respect to the changes of the parameters of both objects and the connection system, and also with always present disturbances.

One of the most important classes of synchronization problems is *problems of synchronization of self-vibratory objects*. This means objects of, usually, similar type, each of which, while being isolated from the others [coupling functions, $F^{(s)}$ and $\Phi^{(s)}$ respectively, are absent in equations (1.3) and (1.4)] at certain conditions, may perform motions of the (1.1) type, which are characterized by a certain frequency (angular velocity) ω_s. The value ω_s may naturally be called the *partial frequency* (velocity) of the object.* The problem of synchronization in this case consists of determining conditions at which, after uniting all the objects into a unified system, the latter may perform motions of the same type, but with the same frequency (velocity) ω or with frequencies (velocities) $\tilde{n}_s\omega$.

Depending on the character of the problem formulation on the synchronization of self-oscillating objects or the systems containing them, we need to distinguish the problem of *internal (mutual, autonomous) synchronization* and the problem of *external (nonautonomous) synchronization*.

In the first, most general, case, which included the above formulated synchronization problem, all objects being synchronized are considered to be elements of the same status of an integral autonomous dynamic system. In this case, the frequency of synchronous motion ω is developed as a result of interaction of all the elements in the system. In such cases, the right-hand sides of equation (1.3) do not contain t in explicit form, and the value of synchronous frequency ω is not known beforehand and has to be determined in the process of problem solution.

In the second case, it is assumed that one of the self-oscillating objects being synchronized is considerably more powerful than the others, and therefore its motion is considered to be independent from the character of the motion of the other elements. The influence of the aforementioned object on the other elements of the system, and therefore on the frequency (or angular velocity) of synchronous motion, is assumed to be given beforehand and to be unchangable. The initial system (1.3) at such an approach is transformed into a nonautonomous one and its order is decreased.

* In application to some types of objects, the concepts of partial frequencies (velocities) will be detailed below and specified (see § 5, chap. 3; § 1, chap. 5 and § 2, chap. 6).

A particular case of an external synchronization problem is one of *catching* when it is considered synchronization under the influence of a given external periodical disturbance of one and only one self-oscillating object. It should be noted that when synchronization is mentioned in the literature, it simply means entrainment.

The synchronization concept assumes the possibility (and usefulness) of the selection of several objects in the united system whose condition is characterized by their "natural" phase coordinates. However, such splitting of the system into "objects" and "connections" is quite conditional. It follows, for example, from the fact that if no reservations on the character of connection functions $F^{(s)}$ and U are made, then all variables enter into the system of equations (1.3) in quite the same status.

However, when considering a specific problem of synchronization, usually no question of the isolation of individual objects arises. The question here is not only (and not as much) that these objects represent isolated physical or biological systems (pendulums, electrical generators, insects, etc.), but that the connections between individual objects in these cases are considerably weaker than between the elements of the objects themselves (see § 2).

In the presence of strong connections between objects, their isolation (at least in the course of the mathematical solution of a problem) is often not recommended, and the phenomenon of mutual synchronization does not differ at all from the class of self-oscillating phenomena.

In technology, *self-synchronization* and *forced synchronization* are sometimes distinguished from one another. In the first case, synchronization and the required correlations between the phases of oscillations and rotations are realized in a natural way, i.e., by means of interactions already existing in the connection system. Thus, for example, synchronization of electrical or mechanical vibration generators (vibro-exciters) often takes place because of properties of the system generators-load itself. In the second case, to achieve the synchronization effect or the required phasing, one has to introduce additional connections.

It should be noted that when considering some specific problems, expressions (1.1), which characterize synchronous motions, are better to be written in the following form:

$$x_j^{(s)} = \sigma_j^{(s)}\left[q_j^{(s)} n_j^{(s)} \omega t + y_j^{(s)}\left(m_j^{(s)} \omega t\right)\right] \qquad (j = 1, \ldots, r_s;\ s = 1, \ldots, k)$$

$$u_\rho = \sigma_\rho\left[q_\rho n_\rho \omega t + v_\rho\left(m_\rho \omega t\right)\right] \qquad (\rho = 1, \ldots, \nu). \tag{1.6}$$

Here, in contrast to (1.1), the integers $n_j^{(s)}$ and n_ρ are assumed to be positive; $\sigma_j^{(s)}$ are numbers, either of which may be equal to +1 or –1; $q_j^{(s)}$ and q_ρ are numbers equal either to 1 or 0 (in the first case, according to the definition given above, the corresponding coordinate is rotational; and in the second case is a vibratory one). The other notations in the formulas (1.6) have the same meaning as in (1.1).

The term "synchronization" is often used side by side with the terms "resonance" or "commensurability of averaged motions" in the field of celestial mechanics. In this case, one understands by these terms the existence of the following linear homogeneous correlations between the average angular velocities ω_s° of the objects:

$$n_{1,p}\overset{\circ}{\omega}_1 + n_{2,p}\overset{\circ}{\omega}_2 + \ldots + n_{k^*,p}\overset{\circ}{\omega}_{k^*} = 0$$
$$(p = 1, \ldots, q;\ q \leqslant k^* - 1) \quad (1.7)$$

Coefficients $n_{j,p}$ are integers; the number of such correlations (assumed to be linear or independant) does not exceed $k^* - 1$, where k^* is the number of rotational motions under consideration.

It is easy to be convinced that the definition of synchronous motions in the form (1.1) envelops the definition based on the "resonance correlations" (1.7). If the correlations (1.1) are correct, then according to (1.2) $\overset{\circ}{\omega}_j = \tilde{n}_j\omega$ where $\tilde{n}_j$ are integers and therefore among the values ω_j there exist no more than $k^* - 1$ independent equations of the (1.7) type. On the other hand, as is easy to find out, the (1.7) type correlations can always be satisfied by expressions like $\overset{\circ}{\omega}_j = \tilde{n}_j\omega$. At the so-called "complete resonance" when $q = k^* - 1$, it can be done in a unique way* (with an accuracy up to a nonessential multiplier), and at the "incomplete resonance" ($q < k^* - 1$) by an infinite number of methods. Thus, the known correlation [see, for example, (100)]

$$\overset{\circ}{\omega}_1 - 3\overset{\circ}{\omega}_2 + 2\overset{\circ}{\omega}_3 = 0$$

between the "average motions" (average angular velocities) of Jupiter's satellites, Io, Europa and Ganymede ($\overset{\circ}{\omega}_1$, $\overset{\circ}{\omega}_2$ and $\overset{\circ}{\omega}_3$ respectively), could be realized, in principle at $\overset{\circ}{\omega}_1 = \overset{\circ}{\omega}_2 = \overset{\circ}{\omega}_3 = \omega$, at $\overset{\circ}{\omega}_1 = -\omega$, $\overset{\circ}{\omega}_2 = \omega$, $\overset{\circ}{\omega}_2 = 2\omega$, at $\overset{\circ}{\omega}_1 = 4\omega$, $\overset{\circ}{\omega}_2 = 2\omega$, $\overset{\circ}{\omega}_3 = \omega$, etc.

Thus, from the formal point of view, there is no principal difference between the cases of frequency commensurability (multiple synchronization) and occurrence of resonance correlations (1.7). It should, however, be considered that the cases with the most application interest are only where integers $|\tilde{n}_j^{(s)}|$, $|\tilde{n}_\rho|$, $m_j^{(s)}$ and m_ρ, as well as the numbers $n_{j\rho}$ are relatively small. Small regions of existence and stability usually correspond to the larger values of these integers, as well as to small regions of attraction of the respective regimes (see § 3 of the Introduction). Taking this circumstance into account, differentiation between multiple synchronization and synchronization in the sense of the presence of resonance correlations being present, might be useful. For example, the case $\overset{\circ}{\omega}_1 = 100\omega$, $\overset{\circ}{\omega}_2 = 102$, $\overset{\circ}{\omega}_3 = \omega$ can be naturally considered as synchronization at the presence of the resonance correlation $\overset{\circ}{\omega}_2 - \overset{\circ}{\omega}_1 = 2\overset{\circ}{\omega}_3$.

In radiophysics, one sometimes speaks of the *synchronization on combination frequencies*, having in mind establishing whole-number ratios (commensurability) not between the frequencies ω° themselves, but between their linear combinations with integer coefficients, such as between the differences $\omega_s - \overset{\circ}{\omega}_j$.

For certain applications, it may be useful to broaden the above formulated synchronization concept, spreading it to objects of a more general nature and enveloping more general motion types. Thus, for example, the number of requirements can

---

* For example, in the case of $q = k^* - 1$, substituting expressions $\tilde{n}_j\,\omega$ in (1.7) instead of ω_j° and canceling ω, we find from the remaining $k^* - 1$ linear equations $\tilde{n}_1 = -\Delta_1\tilde{n}_{k^*}/\Delta, \ldots, \tilde{n}_{k^*-1} = -\Delta_{k^*-1}\tilde{n}_{k^*}/\Delta$. Here Δ is the determinant of matrix $\|n_{j\rho}\|$ ($j,\rho = 1, \ldots, k^* - 1$), which, due to the assumption of the independence of correlations (1.7), can always be considered to be nonzero; Δ_j is the determinant of the same matrix where elements of the j-th column are replaced with $\tilde{n}_{k^*1}, \ldots, \tilde{n}_{k^*,k^*-1}$. Both Δ and Δ_j are integers. Thus, the values $\tilde{n}_1, \ldots, \tilde{n}_{k^*}$ will be integers, at least at $n_{k^*} = n\Delta$ where n is an arbitrary integer number. The latter statement had to be proven.

be limited to a single one, that the correlations (1.1) or (1.6) should only contain expressions for at least one of the objects of the coordinates $x_j^{(s)}$. It may also be assumed that the functions $y_j^{(s)}$ and ν_ρ are quasi periodical. Finally, one can start from a considerably more general definition of synchronization, considering synchronization as the equality of certain functionals from the coordinates of the objects, such as coinciding time moments when these coordinates vanish or reach extremal values. As a rule, it is enough to limit ourselves to the above less general interpretation of synchronization when the abovementioned functionals are assumed to be frequencies (or average velocities of the object coordinates).

We will note that the definition of synchronization which is given above and used throughout this book, and which at the start assumes the selection in the system of the objects of the same type, could conditionally be called *physicotechnological*. There are more abstract definitions that do not assume such a selection of objects [see, for example, (109)]. The latter can be called mathematical.

We will also note that for a number of applications, the synchronization problem in the systems with distributed parameters is of particular interest. In this case, some of the equations (1.3) are equations in partial derivatives.

§ 2) Synchronization of weakly coupled objects. Synchronization problems permitting the application of perturbation methods.

The differential equations of synchronization problems, as a rule, are essentially nonlinear, and therefore their solution, generally speaking, is associated with considerable difficulties. However, at least three classes of problems can be shown to allow the natural insertion of a small parameter into the equations and further utilization of the appropriate effective solution methods. At the same time, these classes of problems are of the highest practical and principal interest. We will give a brief characterization.

1) *Synchronization problem of weakly coupled objects.* It is evident that such problems are actually the most interesting because, on the one hand, synchronization can be realized in the simplest and most economical way by using "weak" connections and, on the other hand, at the "strong" connections between objects, they can be considered as a united system for which there are no practical synchronization problems. For example, two mechanical debalanced vibro-exciters whose shafts are connected by toothed gears effectively represent a two-shaft vibro-exciter.

The problems of the synchronization of objects with weak mutual connections include the majority of synchronization problems for vibro-exciters, synchronization problems of the motions of celestial bodies, Huygens' problem of clock synchronization, the synchronization of self-excited generators and many others.

The mathematical reflection of the fact of the weak mutual connection between the objects is the smallness of the coupling function $F^{(s)}$ in comparison to $X^{(s)}$ in the differential equations (1.3) and (1.4). Otherwise, in this case, the equations (1.3) may be presented in the following form:

$$\dot{x}^{(s)} = X^{(s)}\left(x^{(s)}\right) + \mu \tilde{F}^{(s)}\left(x^{(1)}, \ldots, x^{(k)}, u, \mu\right) \qquad (s = 1, \ldots, k), \tag{2.1}$$

$$\dot{u} = U^*\left(x^{(1)}, \ldots, x^{(k)}, u, \mu\right).$$

and corresponding to them equations (1.4) in variables $y^{(s)}$, v and $\tau = \omega t$ in the form:

$$\frac{dy^{(s)}}{d\tau} = Y^{(s)}\left(y^{(s)}, \tau\right) + \mu\tilde{\Phi}^{(s)}\left(y^{(1)}, \ldots, y^{(k)}, v, \tau, \mu\right) \qquad (s = 1, \ldots, k),$$
$$\frac{dv}{dt} = V^*\left(y^{(1)}, \ldots, y^{(k)}, v, \tau, \mu\right). \tag{2.2}$$

Here, $\mu \geq 0$ is the small parameter, and $\tilde{F}^{(s)}$ and U^*, $\tilde{\Phi}^{(s)}$ and V^* are vector functions of the same type as functions $F^{(s)}$ and U, $\Phi^{(s)}$ and V in equations (1.3) and (1.4). In addition $\tilde{F}^{(s)}$ and U^*, $\tilde{\Phi}^{(s)}$ and V^* are also functions of the parameter μ which can be considered as analytical by μ at $\mu < \mu_0$, here $\mu_0 > 0$.

It should be noted that in radiophysics, the concept of "weak coupling" sometimes has a somewhat different meaning.

A typical class of problems is that of the synchronization of identical or quasi-identical objects weakly interacting with one another. In this case, function $X^{(s)}$ in equation (2.1) and function $Y^{(s)}$ in equation (2.2) do not depend on indices, which also leads to some simplifications.

2) *Problems for objects' synchronization close to those already studied.* In many cases, the objects whose synchronization is analyzed are close to some "standard" model objects whose properties were already investigated. Such standard objects may be, for example, linear oscillators, conservative objects with one or several degrees of freedom or the simplest self-oscillating objects.

Functions $X^{(s)}$ in equations (1.3) and functions $Y^{(s)}$ in equations (1.4) in the considered case may be presented as follows:

$$X^{(s)} = X_0^{(s)}\left(x^{(s)}\right) + \mu X_1^{(s)}\left(x^{(s)}, \mu\right),$$
$$Y^{(s)}\left(y^{(s)}, \tau\right) = Y_0^{(s)}\left(y^{(s)}, \tau\right) + \mu Y_1^{(s)}\left(y^{(s)}, \tau, \mu\right). \tag{2.3}$$

Here, as before, $\mu \geq 0$ is the small parameter, and functions $X_1^{(s)}$ and $Y_1^{(s)}$ are analytical functions by μ at $\mu < \mu_0$ where $\mu_0 > 0$. In this case equations:

$$\dot{x}_0^{(s)} = X_0^{(s)}\left(x_0^{(s)}\right) \qquad (s = 1, \ldots, k), \tag{2.4}$$

following from (1.3), when $\mu = 0$ and $F^{(s)} = 0$ are the equations of the mentioned standard model objects. Usually, solutions of these equations or their character are known, which simplifies the solution of the corresponding synchronization problems. Considerable simplification is achieved, especially in cases where the synchronization problem is considered in the presence of weak connections between objects.

3) *Synchronization problems of objects with an approximately known character of synchronous motions.* There are often cases when one may assume beforehand that in the synchronous regime, if it exists, the objects perform motions which are close to some specific motions. In other words, one can consider that for the motion being studied

$$x^{(s)}(t) = x_0^{(s)}(t) + \mu x_1^{(s)}(t, \mu) \tag{2.5}$$

or

$$y^{(s)}(\tau) = y_0^{(s)}(\tau) + \mu y_0^{(s)}(\tau, \mu), \tag{2.6}$$

where $x_0^{(s)}$ and $y_0^{(s)}$ are functions of a known type determined with accuracy up to certain constant parameters. Thus, for example, when solving synchronization problems of mechanical exciters and celestial bodies, one may often assume that their motions in certain coordinates differ insignificantly from uniform rotations with some initial phases which were not known beforehand.

In similar cases, it is possible to insert the small parameter into equations of the problem (1.3) or (1.4), in spite of its formal absence in these equations. For this purpose, on the right-hand side of the equations, one chooses such terms that if all the other terms were canceled, the equations would have solutions $x_0^{(s)}$ (or $y_0^{(s)}$). These other terms are assumed to be small, i.e., containing a small parameter μ as a multiplier.

This method for introducing a small parameter is described in more detail in the books (54, 57, 61). It was widely used by us for the solution of a number of synchronization problems; see, for example, (32, 36, 57); also § 4, chap. 3 and § 4, chap. 12.

It goes without saying that introducing a small parameter and applying the corresponding methods are effective only in cases when a so-called *generating system*, i.e., a system obtained from the initial one at $\mu = 0$, becomes simpler than the initial one, so that it is possible to find the required family of its solutions (not necessarily the general solution!). It can be seen from the aforesaid that for the three mentioned classes of synchronization problems, this is the case: the system at $\mu = 0$ is always simpler than the initial one and, at least for the second and third classes of problems, it has solutions for the coordinates $x_0^{(s)}$ (or $y_0^{(s)}$), whose character was known beforehand.

Chapter 2

GENERAL CHARACTERIZATION OF MATHEMATICAL APPARATUS FOR THE THEORY OF SYNCHRONIZATION OF WEAKLY COUPLED DYNAMIC OBJECTS*

§ 1) Small parameter (perturbation) methods by Poincare and Lyapunov in synchronization problems of weakly coupled objects.

The methods of Poincare and Lyapunov are now some of the most effective ways of studying and constructing periodical solutions for nonlinear differential equations containing a small parameter. These methods have been largely developed through works by Russian scientists L.I. Mandelshtam, N.D. Papaleksy, A.A. Andronov, A.A. Witte, B.V. Bulgakov, and I.G. Malkin.

A systematic description of Poincare and Lyapunov's methods can be found, for example, in the books (74, 184, 188, 337), and in connection with the problems of synchronization, in the book (57). Here we will dwell only on basic ideas of the methods and on the specifics of their applications to the solution of synchronization problems for weakly coupled objects. We'll discuss these methods in more detail in § 1, chap. 10.

One of the basic ideas of Poincare's method consists of analyzing, instead of the initial system of equations, the usually simpler one which is obtained from the initial system at $\mu = 0$, i.e., the *generating system*. After a periodic solution of the generating system is found, there arises the important question as to whether this solution corresponds to the periodical solution of the initial system for $\mu = 0$. Only if this condition is satisfied can the periodical solution of the generating system be considered as an approximate solution for the initial system, and further approximations, if necessary, be calculated as polynomial series by the small parameter. After this the stability of solutions can be investigated, for which A.M. Lyapunov's stability theory is frequently used. Naturally, the character of the investigation depends largely on the character of the generating system and its solution.

In synchronization problems for weakly coupled objects, the generating system in terms of initial variables $x^{(s)}$ and u is written as [see equations (2.1), chap. 1]

$$\dot{x}_0^{(s)} = X^{(s)}\left(x_0^{(s)}\right) \qquad (s = 1, \ldots, k),$$
$$\dot{u}_0 = U^*\left(x_0^{(1)}, \ldots, x_0^{(k)}, \quad u_0, 0\right). \tag{1.1}$$

and in terms of $y^{(s)}$ and v, is written as [see equations (2.2), chap. 1]

$$\frac{dy_0^{(s)}}{d\tau} = Y^{(s)}\left(y_0^{(s)}, \tau\right) \qquad (s = 1, \ldots, k), \tag{1.2}$$
$$\frac{dv_0}{d\tau} = V^*\left(y_0^{(1)}, \ldots, y_0^{(k)}, v_0, \tau, 0\right).$$

* The mathematical apparatus of synchronization is considered in a more detailed fashion in Part 3 of this book.

Let us assume that system (1.1) has a synchronous solution, i.e., the solution characterized by equations (1.1) in chap. 1. Then, since time does not show explicitly in the righthand sides of equations (1.1), or, in other words, due to the autonomous character of these equations, the latter must also have a synchronous solution in the following format:

$$x_{j0}^{(s)}(t+\alpha_s) = \tilde{n}_j^{(s)}\omega t + \tilde{n}_j^{(s)}\omega\alpha_s + y_j^{(s)}\left[m_j^{(s)}\omega(t+\alpha_s)\right]$$
$$(j = 1, \ldots, r_s;\ s = 1, \ldots, k). \tag{1.3}$$
$$u_{\rho 0} = u_{\rho 0}(t, \alpha_1, \ldots, \alpha_k) \qquad (\rho = 1, \ldots, \nu).$$

where α_s are arbitrary constants. The solution of equation (1.3) is obtained from the original one by substituting t for $t + \alpha_s$ in the solutions $x_{j0}^{(s)}$, corresponding to the s-th object and the subsequent solution of the equation for $u_{\rho 0}$. In this case, it is easy to conclude that if system (2.2) allows for a certain periodic solution of $y_{j0}^{(s)}(\tau)$, $v_{\rho 0}(\tau)$, then it would positively allow for a family of periodic solutions in the form:

$$y_{j0}^{(s)} = \tilde{n}_j^{(s)}\omega\alpha_s + y_j^{(s)}\left[m_j^{(s)}\omega(t+\alpha_s)\right], \tag{1.4}$$
$$v_{\rho 0} = v_{\rho 0}(\tau, \alpha_1, \ldots, \alpha_k),$$

depending on k arbitrary constants $\alpha_1, \ldots, \alpha_k$. These constants play the roles of *initial phases of objects* in the *synchronous motions*.

Thus, problems of synchronization for weakly coupled objects are characterized by the generating system having a family of periodic solutions that depend on at least k arbitrary parameters (the number of such parameters may be even larger—see § 2). This case in the Poincare theory of periodic solutions, is, first of all, a singular one requiring a special analysis, and secondly, is the most interesting case from the principal point of view.

§ 2) Resultative correlations for the theory of synchronization of weakly coupled objects. Integral stability criterion (extremal property) of synchronized motions.

As was shown by the corresponding investigation (184), if the generating system has a family of periodic solutions depending on a certain number of arbitrary parameters $\alpha_1, \ldots, \alpha_k$, then, generally speaking, there is no complete qualitative correspondence between the periodic solutions of the initial and the generating systems of equations. It was found that synchronous solutions of the initial system, i.e., system (2.1) in chap. 1, decreased at $\mu = 0$ into the synchronous solutions (1.3) of the generating system (1.1). These solutions may not correspond to all values of the constants α_s, but only to values conforming to a certain system of finite equations

$$P_s(\alpha_1, \ldots, \alpha_k) = 0 \qquad (s = 1, \ldots, k), \tag{2.1}$$

arranged according to a certain rule (see below).

Further on, generally speaking, not all solutions of (2.1) would correspond to stable synchronous motions. The stability investigations (33, 34, 57, 184, 216) show that for a sufficiently large class of systems, the basic role in the selection of the stable solution is played by the requirement that, for a specific solution of equation (2.1), all the roots χ (excluding, perhaps, one zero root) of the algebraic equations of k-power

$$\begin{vmatrix} \frac{\partial P_1}{\partial \alpha_1} - \varkappa & \frac{\partial P_1}{\partial \alpha_2} & \cdots & \frac{\partial P_1}{\partial \alpha_k} \\ \frac{\partial P_2}{\partial \alpha_1} & \frac{\partial P_2}{\partial \alpha_2} - \varkappa & \cdots & \frac{\partial P_2}{\partial \alpha_k} \\ \cdots & \cdots & \cdots & \cdots \\ \frac{\partial P_k}{\partial \alpha_1} & \frac{\partial P_k}{\partial \alpha_2} & \cdots & \frac{\partial P_k}{\partial \alpha_k} - \varkappa \end{vmatrix} = 0 \tag{2.2}$$

should have negative real parts. For many systems, the formulated requirement is not only necessary but also a sufficient condition of stability (local, and at sufficiently small values for parameter μ). Besides, if this requirement is compared with the corresponding solution to equation (2.1), this solution would actually conform to the only solution of the initial system for the system of differential equations reduced at $\mu = 0$ to the generating solution (1.3).

Thus, if the constants α_s are found from equations (2.1) and all the roots χ of equation (2.2), except, perhaps, for one, conform to the conditions

$$\operatorname{Re} \varkappa_j < 0 \qquad (j = 1, \ldots, k), \tag{2.3}$$

then expressions (1.3) may be considered totally acceptable first approximations for the solution to the problem.

The conditions of equations (2.1), having real solutions with respect to the constants $\alpha_1, \ldots, \alpha_k$, may be considered as the necessary conditions for the object synchronization effect to exist (we consider the case when constants α_s must be real in their physical sense).

The solution of the original system, i.e., system (2.1) in chap. 1, may be found as a polynomial series by the powers of a small parameter (see § 1, chap. 10). Fortunately, however, in the majority of synchronization problems, it is quite sufficient to find the first members of a series determined by approximation (1.3) under the condition, of course, that the constants are found from equations (2.1). It should be noted that these constants determine the phasing of object motion in synchronous regimes of the system.

Thus, equations (2.1) play a very important role in the solution of synchronization problems. They are required for the calculation of phases of object motions in the initial approximation; for determining the existence of the conditions for synchronous motions in the system; to express the stability conditions of synchronous motions of these equations by means of the left-hand sides $P_s(\alpha_1, \ldots, \alpha_k)$. We will call the P_s functions the *generating functions* and the equations in (2.1) the *basic equations of the synchronization problem* of weakly coupled objects. Note that these equations cannot be constructed only on the basis of information contained in the generating system (1.1). It is also necessary to consider the members in the initial equations (2.1) in chap. 1, which contain the small parameter and, in particular, the members corresponding to the coupling functions $\mu F^{(s)}(x^{(1)}, \ldots, x^{(k)}, u, \mu)$. However, for constructing functions $P_s(\alpha_1, \ldots, \alpha_k)$ it is usually sufficient to have expansion terms of functions $\mu F^{(s)}$ which are linear with respect to μ, i.e,. expressions $\mu F^{(s)}$ $(x_0^{(1)}, \ldots, x_0^{(k)}, u_0, 0)$.

It is very important that the difficulties of deriving the expressions for generating functions P_s in an explicit form for the synchronization problems of weakly coupled objects, are due not to the degree of complexity of the system as a whole but only to the degree of complexity of some individual selected objects and the connecting system. This fact considerably facilitates the solution of the synchronization problems for objects with weak couplings.

The considered resultative correlations of the problem of synchronization of weakly coupled objects can often be written in a more convenient form (40, 41, 43, 57, 81, 164, 209–211, 282–286). Let us assume, for example, that such a function $D = D(\alpha_1, \ldots, \alpha_k)$ exists that satisfies the following relations.

$$\partial D/\partial \alpha_s = -P_s(\alpha_1, \ldots, \alpha_k) \qquad (s = 1, \ldots, k). \tag{2.4}$$

The function D (the latter, as well as P_s, are supposed to be real) will be called *potential function.*

Analyzing the relations (2.1)–(2.4) together, it is easy to come to the conclusion that the stable synchronous motions of the system may correspond to those stationary points $(\alpha_1^*, \ldots, \alpha_k^*)$ of the potential function $D(\alpha_1, \ldots, \alpha_k)$ in which this function has a minimum. The minimum can be discovered on the basis of analysis of terms of not higher than the second order in the expansion of the function near its stationary points. A proof and also a more correct formulation of this statement are given in § 8, chap. 10.

Thus, the function D in the problems of synchronization considered, plays the same role as the potential energy of a system in the problem of finding and analyzing the stability of equilibrium states in similar systems.

The formulated stability criterion of synchronous motions is called *integral* (40, 41, 43), since the function D, as well as the generating functions P_s, depend on the average for the period characteristics of system motion calculated for the generating solution. The importance of this criterion is determined by the fact that the potential function D in many actual synchronization problems has a distinctive physical sense which permits one to consider the question of the presence or absence of a tendency toward synchronization in quite a general form, and also facilitates the solution of the problem of the synthesis of systems with synchronizing objects. The other advantage of the integral criterion is the fact that it opens possibilities of the application of well-developed algorithms to the extremum searching for the function of many variables, thus allowing the application of computers. When this criterion is available, the motion stability conditions may also be expressed in explicit form by using the Sylvester theorem.

Highly essential, in particular, is that the existence of stable synchronous motions, i.e, the tendency toward synchronization, can be proved by sufficiently general assumptions in the case of the periodicity of the potential funciton D by $(\alpha_1, \ldots, \alpha_k)$, as well as in cases when D is close to being periodic or functions P_s are close to those for which a quasi-periodic function D exists. (In these cases, we will speak of the existence of quasi-periodic, quasi-potential function; see § 9, chap. 10).

In spite of the fact that the requirement of the existence of a potential (or quasi-potential) function D seemingly puts sufficiently strict conditions on the system under consideration, this function exists in many important synchronization problems and also in other problems of nonlinear oscillations, for which the generating system allows for a family of periodic solutions depending on a certain number of arbitrary parameters. Expanding the class of systems for which the integral criterion applies is helped by the following. As can be shown, for this criterion to be applicable, the existence of function D, with less strict conditions than (2.4) is required, namely,

$$b_{s1}\frac{\partial D}{\partial \alpha_1}+b_{s2}\frac{\partial D}{\partial \alpha_2}+\ldots+b_{sk}\frac{\partial D}{\partial \alpha_k}=-P_s(\alpha_1,\ldots,\alpha_k) \qquad (2.5)$$

$$(s=1,\ldots,k),$$

where $b_{sj}=b_{js}$ are arbitrary constants conforming to the requirement that the quadratic form $B(z,z)=\sum_{s=1}^{k}\sum_{j=1}^{k} b_{sj}z_s z_j$ be positive.

The situation above was intentionally simplified in order to not overload the discussion with minor details and reservations. From the omitted details, we will mention only that in an autonomous case, i.e., in a problem of internal synchronization, the initial phases α_s are not determined from the equations (2.1), but only the phase shifts $\alpha_s - \alpha_k$. However, one usually also finds the first approximation of the synchronous frequency ω. Equation (2.2) should have only $k-1$ roots with negative real parts, since the zero root, always present in this case, does not play a role. Furthermore, the number of arbitrary constants in the generating solution (1.3) may be more than the number of objects k since the appearance of such constants is not necessarily connected with arbitrariness in the selection of the starting time points for the motion of each of the selected autonomous objects. In these, as well as in more complicated cases, the stability conditions expressed by means of equations of the (2.2) type and the integral stability criterion are not sufficient, but only necessary. They, however, play a basic role in the selection of the stable synchronous solutions. Below, in chap. 10, we give specific expressions for generating functions P_S for certain types of systems of differential equations containing a small parameter, which are of interest for the synchronization theory. Expressions for these functions, as well as for the potential or quasi-potential function D, are presented in chaps. 3, 4, 7 and 12. There they are given in more detail in the application to synchronization problems for specific classes of weakly interacting objects.

Concluding this section, we will touch on the important and unfortunately often incorrectly treated issue of how rigorous are the Poincare and Lyapunov methods when they are being used for solving application-related problems.

The fact is that the basic apparatus of these methods permits one to determine the facts of existence and the stability of a solution as well as the convergence of corresponding series at sufficiently small values of the parameter μ and at sufficiently small initial displacements. It is more difficult to determine the boundaries of the areas of the small parameter's variation as well as the initial displacements where these facts take place. In spite of a number of interesting available results [see, for example, (254)], there are still no universally effective methods available. Accordingly, in the majority of applied studies (and the synchronization problems are no exception), the authors limit themselves to the determination of "local" results, not determining either radii of convergence of the series or spheres of attraction for the regimes found in the phase space of the system. However, in practical utilization of the analytical results, one has to deal with the finite values of both the parameter μ and the initial displacements. In other words, the results of local analyses are not used locally.

Of course, such an application of the methods considered is not rigorously correct and may lead to errors. In other words, local results obtained by using the Poincare and Lyapunov methods are not absolute but only trustworthy in applications to practical problems, and then only verisimilar. Such views were already expressed by A.A. Andronov and representatives of his scientific school (19, 102). However, experience using the Poincare and Lyapunov methods shows that the "degree of trustworthiness" with not rigorous but cautious application of local results becomes, in practice, very high. As a rule, these results are in good agreement with experiments which are certainly not redundant here. It seems that the reason for this is a good "correlation" between mathematical and intuitive concepts of "sufficient smallness." Due to this and also some other reasons, the actual chances of getting into the area where the local results are not applicable are much less than the chances of getting into the area where they are correct. Detailed considerations of these concepts are given in chap. 4 of the book (57) and also in the book (61).

§ 3) Other mathematical methods for the solution of synchronization problems for weakly coupled objects.

For the solution of synchronization theory problems for weakly coupled dynamic objects, methods other than those of Poincare and Lyapunov are also used, including those based on the use of a small parameter. The widest applications have asymptotic methods and related principles of averaging and separation of motions. Also used are the methods of point transformations, stroboscopic and variation. Brief descriptions of these methods are given in chap. 11.

We will just mention here that for analysis of just-developed steady-state synchronous motions and their stability, those methods have no special advantages with respect to the Poincare and Lyapunov methods. An advantage of those methods, though, is in the fact that they also allow us to effectively study the "slow" processes of establishing the synchronous regimes, as well as some other more complicated motions. It is noteworthy that, in many cases, equations of the following type are obtained for the description of the slow processes:

$$a_s\ddot{\alpha}_s + b_s\dot{\alpha}_s = c_s P_s(\alpha_1, \dots, \alpha_k) \quad (s = 1, \dots, k), \tag{3.1}$$

where a_s, b_s and c_s are certain constants (a_s and b_s are usually positive) and P_s are the same generating function as mentioned in § 2. At α_s = const., the basic equations (2.1) are naturally obtained from (3.1). From these equations the phase values α_s are determined in stable synchronous motions. Stability analysis of these regimes leads, as a rule, to the requirement of negative real parts for the roots of equation (2.2). Thus, in cases of steady-state regimes, the results of the application of the abovementioned methods are completely in accordance with the results of the Poincare and Lyapunov methods.

Naturally, however, additional information obtained by using these methods also requires, as a rule, some additional work, since application of the Poincare and Lyapunov methods usually amounts to application of the theorems given in chap. 10.

PART TWO

SYNCHRONIZATION OF TECHNOLOGICAL AND NATURAL OBJECTS

Chapter 3

SYNCHRONIZATION OF MECHANICAL VIBRO-EXCITERS

§ 1) Vibro-machinery and devices.

Vibration was, for a long time, considered mostly as a harmful side effect which accompanied the operation of machines and apparatus. Thus, the basic efforts of engineers and scientists were directed toward searching for methods to fight vibration.*

Only in the last several decades has vibration been applied to obtaining useful effects, and have the so-called vibratory machines and devices appeared. Now such machines and devices are widely used in different areas of engineering and technology; a considerable economic and technological effect is achieved with their application. For example, not a single modern hydrotechnical construction site can manage without vibro-compactors; not a single plant for manufacturing of building elements exists without stands for the shaping and compaction of concrete mixtures; and there would be no ore-enrichment mills without vibro-shakers. Miniature microvibratory motors are used to provide highly accurate measured translation or rotation of parts, and vibro-unloaders are used for railroad cars and for vibro-crushers with the mass of the vibrating organ weighing up to several dozen tons. Now numerous monographs and journal papers are published discussing issues related to vibratory processes and machines (for example, 57, 78, 79, 166). The flow of inventions in the field of vibratory technology is continually growing.

The great majority of vibrating machines and devices may be idealized as a system (see Fig. 5) consisting of a certain number of solid or deformable bodies $B_{01}, \ldots, B_{0n}$ connected to each other and with a stationary structure *1* by some "concentrated" geometric connections *2*, and also by elastic *3* and damping *4* elements. Some of these bodies may represent working organs of the machine, the others, together with elastic and damping elements, can be used for modeling the reaction on the working organs of the material being processed.

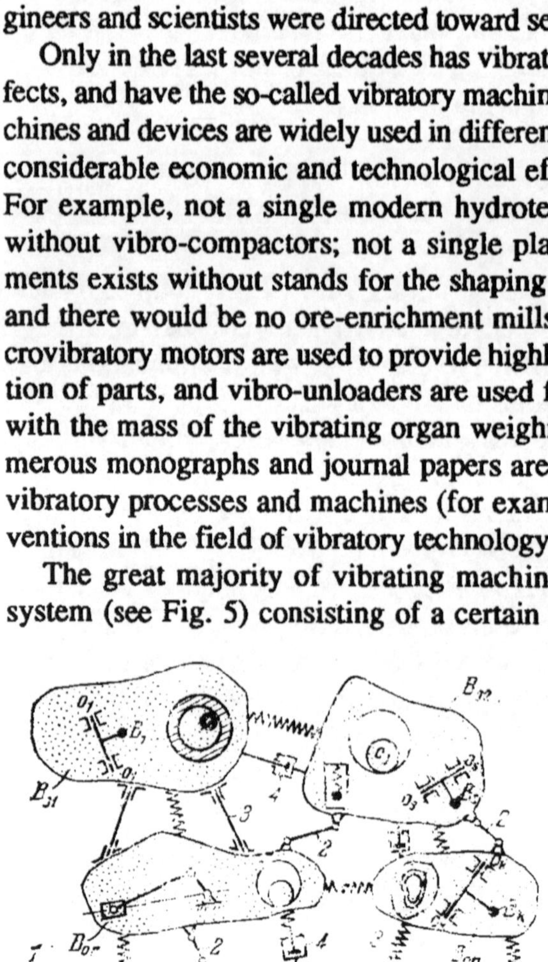

FIG. 5.

As vibration exciters for bodies $B_{01}, \ldots, B_{0n}$, mostly so-called *mechanical vibro-exciters* $B_1, \ldots, B_k$ are used. In simple cases, they consist of unbalanced

*By *vibration*, here and below, we mean mechanical oscillations in which the amplitude is considerably less than a characteristic size of the oscillating body, and the period is less than a certain characteristic time interval.

rotors with the rotational axis fixed in the bodies $B_{01}, \ldots, B_{0n}$, and are driven by induction motors *supporting bodies*. Similar vibro-exciters are often called *debalanced* or *inertia-type machines*. Other types of mechanical vibro-exciters are schematically shown in Fig. 5 (see subsections 3 and 5, § 8). The generalization of the term mechanical vibro-exciter is given in subsection 1, § 8 of this chapter and in § 6, chap. 12.

§ 2) The problem of synchronization and phasing of rotor rotation for mechanical vibro-exciters.*

Many vibratory machines and apparatus designs are using not one but several mechanical debalanced exciters. Thus, for example, application of several relatively low power exciters, instead of one more powerful, may be required in order to distribute the exciting force along a vibrating working organ of a considerable size. In this way, by complying with certain conditions, one can provide an oscillation of the working organ of a vibratory machine that is only slightly different from its oscillation as an absolutely solid body. This can be achieved in spite of the fact that the frequency of the generating force exceeds several natural frequencies of elastic oscillations of the operating organ, i.e., in spite of its "nonrigidity."

In other cases, we use several vibro-exciters instead of one because we wish to decrease the load on the exciter's bearings, thus distributing it over a greater number of bearings. This is typical for heavy industrial machines whose maximum size (when using one exciter) is determined by the maximum capacity of available bearings.

Use of two (or any even number) debalanced exciters whose rotors rotate with the same angular velocity, but in opposite directions, allows the generation of a disturbing force of constant direction which is utilized in many vibratory machines and devices.

A required condition of normal operation for most vibratory machines and units with several exciters is synchronization of the rotation, as well as certain correlations between phases of rotors of individual exciters, e.g., synphasing or opposite phasing. In some cases, there is a problem with generating so-called multisynchronous motion, when the rotors of exciters rotate with angular velocities which are multiples to a certain given velocity, and with specific phase differences.

One may usually be satisfied with inexact but approximate phase correlations between the rotors. The small oscillations which superimpose on the rotation do not lead to distortions of the working process. Similarly, it is almost adequate to provide the equality not of instant but of averaged one per rotation values of angular rotation velocity.

Recently, the only widespread method for coordinating rotor rotations for vibro-exciters was by means of kinematic connections between rotors in the form of gears, chain drives, etc. (see, for example, Fig. 2a). Such forced kinematic synchronization has, however, a number of essential shortcomings. The principal ones are considerable noise and the rapid wear of gear or chain drives due to the presence of alternating or pulsating loads. Besides synchronization, such a method hampered development of vibration technology, since it could not be reasonably used in many practical and important cases, for example, at considerable distances between exciters.

---

* This problem is discussed in a broader and more formal way in § 11.

In recent years, a number of methods for the forced electric synchronization of rotors for vibro-exciters have been offered [see, for example, the review in the book (57)]. However, this method has not yet found a widespread practical application, mostly because of its relative complexity and the nonavailability of the necessary hardware. Therefore, an important role in the development of vibratory technology was played by the discovery of the *self-synchronization of the mechanical vibro-exciter phenomenon*, which allows for the automatic coordination of rotor rotation between vibro-exciters. If certain conditions are satisfied, equality of integer ratios between average angular velocities and the presence of definite phase relations at rotor rotation are realized, in this case by means of the vibration of supporting structures on which the rotors are mounted, i.e., due to internal properties of the vibratory system itself.

§ 3) Discovery, study and implementation of synchronization phenomena to mechanical vibro-exciters.

To the best of our knowledge, the synchronization phenomenon for mechanical debalanced vibro-exciters was discovered in the "Mekhanobr" Institute in Leningrad in 1947–48 in the course of a research project under the guidance of D.A. Pliss. The interest in this phenomenon and the understanding of its practical implications were due to the following observation. There was a protracted test of a vibratory machine with two nominally identical independent debalanced exciters which were driven by individual induction motors (the machine schematic is similar to Fig. 2b). The exciters' rotors, if mounted on the stationary foundation, would inevitably rotate with somewhat different angular velocities and random phases. This was due to manufacturing inaccuracies in the motors and exciters. However, having been installed on a common vibratory organ of the machine, both exciters, in average, ran in a synchronous and synphaseous regime. The tendency toward synchronous rotor rotation was so strong that their rotation was not disturbed, even by an occasional rupture of the electric supply cable for one of the motors. The ruptures were only discovered after several hours, during which the machine was operating more or less normally (and continued to work until the second motor was switched off). In other words, what took place here was the *vibratory maintenance of the rotation of a debalanced rotor*. This phenomenon is a special case of synchronization which will be discussed later in § 12.

The observation was recorded in the "Mekhanobr" internal report (232) in 1948. There, attention was brought to the potential importance of this phenomenon for vibratory technology.

The first publication of this phenomenon abroad was made later—in 1950—as a patent for specific simplistic vibratory devices, and is attributed to Signul, a Swedish inventor (327, 328). The patent was, however, filed in Sweden in 1946.

In 1953 the author gave a physical explanation and mathematical description of self-synchronization effects on mechanical vibro-exciters (31). Later, the conclusions of this paper were further developed by both the author and other Russian researchers, see (5, 14–17, 35–38, 40–43, 45–49, 51, 53, 55–59, 63, 64, 79, 91, 99, 104, 105, 114–118, 130–132, 140, 148, 159, 230, 231, 235, 241, 242, 246, 249, 256, 275, 283, 284, 293, 300, 306). The following foreign publications on the self-synchronization of vibro-exciters, which appeared after 1965, are worth mentioning: (307), (320), (323), (329, 330).

The principal results of the enumerated investigations are either presented in this chapter with appropriate references, or are mentioned in the short review in subsection 7, § 8. A more detailed review of the pre-1970 results is given in the book (57).

The development of the theory and of the computational methods for devices with self-synchronizing vibro-exciters resulted in development and implementation of a new class of vibratory machines—conveyors, feeders, crushers, screens, mills, concentration tables, special stands, etc. Now there are more than a hundred Soviet inventions based on the utilization of the self-synchronization effect. Note that many of these inventions could evidently not have been made without the corresponding theoretical studies and computational methods. The fact is, as will be shown below, that a number of features of the self-synchronization phenomenon for vibro-exciters can hardly be predicted on the basis of purely intuitive considerations or be discovered on the basis of nontargeted experimentation. Descriptions of a number of devices with self-synchronizing vibro-exciters are given in § 15.

It should be noted that the majority of both the theoretical investigations and prototype designs of self-synchronizing vibro-exciters were made at the "Mekanobr" Institute.

The discovery of the self-synchronization phenomenon for vibro-exciters allows one to approach the problem of group foundations for unbalanced machines in a new fashion. This is the case when several machines with rotating or oscillating unbalanced masses are located on the same foundation. In this special case, the self-synchronization may become harmful and even dangerous. If, in the ensuing motions, the phases of rotor rotations are such that imbalances of individual machines are summed up, then this may cause excessive foundation oscillations and even destruction of building foundations. There are several recorded cases of such mishaps. On the other hand, designing a group foundation using the results of the self-synchronization theory, often results in the cancellation of the unbalanced forces generated by the individual machines, thus leading to a considerable reduction in foundation vibrations.

An additional consequence of the efforts to develop the theory of synchronization for vibro-exciters was a stimulus toward the development of a general synchronization theory. These efforts also led to the consideration of the tendency toward synchronization as a general property of material objects of very different origins. In particular, it was found that synchronization of natural and artificial celestial bodies follows the same laws as the synchronization of unbalanced rotors. Thus, for example, in both cases, the so-called integral stability criterion (extremal property) of synchronous motions turns out to be correct (see chap. 7).

§ 4) Self-synchronization of debalanced vibro-exciters in the simplest oscillating system (basic correlations).

The basic specifics of both the formulation and solution of the synchronization problem for mechanical vibro-exciters, as well as many features of the phenomenon, may be discovered by using the simplest model related to the self-synchronization of debalanced vibro-exciters located on an absolutely rigid platform with one degree of freedom (31, 57) (see Fig. 6). Rigid platform *1* (*the supporting body*) is movable in relation to the stationary foundation *2* in the fixed direction *Ox*. The platform is con-

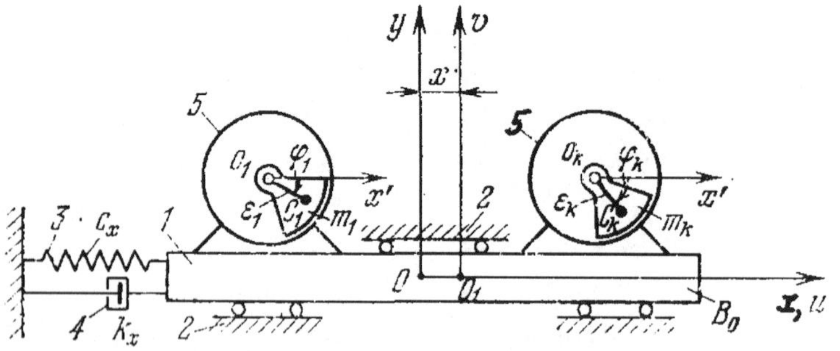

FIG. 6.

nected with the foundation by elastic elements *3* (stiffness c_x) and a linear damping element *4* (resistance coefficient k_x). Mounted on the platform are a certain number k of debalanced vibro-exciters *5* (unbalanced rotors) whose axes are perpendicular to the direction of the platform oscillations and which are driven by induction motors.

The system described presents a specific case from a wide range of mechanical systems considered in § 6, chap. 12—the so-called *orbital systems.*

Displacement x of the platform from a position corresponding to the nonstrained elastic elements, and rotor rotation angles φ_s ($s = 1, \ldots, k$), counted from the Ox-axis direction in a clockwise direction, are assigned as the generalized coordinates of the system. Then expressions for kinetic and potential energy of the system will be as follows:

$$T = \frac{1}{2}\mathcal{M}\dot{x}^2 + \frac{1}{2}\sum_{s=1}^{k}\mathcal{I}_{cs}\dot{\varphi}_s^2 + \frac{1}{2}\sum_{s=1}^{k} m_s\left(\dot{x}_{cs}^2 + \dot{y}_{cs}^2\right), \qquad (4.1)$$

$$\Pi = \frac{1}{2}c_x x^2 + \sum_{s=1}^{k} m_s g \varepsilon_s (1 - \sin\varphi_s). \qquad (4.2)$$

Here $\mathcal{M}$ is the mass of the platform, m_s and $\mathcal{I}_{cs}$ are the mass and moment of inertia of the s-th vibro-exciter's rotor about the axis passing through its center of gravity C_s, respectively; g-acceleration due to gravity, ε_s is the eccentricity and :

$$x_{cs} = u_s + x + \varepsilon_s \cos\varphi_s, \quad y_{cs} = v_o - \varepsilon_s \sin\varphi_s \qquad (4.3)$$

are coordinates of the center of gravity of the s-th rotor in the system of stationary axes xOy (uO_1v are axes rigidly connected with the platform and coinciding with axes xOy at $x = 0$; u_s and v_o are constants representing the coordinates of axes 0_s of rotor rotation in the frame uO_1v).

Considering formulas (4.3), expression (4.1) for kinetic energy can be reduced to the form:

$$T = \frac{1}{2}M\dot{x}^2 + \frac{1}{2}\sum_{s=1}^{k} I_s\dot{\varphi}_s^2 - \dot{x}\sum_{s=1}^{k} m_s\varepsilon_s\dot{\varphi}_s \sin\varphi_s. \qquad (4.4)$$

The notations here are:

$$M = \mathcal{M} + \sum_{s=1}^{k} m_s, \quad I_s = \mathcal{J}_{cs} + m_s \varepsilon_s^2. \tag{4.5}$$

Composing Lagrange equations

$$\frac{d}{dt}\frac{\partial T}{\partial \dot{q}_j} - \frac{\partial T}{\partial q_j} = -\frac{\partial \Pi}{\partial q_j} + Q_j.$$

where q_j are the generalized coordinates of the system and Q_j are the generalized nonconservative forces, the following differential equations of motion for the system can be obtained:

$$I_s \ddot{\varphi}_s = L_s(\dot{\varphi}_s) - R(\dot{\varphi}_s) + m_s \varepsilon_s (\ddot{x} \sin \varphi_s + g \cos \varphi_s)$$
$$(s = 1, \ldots, k), \tag{4.6}$$

$$M\ddot{x} + k_x \dot{x} + c_x x = \sum_{j=1}^{k} m_j \varepsilon_j (\ddot{\varphi}_j \sin \varphi_j + \dot{\varphi}_j^2 \cos \varphi_j). \tag{4.7}$$

In these equations the nonconservative forces $Q_x = -k_x \dot{x}$ and $Q_s = L_s(\dot{\varphi}_s) - R_S(\dot{\varphi}_s)$ are considered. They represent, respectively, the force of viscous resistance to the platform oscillation and the torque acting on the rotor of s-th vibro-exciter. In this case, $L_s(\dot{\varphi}_s)$ is the driving torque of the induction motor (its so-called static characteristic*) and $R_s(\dot{\varphi}_s)$ is the resistance torque which is usually determined by the bearing friction. In this case:

$$R_s(\dot{\varphi}_s) = R_s^{\circ}(|\dot{\varphi}_s|) \operatorname{sgn} \dot{\varphi}_s, \tag{4.8}$$

where R_s° is the magnitude of the resistance torque.

The equations in (4.6) describe the motion of the exciter rotors, and equation (4.7) is the equation for platform oscillation. By their structure, these equations completely correspond to the general equations for the synchronization problem for dynamic objects (see equations (1.3) in chap. 1). It is easy to distinguish in them phase coordinates of the objects (φ_s and $\dot{\varphi}_s$) and the coupling systems (x and $\dot{x}$) as well as the coupling functions.

*Usually, when considering the problem of the synchronization of vibro-exciters, it is enough to consider the motor's *static characteristics*, since the time constants of the mechanical transients are usually considerably larger than the time constants of the electric transients (this is treated in (57) in more detail). The somewhat schematic typical static characteristic of an induction motor is shown in Fig. 7, where $L_{st}^{(a)}$, $L_{max}^{(a)}$, and $L_{nom}^{(d)}$ designate respectively starting maximum and nominal torques; ω_{cz}, ω_{nom} and ω_{syn} represent the critical, nominal and synchronous frequencies of the rotor rotation.

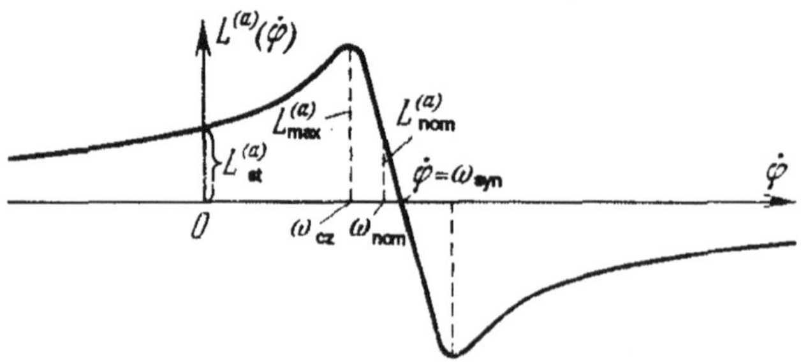

FIG. 7.

As a whole, the system (4.6) and (4.7) is essentially nonlinear. Only if we assume that the rotors of the exciters rotate uniformly, i.e., $\varphi_s = \sigma_s(\omega t + \alpha_s)$, where α_s are constants and $\sigma_s = 1$, then equation (4.7) is reduced to a standard linear equation for small oscillations of the platform under the action of harmonic disturbing forces. The equation complies with the assumption that the platform oscillations have no influence on the motion of the exciters. In many cases, such a simplified formulation of a problem is quite adequate.

When considering synchronization problems for vibro-exciters, one cannot ignore the presence of coupling between the oscillating and rotating coordinates of the system, since these coupling terms specifically determine the pattern of motion. Equations of motion which are adequate in this case are naturally much more complicated. The active influence of platform oscillations on the motion of exciter rotors is expressed by the terms $m_s\varepsilon_s x \sin\varphi_s$ in equations (4.6). As will become clear from the following, these terms specifically predetermine the synchronization of vibro-exciters, as well as other effects in the considered system.

The principal problem of the self-synchronization of vibro-exciters consists of the formulation conditions, compliance with which would lead the rotors of all exciters to rotate with equal, by magnitude, average angular velocities, in spite of the absence of any direct connections between them and in spite of differences in the parameters characterizing the exciters and the torques acting on them. In other words, it is the determination of the conditions of existence and the stability of solutions for the systems (4.6) and (4.7) which have the form:

$$\varphi_s = \sigma_s\,[\omega t + \alpha_s + \psi_s(\omega t)] \quad (s = 1, \ldots, k), \quad x = x(\omega t), \tag{4.9}$$

where ω is the magnitude of the average rotor rotational velocity, α_s are constants (initial rotation phases), ψ_s and x_s are periodic functions of time t with the period $T = 2\pi/\omega$ and each of values σ_s is equal to either 1 or −1. The first case corresponds to the positive and the second to the negative direction of rotor rotation in the s-th exciter. The value of synchronous angular velocity ω is not known beforehand, and it has to be determined during the solution process.

The motions of (4.9), in accordance with the terminology assumed in chap. 1, are *simple synchronous*. Sometimes the *multiple-synchronous* motions are also of interest. Since the latter case is more complicated, it requires, as a rule, a special consideration [see (57) and also the review in subsection 7, § 8].

In addition to finding the conditions of existence and stability of synchronous motions, it is also of interest to determine at least the approximate motion character of rotors and the platform in stable synchronous motions, as well as the solution of the inverse problem ("*problem of synthesis*") consisting of such a selection of system parameters at which the existence and stability of the synchronous motions of a specified character are assured.

As was shown by practical experience and analytical evaluation in the considered synchronous motions of the system, rotation of the rotors differs only slightly from the uniform rotation. Therefore, functions $L_s(\dot{\varphi}_s)$ and $R_s(\dot{\varphi}_s)$ may be linearized in the vicinity of $\dot{\varphi}_s = \sigma_s\omega$ and assuming the following correlations considering equality (4.8):

$$L_s(\dot{\varphi}_s) = L_s(\sigma_s\omega) - \overset{*}{k}_s(\dot{\varphi}_s - \sigma_s\omega),$$
$$R_s(\dot{\varphi}_s) = \sigma_s \overset{\circ}{R}_s(\omega) + \overset{\circ}{k}_s(\dot{\varphi}_s - \sigma_s\omega). \tag{4.10}$$

where

$$\overset{*}{k}_s = -\left(\frac{dL_s}{d\dot{\varphi}_s}\right)_{\dot{\varphi}_s=\sigma_s\omega}, \quad \overset{\circ}{k}_s = \left(\frac{d\overset{\circ}{R}_s}{d|\dot{\varphi}_s|}\right)_{\dot{\varphi}_s=\sigma_s\omega} = \frac{d\overset{\circ}{R}_s(\omega)}{d\omega} \tag{4.11}$$

are, respectively, *coefficients of electrical and mechanical damping*. Both these coefficients are usually positive, and this will always be assumed below.

It can be noted, however, that the final result would be the same if relation (4.10) were not used, but it would be achieved in a more complicated way.

Considering the solution of the problem by the Poincare method, we will introduce the small parameter μ into system (4.6) and (4.7), thus presenting it in the form:

$$I_s\ddot{\varphi}_s + k_s(\dot{\varphi}_s - \sigma_s\omega) = \mu\Phi_s(\varphi_s, \ddot{x}) \qquad (s = 1, \ldots, k),$$
$$M\ddot{x} + c_x x = \sum_{j=1}^{k} m_j\varepsilon_j(\ddot{\varphi}_j \sin\varphi_j + \dot{\varphi}_j^2 \cos\varphi_j) - \mu k'_x \dot{x}, \tag{4.12}$$

where *

$$k_s = \overset{*}{k}_s + \overset{\circ}{k}_s, \quad k_x = \mu k'_x, \tag{4.13}$$
$$\mu\Phi_s = L_s(\sigma_s\omega) - \sigma_s\overset{\circ}{R}_s(\omega) + m_s\varepsilon_s(\ddot{x}\sin\varphi_s + g\cos\varphi_s).$$

Such a method for introducing the small parameter corresponds to an assumption of quasi-uniform rotor rotation in the analyzed synchronous motion, and also to an assumption that the *motion is analyzed far from resonance when the friction force $k_x\dot{x}$ in the vibratory part of the system may be considered as small compared to the elastic or inertia forces, as well as the exciting force*. We should note, by the way, that the assumed method of introducing the small parameter is also supported by the realistic estimation of the orders of magnitude for the terms in the equations. This is discussed in more detail in § 3, chap. XIII and in chap. IV of (57). Also shown there is the "stability," of a sort, of the results obtained in spite of the method of introducing the small parameter. For example, these results would not change significantly if $k_s = \mu k'_s$ is also assumed or the assumption that $k_x = \mu k'_x$ is dropped.

* Parameter $k_s = \overset{*}{k}_s + k_s^{\circ} > 0$ will be called the *total damping coefficient*.

In solving the considered problem we may directly use either the theorem in § 2, chap. 10, or the result of the solution for the case of the synchronization of objects with almost uniform rotational motions (see § 4, chap. 12). We will focus on the latter approach.

The generating system corresponding to the equations (4.12), which is as follows,

$$I_s\ddot{\varphi}_s^0 + k_s(\dot{\varphi}_s^0 - \sigma_s\omega) = 0 \qquad (s = 1, \ldots, k),$$
$$M\ddot{x}^0 + c_x x^0 = \sum_{j=1}^{k} m_j\varepsilon_j\left[\ddot{\varphi}_j^0 \sin\varphi_j^0 + (\dot{\varphi}_j^0)^2\cos\varphi_j^0\right] \tag{4.14}$$

permits the family of synchronous solutions

$$\varphi_s^0 = \sigma_s(\omega t + \alpha_s), \quad x^0 = -\frac{\omega^2}{\omega^2 - p_x^2}\sum_{j=1}^{k}\frac{m_j\varepsilon_j}{M}\cos(\omega t + \alpha_j), \tag{4.15}$$

which depends on k arbitrary parameters α_s. The basic equations (see equations (4.6), chap. 12) from which the parameters $\alpha_1, \ldots, \alpha_k$ of the generating solutions are determined are expressed in this case as:*

$$P_s(\alpha_1, \ldots, \alpha_k) \equiv \frac{\sigma_s}{k_s}\langle(\mu\Phi_s(\dot{\varphi}_s^0, \ddot{x}^0))\rangle = 0$$
$$(s = 1, \ldots, k). \tag{4.16}$$

Averaging expressions (4.13) and (4.15) as well as equations

$$\sigma_s = 1/\sigma_s, \quad \sin\sigma_s\alpha_s = \sigma_s\sin\alpha_s, \quad \cos\sigma_s\alpha_s = \cos\alpha_s;$$
$$\langle\sin(\omega t + \alpha_s)\cos(\omega t + \alpha_s)\rangle = \frac{1}{2}\sin(\alpha_s - \alpha_j),$$
$$\langle\sin(\omega t + \alpha_s)\sin(\omega t + \alpha_j)\rangle = \langle\cos(\omega t + \alpha_s)\cos(\omega t + \alpha_j)\rangle =$$
$$= \frac{1}{2}\cos(\alpha_s - \alpha_j), \tag{4.17}$$

we obtain

$$\langle L_s(\sigma_s\omega) - \sigma_s R_s(\omega)\rangle = L_s(\sigma_s\omega) - \sigma_s R_s(\omega).$$

$$\langle\ddot{x}^0\sin\varphi_s^0\rangle =$$
$$= \frac{\omega^4}{\omega^2 - p_x^2}\sum_{j=1}^{k}\frac{m_j\varepsilon_j}{M}\langle\cos(\omega t + \alpha_j)\sin\sigma_s(\omega t + \alpha_s)\rangle =$$
$$= \frac{1}{2}\frac{\sigma_s\omega^4}{\omega^2 - p_x^2}\sum_{s=1}^{k}\frac{m_s\varepsilon_s}{M}\sin(\alpha_s - \alpha_j).$$
$$\langle g\cos\varphi_s^0\rangle = g\langle\cos(\omega t + \alpha_s)\rangle = 0, \tag{4.18}$$

---

* Here and below the angle brackets< > show averaging of the expression within them for one period by the variable t or τ . The additional round parentheses (which do not play their conventional role), which embrace function designations, show that these functions are calculated for the generating solution.

Consequently, the basic equations (4.16) will be written as

$$P_s(\alpha_1, \ldots, \alpha_k) \equiv \frac{1}{k_s}\left[Z_s(\sigma_s, \omega) - W_s^{(K)}(\alpha_1, \ldots, \alpha_k)\right] = 0$$
$$(s = 1, \ldots, k). \tag{4.19}$$

Here

$$W_s^{(K)}(\alpha_1, \ldots, \alpha_k) = -m_s\varepsilon_s\langle\ddot{x}^0 \sin\varphi_s^0\rangle =$$
$$= -\frac{1}{2}\frac{\omega^4}{\omega^2 - p_x^2}\frac{m_s\varepsilon_s}{M}\sum_{j=1}^{k} m_j\varepsilon_j \sin(\alpha_s - \alpha_j). \tag{4.20}$$
$$Z_s(\sigma_s\omega) = \sigma_s L_s(\sigma_s\omega) - R_s(\omega).$$

The values Z_s and W_s^K, having dimensions of moments (torques), play a considerable role in the further analysis. We will call them, respectively, excessive *moments (torques) of exciters* and *vibratory moments (torques).* The physical sense of these values is discussed in detail in § 5; the sense of index K at W_s will become clear from what is said in subsection 2, § 8.

As it is supposed to be for the solution of internal synchronization problems, equation (4.19) contains only phase differences $\alpha_s - \alpha_j$. Accordingly, one of the phases may be considered as arbitrary, assuming for example, $\alpha_k = 0$. These equations, therefore, contain only $k - 1$ independent unknown phase differences $\alpha_1 - \alpha_k$, $\ldots$, $\alpha_{k-1} - \alpha_k$. At the same time, these equations, generally speaking, permit one to determine beforehand the unknown first approximation of the angular velocity ω magnitude for the synchronous rotation. One independent solution may easily be obtained for this value from the system (4.19), if it is noted that

$$\sum_{s=1}^{k} W_s^{(K)}(\alpha_1, \ldots, \alpha_k) =$$
$$= -\frac{1}{2M}\frac{\omega^4}{\omega^2 - p_x^2}\sum_{s=1}^{k} m_s\varepsilon_s \sum_{j=1}^{k} m_j\varepsilon_j \sin(\alpha_s - \alpha_j) =$$
$$= -\frac{1}{2M}\frac{\omega^4}{\omega^2 - p_x^2}\left(\sum_{s=1}^{k}\sum_{j=1}^{k} m_s\varepsilon_s m_j\varepsilon_j \sin\alpha_s \cos\alpha_j - \right.$$
$$\left. - \sum_{s=1}^{k}\sum_{j=1}^{k} m_s\varepsilon_s m_j\varepsilon_j \cos\alpha_s \sin\alpha_j\right) \equiv 0 \tag{4.21}$$

(the identity of both double summations in the last equation becomes evident if we change in one of them, the index s for j and j for s). Summing up all of the equations (4.19), first multiplied by k_s, and taking into account (4.21), we obtain:

$$\sum_{s=1}^{k} Z_s(\sigma_s, \omega) = 0, \tag{4.22}$$

or according to (4.20):

$$\sum_{s=1}^{k} \sigma_s L_s(\sigma_s\omega) = \sum_{s=1}^{k} R_s^0(\omega). \tag{4.23}$$

These relations express the condition of energy balance in the system and are used to determine ω.

According to what has been said in § 4, chap. 12, *for the existence of synchronous motions of the type analyzed, it is necessary that the equations (4.19) would allow a real solution relative to the phase differences* $\alpha_1 - \alpha_k, \ldots, \alpha_{k-1} - \alpha_k$ *and equation (4.22) would allow positive solutions relative to* ω. *For each such solution there is an actual corresponding unique orbital asymptotically stable synchronous motion which transforms at* $\mu=0$ *into the generating solution (4.15), if for this solution all the roots* $\varkappa$ *of the k-th order algebraic equation*

$$\begin{vmatrix} \frac{\partial (P_1 - P_k)}{\partial \alpha_1} - \varkappa & \frac{\partial (P_1 - P_k)}{\partial \alpha_2} & \cdots & \frac{\partial (P_1 - P_k)}{\partial \alpha_{k-1}} \\ \frac{\partial (P_2 - P_k)}{\partial \alpha_1} & \frac{\partial (P_2 - P_k)}{\partial \alpha_2} - \varkappa & \cdots & \frac{\partial (P_2 - P_k)}{\partial \alpha_{k-1}} \\ \cdots & \cdots & \cdots & \cdots \\ \frac{\partial (P_{k-1} - P_k)}{\partial \alpha_1} & \frac{\partial (P_{k-1} - P_k)}{\partial \alpha_2} & \cdots & \frac{\partial (P_{k-1} - P_k)}{\partial \alpha_{k-1}} - \varkappa \end{vmatrix} = 0 \tag{4.24}$$

would have negative real parts. If even a single root with a positive real part is present, the corresponding solution is unstable. In cases of zero or imaginary roots, there is a need, generally speaking, for an additional investigation.

It can be noted that, besides the abovementioned stability conditions for the considered problem, there is an additional condition. This refers to the presence of roots $\rho^0 = e^{\pm i\lambda_x}$ with unity magnitude in the characteristic equation in variations of the system, which corresponds to the generating system and its (generating) solution (31, 57). However, this additional condition is usually satisfied at a sufficient distance from the resonance zone $\lambda_x \approx \omega$ and besides is independent from the phases $\alpha_1, \ldots, \alpha_k$.

Therefore, equations (4.19), (4.22) and (4.24) are *basic resultative equations of the self-synchronization theory for vibro-exciters in the considered case.*

§ 5) General analysis and different formats for basic expressions for the problem of self-synchronization of vibro-exciters. Vibratory torques, partial angular velocities.

Let us consider in more detail the meaning of basic equations (4.19). First of all, these equations may be formally derived if an assumption is made that that motion of the system follows the law (4.15) corresponding to the generating solution, and if the equations of rotor motion (4.6) are averaged for period $T = 2\pi/\omega$ after having been multiplied by σ_s/k_s. Thus, *the basic equations (4.19) are approximate equations for the equilibrium of the average moments of forces acting on the rotors of vibro-exciters in a steady synchronous motion.*

Referring to analysis of the individual members of equations (4.19), we note that the value $\sigma_s L_s(\sigma_s\omega)$ represents the value of the motor torque, which is calculated in

*It is necessary to consider meaning of the value σ_s as explained in § 4 (see also the footnote on page 39.)

the generating (fundamental) approximation $\dot{\varphi}_s = \dot{\varphi}_s^0 = \sigma_s\omega$. In this case positive direction of the torque $\sigma_s L_s(\sigma_s\omega)$ coincides with the direction of rotation of the s-th rotor* in contrast to torque $L_s(\sigma_s\omega)$. $R_s(\omega)$ resistance torque opposing the rotor motion is also calculated in the fundamental approximation and measured in the direction of rotation of the s-th rotor.

The values $W^{(K)}$ above were called *vibratory torques* (see § 4). *These torques could be defined as additional average torques acting on the unbalanced rotors due to vibrations of their supporting structures.** The vibratory torques can be considered as present as additional average loads for the rotors if $W_s^{(K)} > 0$ or as additional driving torques if $W_s^{(K)} < 0$). The self-synchronization of vibro-exciters, vibratory maintenance of rotation, as well as a number of peculiar phenomena accompanying forced electric or mechanical synchronization processes, are explained by the presence of vibratory torques.

The physical meaning of these effects (which will be described in detail below), consists of an essential "*vibratory coupling*" between the individual rotors, which is caused by vibrations of their common supporting structure. Development of vibratory torques is a manifestation of this coupling. Sometimes the coupling is sufficiently strong and has such a character that the motion of vibro-exciters with a desired combination of phase shifts exists and is stable by itself. In this case, there is no need for artificial connections between rotors, since the self-synchronization effect can be used. In other cases, to assure the existence and stability of the desired motion, either the character of "natural" vibratory coupling has to be changed, or an artificial connection (elements of forced synchronization) has to be introduced.

According to (4.20), the vibratory torque $W_s^{(K)}$, acting on the s-th rotor, may be considered as a sum

$$W_s^{(K)} = \sum_{j=1}^{k} w_{sj}^{(K)}, \tag{5.1}$$

the terms of which

$$w_{sj}^{(K)} = -w_{js}^{(K)} = -\frac{1}{2}\frac{m_s\varepsilon_s m_j\varepsilon_j\omega^4}{M\left(\omega^2 - p_x^2\right)}\sin\left(\alpha_s - \alpha_j\right) \tag{5.2}$$

present "*individual (partial)" vibratory torques* characterizing the action of the s-th exciter on the j-th exciter. In this case, the reciprocity $w_{sj}^{(K)} = -w_{js}^{(K)}$, $w_{ss}^{(K)} = 0$ takes place. As a result (as was mentioned in § 4), the sum of all vibratory torques is equal to zero:

$$\sum_{s=1}^{k} W^{(K)} = \sum_{s=1}^{k}\sum_{j=1}^{k} w_{sj}^{(K)} = 0. \tag{5.3}$$

For exactly this reason, the vibratory torques $W_s^{(K)}$ do not enter into the energy balance equation (4.22): vibratory torques do not change the total energy balance in the system but only redistribute the energy input to the system among individual exciters, thus speeding the "slow" and slowing down the "fast" exciters just to provide synchronous rotation. (The meaning of the terms "fast" and "slow" exciter will

*The concept of vibratory torque was initially introduced by P. L. Kapitsa while solving the problem of a pendulum with a vibrating axis (135).

become somewhat clearer below, after introducing the concept of partial angular velocities for rotors of the exciter.) It should be noted, however, that equalities like (5.1) and (5.2) are not correct for all types of vibro-exciters and not for every type of the system idealization (see § 8).

The largest values of $\widetilde{W}_s^{(K)}$ of vibratory torques $W_s^{(K)}$ (we call them *moduli of vibratory torques*) are a measure of the vibratory coupling between the rotors. From (5.1), it is easy to come to the conclusion that the vibratory torque modulus $\widetilde{W}_s^{(K)}$ is equal to the product of the disturbing force $F_s = m_s\varepsilon_s\omega^2$ generated by the *s*th exciter and the "arm" of this force A_s, which will be called the *effective vibratory amplitude of the rotor axis* of this exciter caused by the action of all other exciters:

$$\widetilde{W}_s^{(K)} = F_s A_s. \tag{5.4}$$

A specific expression for effective vibratory amplitude will be given in § 6 for the case of two vibro-exciters. A more general expression is given in § 12. If static moments of the rotors for all exciters are increased by a certain factor ρ, then according to (4.20), the moduli of vibratory torques $W_s^{(K)}$ would increase by ρ^2 times. For the same frequency shift $\omega^2 - p_x^2$, values $\widetilde{W}_s^{(K)}$ are increasing as ω^4, and at $p_x \approx 0$ (a case of soft suspension) as ω^2 magnitudes of vibratory torques increase with increasing intensity of vibratory motion of the body on which the rotors of exciters are installed. The values $\widetilde{W}_s^{(K)}$ decrease with increasing mass M of the body while maintaining the same frequency shift. At $M \to \infty$, as well as when the rotors are completely balanced ($m_s\varepsilon_s = 0$) corresponding to the stationary body, the values $\widetilde{W}_s^{(K)} = 0$. Thus the vibratory coupling between the rotors is absent. The action of partial vibratory torques $W_s^{(K)}$, i.e., the action of vibratory coupling between the rotors, is similar to the action of a spring between the *s*-th and *j*-th rotors, which generate a torque proportional to sin $|\alpha_s - \alpha_j|$, where $|\alpha_s - \alpha_j|$ is magnitude of the phase shift $|\alpha_s - \alpha_j|$ between angular positions of the rotors. Magnitude of angular stiffness c_w of such an imaginary spring at $|\alpha_s - \alpha_j| = 0$ in the simplest case considered, is equal to

$$\widetilde{w}_{sj}^{(K)} = \left| \frac{\partial w_{sj}^{(K)}}{\partial(\alpha_s - \alpha_j)} \right|_{\alpha_s - \alpha_j = 0} = \frac{1}{2} \frac{m_s\varepsilon_s m_j\varepsilon_j\omega^4}{M\,|\omega^2 - p_x^2|},$$

i.e., to the modulus of a partial vibratory torque. The stable state corresponds, depending on the sign of the difference $\omega - p_s$, either to the angle of phase shift $\alpha_s - \alpha_j = 0$ or the angle $\alpha_s - \alpha_j = \pi$.

The aforementioned simple interpretation of action of the vibratory torques, which at the same time explains the mechanism of self-synchronization, is often quite convenient for qualitative analysis.

Now values ω_s are introduced which are equal to the steady angular velocities of the rotors of vibro-exciters if they are mounted not on a vibrating body but on a stationary foundation, and if for the positive direction of angular velocity, the direction in which the rotor of the *s*-th exciter rotates in the synchronous motion is assumed. We will call these angular velocities ω_s *partial angular velocities of the exciters.*

Angular velocities $\dot{\varphi}_{s0}$ of the rotors of exciters mounted on the stationary foundation are determined from the equations:

$$L_s(\dot{\varphi}_{s0}) = \overset{\circ}{R}(|\dot{\varphi}_{s0}|)\,\mathrm{sgn}\,\dot{\varphi}_{s0}, \tag{5.5}$$

and partial angular velocities, according to the given definition, are:*

$$\omega_s = \dot{\varphi}_{s0}\sigma_s, \tag{5.6}$$

i.e., they comply with equation:

$$\sigma_s L_s(\sigma_s\omega_s) = \overset{\circ}{R}_s(\omega_s)\,\mathrm{sgn}\,\omega_s. \tag{5.7}$$

The partial angular velocity ω_s is positive if the s-th rotor, when mounted on the stationary foundation, rotates in the same direction as it does in the considered synchronous motion.

In the case where the mentioned directions of rotation are opposite, ω_s is negative. In this case, an induction-type motor operates in the generating regime. Partial angular velocities of the vibro-exciters with disconnected motors, or with no motors at all, are equal to zero. It is shown below that the concept of partial angular velocities may be considered as a generalization of the concept of frequencies of self-exciting vibrations.

Let us now assume that all partial angular velocities are equal to each other and positive:

$$\omega_1 = \omega_2 = \ldots = \omega_k = \omega^* > 0. \tag{5.8}$$

Then due to (5.7), the relations

$$L_s(\sigma_s\omega^*) = \sigma_s\overset{\circ}{R}_s(\omega^*) \qquad (s = 1, \ldots, k). \tag{5.9}$$

are correct.

After multiplying by σ_s, we sum all these equations to get:

$$\sum_{s=1}^{k} \sigma_s L_s(\sigma_s\omega^*) = \sum_{s=1}^{k} \overset{\circ}{R}_s(\omega^*). \tag{5.10}$$

The last relation exactly coincides with equation (4.23) for angular velocity ω. Thus, in the case of equal and positive partial angular velocities of the debalanced vibro-exciters, one of the possible values of the angular velocity of synchronous rotation ω is equal to the common partial angular velocity ω^* of vibro-exciters:

$$\omega = \omega^*. \tag{5.11}$$

If, as often happens, equation (4.22) has a unique solution, then the synchronous rotation of debalanced vibro-exciters can occur only with angular velocity equal to the common partial angular velocity ω^*.

*The values $\sigma_s = \pm 1$ in equations (4.4) and many other expressions cannot be considered as parameters of the system in cases of self-synchronization. They only characterize the type of synchronous motion. Thus, if $\sigma_1 = 1, \sigma_2 = 1, \sigma_3 = -1$, etc., then it means that the pattern of the synchronous motion is such that the rotors of the first and second exciters rotate in the positive direction (clockwise), and the rotor of the third exciter rotates in the negative direction, etc.

This result can easily be understood from physical considerations. Vibrations of the structures on which debalanced vibro-exciters are mounted, for all practical purposes, do not cause additional friction losses. Therefore, it is natural to expect that if the rotors of such exciters, while mounted on a stationary foundation, were rotating with angular velocities of the same magnitudes, they would rotate with the same average angular velocities if mounted on the vibrating body. Of course, the situation can change, as in the case of so-called *planetary vibro-exciters* in which energy consumption to overcome rolling resistance for roller rotation depends significantly on the character of the oscillatory motion of the supporting body. (See subsection 5, § 8.)

If equation (5.11) is satisfied, i.e., at equal positive partial velocities of the exciters, then, due to equations (5.9),

$$Z_s(\sigma_s, \omega) = 0 \qquad (s = 1, \ldots, k), \tag{5.12}$$

This means that all additional torques are becoming zero. Thus the basic equations (4.19) take the form:

$$P_s(\alpha_1, \ldots, \alpha_k) \equiv -\frac{1}{k_s} W_s^{(K)}(\alpha_1, \ldots, \alpha_k) \equiv$$

$$\equiv \frac{1}{2k_s} \frac{\omega^4}{\omega^2 - p_x^2} \frac{m_s \varepsilon_s}{M} \sum_{j=1}^{k} m_j \varepsilon_j \sin(\alpha_s - \alpha_j) = 0 \qquad (s = 1, \ldots, k). \tag{5.13}$$

Thus, the generating (fundamental, first approximation) phases in this case are determined from the condition that vibratory torques $W_s^{(K)}$ are equal to zero.

It is easy to see that equations (5.13) allow for the solution in the form:

$$\alpha_s = q_s^* \pi + \alpha_0 \qquad (s = 1, \ldots, k),$$

where each of the numbers q_s^* may be zero or one, and α_0 is an arbitrary constant. Below, in § 6 and § 7, it will be shown that, in many important specific cases, as well as under sufficiently general assumptions, there is one solution among these solutions which corresponds to the stable, at least in a local sense, synchronous motion. It will prove that, in the case of equal and positive partial angular velocities of the vibro-exciters, their synchronization is certainly possible. According to the terminology introduced in § 1, chap. 1, one can say that, in these cases, vibro-exciters have a tendency toward self-synchronization. From continuity considerations, it follows that this conclusion would also apply when there are reasonably small differences in partial velocities. Actually, as will be shown in § 6, self-synchronization under certain conditions would also take place for vastly different values of partial velocities of vibro-exciters.

When the partial velocities ω_s are positive and do not differ very much from one another, the expressions for excess torques $Z_s(\sigma_s, \omega)$ may be considerably simplified. Actually, in this case, the linear relations in (4.10) are quite accurate, where

$$Z_s(\sigma_s, \omega_s) = Z_s(\sigma_s, \omega) - k_s(\omega_s - \omega), \tag{5.14}$$

Here, as before [see equations (4.11) and (4.13)],

$$k_s = -\frac{dZ_s(\sigma_s, \omega)}{d\omega} = -\sigma_s \frac{dL_s(\sigma_s \omega)}{d\omega} + \frac{dR_s^{\circ}(\omega)}{d\omega} =$$

$$= -\frac{dL_s(\dot{\varphi}^0)}{d\dot{\varphi}_s^0} + \frac{dR_s^{\circ}}{d\omega} = k_s^* + k_s^{\circ}.$$

However, at $\omega_s > 0$ ($s = 1, \ldots, k$), according to (4.20), the equation (5.7) for partial velocity ω_s may be written as

$$Z_s(\sigma_s, \omega_s) = 0. \tag{5.15}$$

Therefore, from equations (5.14) it follows:

$$Z_s(\sigma_s, \omega) = k_s(\omega_s - \omega). \tag{5.16}$$

The obtained correlation allows one to write the basic equations (4.19) in the form:

$$P_s(\alpha_1, \ldots, \alpha_k) \equiv (\omega_s - \omega) - \frac{1}{k_s} W_s(\alpha_1, \ldots, \alpha_k) = 0 \qquad (s = 1, \ldots, k) \tag{5.17}$$

and to obtain from the balance energy equation (4.22) the following simple expression for the approximate value of the angular velocity of synchronous rotation ω expressed through partial angular velocities:

$$\omega = \frac{k_1\omega_1 + \ldots + k_k\omega_k}{k_1 + \ldots + k_k}. \tag{5.18}$$

Thus, the angular velocity ω is equal to the weighted mean value of partial velocities ω_s. Positive overall damping coefficients k_s play the role of weighting coefficients. Hence, in particular, it follows that the angular velocity of the synchronous rotation is not lower than the smallest and not higher than the largest of the partial velocities:

$$\mathrm{Inf}\,[\omega_1, \ldots, \omega_k] \leqslant \omega \leqslant \mathrm{Sup}\,[\omega_1, \ldots, \omega_k]. \tag{5.19}$$

Finally, we will exhibit an expression for vibratory torques $W_s^{(K)}$ which follows, for example, from formulas (8.35) of this chapter:

$$W_s^{(K)} = \frac{\partial \Lambda^{(\mathrm{I})}}{\partial \alpha_s} \qquad (s = 1, \ldots, k). \tag{5.20}$$

Here

$$\Lambda^{(\mathrm{I})} = \langle (L^{(\mathrm{I})}) \rangle = \langle (T^{(\mathrm{I})} - \Pi^{(\mathrm{I})}) \rangle \tag{5.21}$$

is averaged for the period $T = 2\pi/\omega$, the value of Lagrange function $L^{(1)} = T^{(1)} - \Pi^{(1)}$

of the supporting body B_0 (vibrating platform), which is calculated for the generating approximation (4.15), and

$$T^{(\mathrm{I})} = \frac{1}{2} M\dot{x}^2, \quad \Pi^{(\mathrm{I})} = \frac{1}{2} c_x x^2 \tag{5.22}$$

are the kinetic and potential energy of the platform, respectively. After calculation using the expressions (4.15) for x^0 and equations (4.17), we find:

$$\begin{aligned}\Lambda^{(\mathrm{I})} &= \left\langle \frac{1}{2} M (\dot{x}^0)^2 - \frac{1}{2} c_x (x^0)^2 \right\rangle = \frac{1}{2} [M \langle (\dot{x}^0)\rangle^2 - c_x \langle (x^0)^2 \rangle] = \\ &= \frac{1}{2} \frac{\omega^4}{\omega^2 - p_x^2} \left[M\omega^2 \left\langle \sum_{s=1}^{k} \sum_{j=1}^{k} \frac{m_s \varepsilon_s}{M} \sin(\omega t + \alpha_s) \sin(\omega t + \alpha_j) \right\rangle - \right. \\ &\left. - c_x \left\langle \sum_{s=1}^{k} \sum_{j=1}^{k} \frac{m_s \varepsilon_s}{M} \frac{m_j \varepsilon_j}{M} \cos(\omega t + \alpha_s) \cos(\omega t + \alpha_j) \right\rangle \right] = \\ &= \frac{M}{4} \frac{\omega^4}{\omega^2 - p_x^2} \sum_{s=1}^{k} \sum_{j=1}^{k} \frac{m_s \varepsilon_s}{M} \frac{m_j \varepsilon_j}{M} \cos(\alpha_s - \alpha_j). \end{aligned} \tag{5.23}$$

This expression allows us to directly check the accuracy of equations (5.20) obtained in § 8 for a considerably more general case. Considering these relations, the basic equations are written as

$$P_s(\alpha_1, \ldots, \alpha_k) \equiv \frac{1}{k_s} \left[Z_s(\sigma_s, \omega) - \frac{\partial \Lambda^{(\mathrm{I})}}{\partial \alpha_s} \right] = 0 \quad (s = 1, \ldots, k), \tag{5.24}$$

and the somewhat less general equations (5.17) as

$$P_s(\alpha_1, \ldots, \alpha_k) = (\omega_s - \omega) - \frac{1}{k_s} \frac{\partial \Lambda^{(\mathrm{I})}}{\partial \alpha_s} = 0 \quad (s = 1, \ldots, k). \tag{5.25}$$

§ 6) Some important particular cases.

1. A case of two vibro-exciters. The phenomenon of vibratory support of rotation. Many important features of self-synchronization of vibro-exciters can be followed on a simple case of two exciters. At $k = 2$, the basic equations in (4.19) are:

$$\begin{aligned} P_1 &\equiv \frac{1}{k_1} \left[Z_1(\sigma_1, \omega) + \frac{1}{2} \frac{\omega^4}{\omega^2 - p_x^2} \frac{m_1 \varepsilon_1 m_2 \varepsilon_2}{M} \sin(\alpha_1 - \alpha_2) \right] = 0, \\ P_2 &\equiv \frac{1}{k_2} \left[Z_2(\sigma_2, \omega) - \frac{1}{2} \frac{\omega^4}{\omega^2 - p_x^2} \frac{m_1 \varepsilon_1 m_2 \varepsilon_2}{M} \sin(\alpha_1 - \alpha_2) \right] = 0. \end{aligned} \tag{6.1}$$

By introducing notations:

$$\begin{aligned} &A_1^{(x)} = \frac{m_2 \varepsilon_2 \omega^2}{M |\omega^2 - p_x^2|}, \quad A_2^{(x)} = \frac{m_1 \varepsilon_1 \omega^2}{M |\omega^2 - p_x^2|}, \quad \alpha_1 - \alpha_2 = \alpha; \\ &A_s = A_s^{(x)}/2, \qquad F_s = m_s \varepsilon_s \omega^2 \qquad (s = 1, 2); \\ &\widetilde{W}^{(K)} = \widetilde{W}_1^{(K)}(\omega) = \widetilde{W}_2^{(K)}(\omega) = \frac{1}{2} \frac{\omega^4}{|\omega^2 - p_x^2|} \frac{m_1 \varepsilon_1 m_2 \varepsilon_2}{M} = F_1 A_1 = F_2 A_2, \end{aligned} \tag{6.2}$$

we will present these equations in the following more compact form:

$$P_1 \equiv \frac{1}{k_1}\left[Z_1(\sigma_1, \omega) + \widetilde{W}^{(K)}(\omega)\ \operatorname{sgn}\left(\omega^2 - p_x^2\right)\sin\alpha\right] = 0,$$
$$P_2 \equiv \frac{1}{k_2}\left[Z_2(\sigma_2, \omega) - \widetilde{W}^{(K)}(\omega)\ \operatorname{sgn}\left(\omega^2 - p_x^2\right)\sin\alpha\right] = 0. \quad (6.3)$$

The value $\widetilde{W}^{(K)}(\omega)$ presents, in the considered case, the largest possible value *(modulus) for vibratory torques* $W_1^{(K)}$ *and* $W_2^{(K)}$. The values A_1 and A_2 are *effective vibration amplitudes of the axes of the first and second rotors respectively*. The value $A_1^{(x)}$ is the amplitude of a steady-state vibration of the axis of the first rotor (together with the platform on which it is mounted) under the influence of centrifugal force $F_2 = m_2\varepsilon_2\omega^2$, which is developed by the second rotor. The value $A_2^{(x)}$ is the vibration amplitude of the axis of the second rotor under the influence of centrifugal force $F_1 = m_1\varepsilon_1\omega^2$ developed by the rotation of the first rotor. Thus, the effective vibration amplitude of a rotor axis in the considered case of rectilinear vibrations of the platform, is equal to one half of the vibration amplitude of the axis of the rotor which is excited by the other rotor.

Let us analyze possible synchronous motions. We find from equations (6.3):

$$Z_1(\sigma_1, \omega) + Z_2(\sigma_2, \omega) = 0,$$

$$\sin\alpha = -\frac{Z_1(\sigma_1, \omega)}{\widetilde{W}^{(K)}(\omega)\operatorname{sgn}\left(\omega^2 - p_x^2\right)} = \frac{Z_2(\sigma_2, \omega)}{W^{(K)}(\omega)\operatorname{sgn}\left(\omega^2 - p_x^2\right)}. \quad (6.4)$$

The first of these equations presents a particular case of equation (4.22) and serves to find the first approximation of angular velocity ω for the synchronous rotation. The second equation serves to determine phase shift $\alpha = \alpha_1 - \alpha_2$ for rotor rotations in possible synchronous motions. If inequality

$$|Z_1(\sigma_1, \omega)| = |Z_2(\sigma_2, \omega)| < \widetilde{W}^{(K)}(\omega), \quad (6.5)$$

is satisfied, i.e., under the condition that the moduli of excessive torques do not exceed the modulus of the vibratory torque, then equation (4.22) has two essentially different real solutions:

$$\alpha = \alpha^{(1)} = \chi, \quad \alpha = \alpha^{(2)} = \pi - \chi, \quad (6.6)$$

where

$$-\frac{\pi}{2} < \chi = -\arcsin\frac{Z_1(\sigma_1, \omega)}{\widetilde{W}^{(K)}(\omega)\operatorname{sgn}\left(\omega^2 - p_x^2\right)} =$$
$$= \arcsin\frac{Z_2(\sigma_2, \omega)}{\widetilde{W}^{(K)}(\omega)\operatorname{sgn}\left(\omega^2 - p_x^2\right)} < \frac{\pi}{2}. \quad (6.7)$$

It is obvious that the solutions of this equation, which differ by $2\pi n$, where n is an integer, can be considered to be identical.

Therefore, inequality (6.5) is the condition at which the self-synchronization effect for vibro-exciters is possible.

Synchronous motion corresponding to the first of the solutions (6.6) will be called *synphase*, and that corresponding to the second solution will be called *antiphase*. Let us consider the stability of these motions. Equation (4.24) is linear in this case. Considering (6.3), the following condition of stability is obtained from (4.24):

$$\varkappa = \frac{\partial (P_1 - P_2)}{\partial \alpha_1} = \frac{\partial (P_1 - P_2)}{\partial \alpha} =$$

$$= \frac{k_1 + k_2}{k_1 k_2} \widetilde{W}^{(K)}(\omega) \operatorname{sgn}(\omega^2 - p_x^2) \cos \alpha < 0. \quad (6.8)$$

Since $k_s > 0$ and $\widetilde{W}^{(K)}(\omega) > 0$, it then follows that *before the resonance* ($\omega < p_x$), *the synphase motion is stable and the antiphase motion is unstable; after the resonance* ($\omega > p_x$), *on the contrary, synphase motion is unstable and the antiphase motion is stable.*

Let us consider in more detail the condition of existence of the self-synchronization effect for vibro-exciters. As in the general case, this condition is certainly satisfied if the partial velocities ω_1 and ω_2 are positive and equal, since excess torques $Z_1(\sigma_1, \omega)$ and $Z_2(\sigma_2, \omega)$ become zero. In the case of positive and not very different partial velocities, the synchronous angular velocity ω according to (5.18) will be:

$$\omega = \frac{k_1\omega_1 + k_2\omega_2}{k_1 + k_2}, \quad (6.9)$$

and therefore expression (5.16) for the excess torque $Z_1(\sigma_1, \omega)$ may be written as:

$$Z_1(\sigma_1, \omega) = k_1(\omega_1 - \omega) = \frac{k_1 k_2}{k_1 + k_2}(\omega_1 - \omega_2). \quad (6.10)$$

As a result, the synchronization condition (6.5) becomes

$$|\omega_2 - \omega_1| < \frac{k_1 + k_2}{k_1 k_2} \widetilde{W}^{(K)}(\omega). \quad (6.11)$$

This relation plays a significant role in the theory of vibratory devices with self-synchronizing vibro-exciters (see § 9). From the condition in (6.11), one can directly see that synchronization is definitely possible if the partial angular velocities ω_1 and ω_2, of vibro-exciters, while being positive, differ little from one another. At the same time, the self-synchronization effect can also take place at the very essential differences in partial angular velocities, such as even in a case where the motor of one exciter is deenergized, i.e., when $L_1(\sigma_1, \omega) = 0$, $Z_1(\sigma_1, \omega) = -R_1(\omega)$, and thus the partial velocity $\omega_1 = 0$. The condition which makes this effect possible is called the *effect of vibratory maintenance of rotation of an unbalanced rotor.** According to (6.5), this will be inequality:

$$\overset{\circ}{R}_1(\omega) < \widetilde{W}^{(K)}(\omega), \quad (6.12)$$

*This effect is used in the well-known game-exercise "hoola-hoop." For more serious applications, see the book (57) and also § 15.

It formulates the requirement that the resistance torque for the rotor of a disconnected exciter should not exceed the magnitude of vibratory torque $\widetilde{W}^{(K)}(\omega)$. The latter torque can be considered as a limiting moment which can be transferred to the rotating unbalanced rotor by means of vibrations of its axis. The value $N(\omega)=\widetilde{W}^{(K)}(\omega)\omega$ is a corresponding limiting power which can be transferred by means of vibrations to assure self-synchronization. *It is remarkable that the value of this power at practically realizable parameters of vibratory devices is very large. Thus, for example, at a static moment of unbalanced rotor* $m\varepsilon = 10$ *kg, where* M *is the effective amplitude* $A_1 = 0.25 \times 10^{-2}$ *m, and frequency* $\omega = \pi n/30 = \pi\ 3000/30 = 314$ *sec*$^{-1}$*, we obtain, according to (6.2):*

$$N_{\max} = \widetilde{W}^{(K)}(\omega)\,\omega = FA_1\omega = m\varepsilon A_1\omega^3 =$$
$$= 10\cdot 0{,}25\cdot 10^{-2}\cdot 314^3 \approx 0{,}8\cdot 10^6 \text{Н}\cdot\text{м/с} \approx 800\text{кВт}\ (!)$$

Jumping ahead, one can note that the similar value in cases of self-synchronization of oscillating objects (for example, pendulums), at other equal conditions, turns out to be considerably less. (See § 3, chap. 5.)

This example essentially explains wide possibilities for the practical utilization of the self-synchronization phenomenon in vibratory machines and devices where rotating unbalanced rotors are used.

Condition (6.12) becomes an especially simple one when resistance to rotor rotation is largely due to friction in bearings. For roller bearings, $R_1(\omega) = {}^1\!/_2 fF_1 d$, where f is the bearing friction coefficient and d the diameter of the inner race, and the condition, considering expression (6.2) for $\widetilde{W}^{(K)}(\omega)$, is reduced to $2A_1/d > f$.

Since usually $0.001 \le f \le 0.01$, it is obvious that rotor rotation can be supported by vibrations with amplitudes considerably less than the size of the rotor bearings.

These results allow us to consider the phenomenon of the vibratory maintenance of rotation as a special case of the self-synchronization effect. This phenomenon is investigated in more detail in § 12, where the problem is solved as a problem of rotation entrainment, and references are given.

2. A case of almost identical vibro-exciters. Let us consider in more detail an important case for applications of vibro-exciters whose partial angular velocities (supposedly positive) differ from one another so little that the basic equations may be written in the form of (5.13). As was already mentioned in § 5, these equations allow for the set of solutions:

$$\alpha_s = \overset{*}{q}_s\pi + \alpha_0 \qquad (s = 1, \ldots, k), \tag{6.13}$$

where $\overset{*}{q}_s$ are numbers, each of which can be zero or one. We will call these solutions *solutions of the first group*, with solutions in which m of the numbers $\overset{*}{q}_s$ that equal one are called solutions of the (m) type. Let us analyze the stability of such solutions for a case when parameters of exciters $m_s\varepsilon_s$ and k_s are almost identical. In this case, equation (4.24) may be presented in the form (We assume

$$m_1\varepsilon_1 = m_2\varepsilon_2 = \ldots = m_k\varepsilon_k = m\varepsilon,\ k_1 = k_2 = \ldots = k_k = k_0,\ \overset{*}{q}_1 =$$
$$= \overset{*}{q}_2 = \ldots = \overset{*}{q}_m = 1,\ \overset{*}{q}_{m+1} = \ldots = \overset{*}{q}_k = 0)$$

$$\begin{vmatrix} \overbrace{\begin{matrix} A_1-u & 2 & \dots & 2 \\ 2 & A_1-u & \dots & 2 \\ \dots & \dots & \dots & \dots \\ 2 & 2 & \dots & A_1-u \\ 0 & 0 & \dots & 0 \\ 0 & 0 & \dots & 0 \\ \dots & \dots & \dots & \dots \\ 0 & 0 & \dots & 0 \end{matrix}}^{m \text{ columns}} & \overbrace{\begin{matrix} -2 & -2 & \dots & -2 \\ -2 & -2 & \dots & -2 \\ \dots & \dots & \dots & \dots \\ -2 & -2 & \dots & -2 \\ A_2-u & 0 & \dots & 0 \\ 0 & A_2-u & \dots & 0 \\ \dots & \dots & \dots & \dots \\ 0 & 0 & \dots & A_2-u \end{matrix}}^{k-m-1 \text{ columns}} \end{vmatrix} = 0, \tag{6.14}$$

where

$$u = -\varkappa \frac{2k_0 M}{(m\varepsilon)^2 \omega^4} \left(\omega^2 - p_x^2\right),$$
$$A_1 = -2m + k + 2, \quad A_2 = 2m - k. \tag{6.15}$$

Equation (6.14) is easy to solve. At $m = 1, \dots, k-1$, it has the $m-1$ multiple root $u_1 = A_1 - 2 = -2m + k$, the $k-m-1$ multiple root $u_2 = A_2 = 2m - k$, and the simple root $u_3 = A_1 + 2(m-1) = k$. At $m = 0$ [and also at $m = k$, since the solution (0) is allowed not to be distinguished from (k)], there is a $k-1$ multiple root $u = -k$. Hence, it follows that the stable solution at the region before the resonance ($\omega < p_x$) is only a solution of the (0) type, i.e., a synchronous motion close to synphaseous. (Positive values of $\varkappa$ are absent only for this solution.) At the region after resonance ($\omega > p_x$) stable motions are only possible in the special case of an even number of exciters $k = 2l$. Specifically, stability is characteristic only for solutions of the (l) type because the $(2l-2)$ multiple root $u = u_1 = u_2 = 0$ and simple root $u = k$ correspond to these solutions. In the case of two exciters ($k = 2$, $l = 1$), the zero root is absent and motion is actually stable at $\omega > p_x$. This is in accordance with the result obtained in subsection 1 of this section. In the case of $k > 2$, due to the presence of zero roots $\varkappa$, one needs an additional investigation to solve the stability problem.

Thus, for almost identical exciters, the tendency toward synchronization takes place at least in the regime before resonance for any number of exciters and in the regime after resonance at least for two exciters.

Returning to the general case of exciters with almost equal positive partial velocities, one can see that equations (5.13) are satisfied if the following equalities are satisfied:

$$\sum_{s=1}^{k} m_s \varepsilon_s \cos \alpha_s = 0, \quad \sum_{s=1}^{k} m_s \varepsilon_s \sin \alpha_s = 0. \tag{6.16}$$

We will call the corresponding solutions of equations (5.13) *solutions of the second group*. By considering them, we see that correlations (6.16) are equivalent to the requirement that k vectors e_s, having magnitudes $m_s \varepsilon_s$ and directed under angles α_s to the axis Ox in the plane xOy, would form a closed polygon (Fig. 8). In this case, due to the arbitrariness of one of the phases α_s, it can be assumed that

$$\alpha_k = 0, \tag{6.17}$$

i.e., the vector e_k is directed along the axis O_x. It is evident that for $k = 2$ and $k = 3$, only one such polygon can exist (not considering permutation of exciter numbers), and for $k > 3$ an unlimited number of polygons* can exist. This means that, for more than three exciters, the phase differences $\alpha_s - \alpha_k$ for the second-group solutions are not uniquely determined from the equation (5.13). In order to find the phase differences, according to remark 8, § 7, chap. 10, one must consider the higher order approximations. In this case, additional investigations would be needed also for the problem of existance and the stability of synchronous motions since, in such a case, equation (4.24) will have zero roots. Such a situation is the consequence of the certain degeneration of the considered system. This degeneration would disappear if the inequality of the partial velocities of exciters and some other factors were considered.

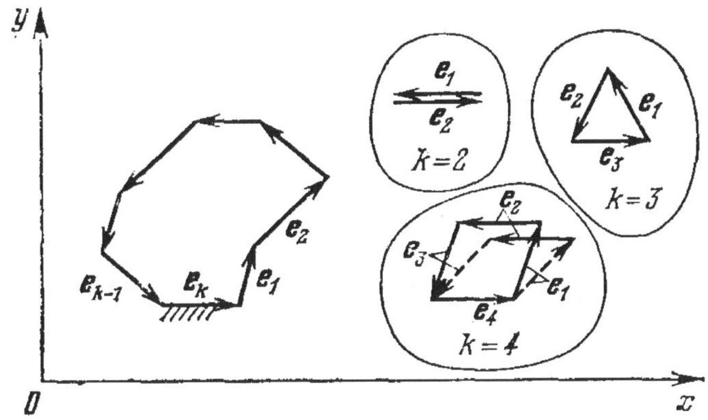

FIG. 8.

In cases of two and three exciters, the problem is completely solved on the basis of equations (5.13) and (4.24). Moreover, in the case of two exciters, the second group solution $\alpha_1 = \pi$, $\alpha_2 = 0$ coincides with solution (1) of the first group. As was shown in subsection 1, this solution corresponds to the antiphase synchronous motion of exciters, is stable in the region after resonance ($\omega > p_x$) and unstable in the region before resonance ($\omega < p_x$).

§ 7) Integral stability criterion (extremal property) of synchronous motions and the trend of vibro-exciters towards synchronization. Some important conclusions.

It follows from the basic equations as written in the format (5.24) that they also can be written as

$$P_s \equiv -\frac{1}{k_s}\frac{\partial}{\partial \alpha_s}\left(\Lambda^{(I)} - \sum_{s=1}^{k} Z_s(\sigma_s, w)\,\alpha_s\right) = 0$$

$$(s = 1, \ldots, k), \tag{7.1}$$

*For the existence of such polygons to be possible, the conditions must be satisfied, in which the values $m_1\varepsilon_1, \ldots, m_k\varepsilon_k$ can be the side lengths of k-polygon. For example, at $k = 2$, it must be $m_1\varepsilon_1 = m_2\varepsilon_2$; at $k \geq 3$, it must be

$$m_s\varepsilon_s \leqslant \sum_{\substack{j=1 \\ j\neq s}}^{k} m_j\varepsilon_j.$$

This means that, in this case, in accordance with § 2 of chap. 2, § 8 of chap. 10, and § 4 of chap. 12, there exists the so-called *potential function*

$$D = \Lambda^{(I)}(\alpha_1, \dots, \alpha_k) - B(\alpha_1, \dots, \alpha_k), \tag{7.2}$$

where

$$B(\alpha_1, \dots, \alpha_k) = \sum_{s=1}^{k} Z_s(\sigma_s, \omega)\,\alpha_s. \tag{7.3}$$

It is characteristic for the function D, that stable synchronous motions would correspond to its rough minimum points. This is similar to correspondence between equilibrium positions of a conservative mechanical system and minima of its potential energy. The "rough minimum" concept is explained in § 8, chap. 10. In (7.2), (7.3) B denotes the so-called *potential of averaged nonconservative forces.* For identical partial velocities of vibro-exciters, when excess torques $Z_s = 0$, there will be

$$D = \Lambda^{(I)}(\alpha_1, \dots, \alpha_k), \tag{7.4}$$

This means that stable synchronous motions are characterized by a minimum of the Lagrange function of the oscillatory part of the system, averaged during the oscillation period $T = 2\pi/\omega$ and computed in the fundamental approximation.

At least three important conclusions can be made from the aforesaid:

1. Since $\Lambda^{(I)}$ is a periodic function of α_s [see (5.23)], then, according to § 2, chap. 2 and § 9, chap. 10, with identical or quasi-identical partial velocities of vibro-exciters, minima of the function D always exist.* Consequently, locally stable synchronous motions of the vibro-exciters also always exist. In other words, under these conditions there exists a trend toward synchronization of vibro-exciters. It was demonstrated in the preceding paragraph for a special case of two exciters.

2. Practical solution of the problem of synchronization of vibro-exciters can be performed by using an integral criterion (extremal property) of stable synchronous motions. It can be done as follows:

a) Equation of oscillatory motion (or, in more complex cases—equations) for the supporting body B_0 (platform) is constructed. The body is acted upon by exciting forces from vibro-exciters when rotors of the latter are uniformly rotating as $\varphi_s = \sigma_s(\omega t + \alpha_s)$, i.e., with the common angular velocity ω and with initially unknown initial phases α_s. This is the last equation in (4.14), in which $\varphi_s = \sigma_s(\omega t + \alpha_s)$ is substituted.

b) Solution x^0 of this equation, corresponding to steady forced vibrations of the platform, is constructed, see (4.15).

c) This solution x^0 is substituted into the Lagrange function expression $L^{(I)} = T^{(I)} - \Pi^{(I)}$, corresponding to vibratory motion of the supporting body, see (5.22). Then averaging along the period $T = 2\pi/\omega$ is performed, that is, function $\Lambda^{(I)}$ is calculated. As a result, an expression similar to (5.23) is obtained.

d) Expression (7.2) for the potential function D is constructed. Phase shifts

*It is easy to show that minima of the function D are rough minima in this case. Exceptions from this statement are some degenerated cases, especially such cases when the minima are nonrigorous, thus nonrough (see subsection 2, § 6). In such cases, stability conditions obtained from the requirement of minimum for the function D would be the only necessary ones. An additional analysis is required here to obtain sufficient conditions.

α_s,which would comply with stable synchronous motions, are found from minima conditions for the function D.

This solution algorithm remains also for much more complex problems of synchronization of vibro-exciters. In many cases, it is more appropriate than other methods (see § 10).

3. Let us assume that elastic connections between the supporting body and the stationary foundation are so soft that the average potential energy $<(\Pi^{(I)})>$ is negligible in comparison with the average kinetic energy $<(T^{(I)})>$. It is easy to show that this assumption is equivalent to the requirement $\omega \gg p_x$, i.e., the motion occurs far in the after-resonance region. With this assumption, for exciters with identical partial velocities their two stable synchronous motions would correspond to a minimum of the averaged motion during one period of kinetic energy of the platform. In other words, phase values of the exciters during their self-synchronization would be such that the platform vibrations are minimized in terms of average kinetic energy of vibrations. As a special case, for equal static moments $m_s\varepsilon_s$ of the exciters, and when there is an even number of exciters, $k = 2k'$, there would always exist a stable synchronous motion for which one half (i.e., k') of the exciters are in opposite phase relative to the other half. In such a case, of course, excitation forces from exciters are mutually compensated, and vibrations of the platform are essentially canceled. These interesting effects, important for some applications, are discussed in subsection 4, § 8, and subsections 6 and 7, § 14.

Another possible assumption is of a very rigid elastic connection, so that kinetic energy is negligible relative to potential energy ($\omega \ll p_x$, before-resonance region, far from the resonance). In this case, for the same conditions, average potential energy for a stable synchronous motion is maximal. As a result, synphaseous motion of all the exciters would be stable, in which case excitation forces are summing up and vibrations of the supporting body are the most intensive. It can be seen that these conclusions are fully equivalent to ones obtained in § 6 for the two-exciters case.

These results are further expanded for both more complex cases of synchronization of vibro-exciters in subsection 4, § 8, and for broad classes of other dynamic objects in §§ 4–7, chap. 12.

§ 8) Expansion of the problem of synchronization of vibroexciters.

1. Description of the system and of the basic assumptions. Features of the self-synchronization of vibro-exciters as considered above for the simplistic system can be generalized for much more complex cases.

As before, the problem of simple (not multiple) synchronization is considered. However, it will be assumed, as in Fig. 5, that the exciters are attached to an arbitrary system of rigid bodies $B_{01}, \ldots, B_{0n}$ (supporting bodies). These are connected between themselves and with stationary foundation *1* by certain "lumped" geometric connections *2*, as well as by elastic *3* and damping *4* connections. The vibratory part of the system may have any number of degrees of freedom and may not necessarily be a linear system. However, it is assumed that nonlinear characteristics are sufficiently smooth, thus excluding, for example, compacting of the supporting bodies (Fig. 9). Vibro-exciters $B_1, \ldots, B_k$ are not assumed to be simple debalanced exciters any more. They can be of any inertial type, representing bodies which can perform periodic motions in relation with one of the supporting bodies. Excitation forces from such exciters (which will be called *generalized mechanical exciters*)

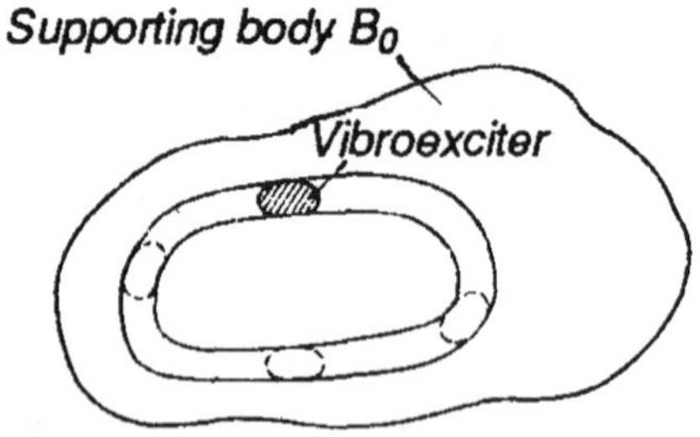

FIG. 9.

are not necessarily pure harmonic, even with the stationary supporting body. However, the majority of commercially available industrial mechanical vibro-exciters generate either harmonic excitation, or excitation consisting of two or three harmonics. Some typical exciters are sketched in Fig. 10. Fig. 10a presents an (already discussed) basic *debalanced exciter*; Fig. 10b, c show *internal planetary exciters*; Fig. 10d shows an *external planetary exciter* (for their descriptions see subsection 5, § 8). Fig. 10e depicts a so-called *piston exciter* in which the excitation force is generated by mass in performing a reciprocal motion in the cavity of the housing. The most frequently used system for generating a unidirectional excitation force is using two debalanced exciters whose rotors are connected by gears with a 1:1 transmission ratio, Fig. 10f. Such an exciter is sometimes called *self-balancing*. However, as will be shown, the same effect can be more effectively achieved by utilizing the self-synchronization effect. Fig. 10g, h show a *biharmonic exciter* of nondirectional (Fig. 10g) and directional (Fig. 10h) action.

The position of the inertial element (rotor, piston, etc.) in each of the exciters shown in Figs. 9 and 10, relative to the supporting body, could be described by just one generalized coordinate φ_s. It will be assumed that this coordinate can be selected in such a manner that its variation during the considered synchronous motion presents a quasi-uniform rotation.

$$\varphi_s = \sigma_s \left[\omega t + \alpha_s + \mu\psi_s^* (\omega t, \mu) \right]. \tag{8.1}$$

Here, as before, ω, α_s are constants, $\sigma_s = \pm 1$ and $\mu\psi_s^*(\omega t, \mu)$ is an additive term characterizing small oscillations in relation to the uniform rotation.

It is also assumed that shafts or inertial elements might be connected by the supported connections (not shown in Fig. 9). They might have the appearance of torsion (elastic) shafts, couplings, electric shaftlike connections, and also so-called centrifugal-inertial connections and couplings (57, 242, 256), as shown in Fig. 11 in several embodiments. In Fig. 11 links *1* are connected with one of the shafts (rotors), and links *2*, with the other one. Figure 11a–f show coaxial, while Fig. 11 g, h show noncoaxial shafts. Designs in Fig 11f–i represent connections with "associated degrees of freedom," while those in Fig. 11a–e represent connection without "associated degrees of freedom." For the former type, inertial components of the connections can be moved while positions of links *1* and *2* are fixed; with the latter type they cannot be moved.

The described system belongs to a class of *constrained orbital systems*, specifically such systems in which supported bodies have one degree of freedom relative to bodies comprising the supporting system (see § 6, chap. 12). In such a system, the problem of self-synchronization of vibro-exciters can be considered as a problem of synchronization in a system with quasi-uniform rotations (see § 4, chap. 12).

Equations of motion with the formulated assumptions can be written in the format of equations (4.2), (4.3) from chap. 12. In this case, equations (4.2) for rotational coordinates $\varphi_1, \ldots, \varphi_k$ are equations of motion for inertial components of vibro-exciters, and equations (4.3) for vibratory coordinates $u_1, \ldots, u_v$ are equations of motion for supporting and supported connections [compare them with equations (4.12) derived for a special case of the considered system].

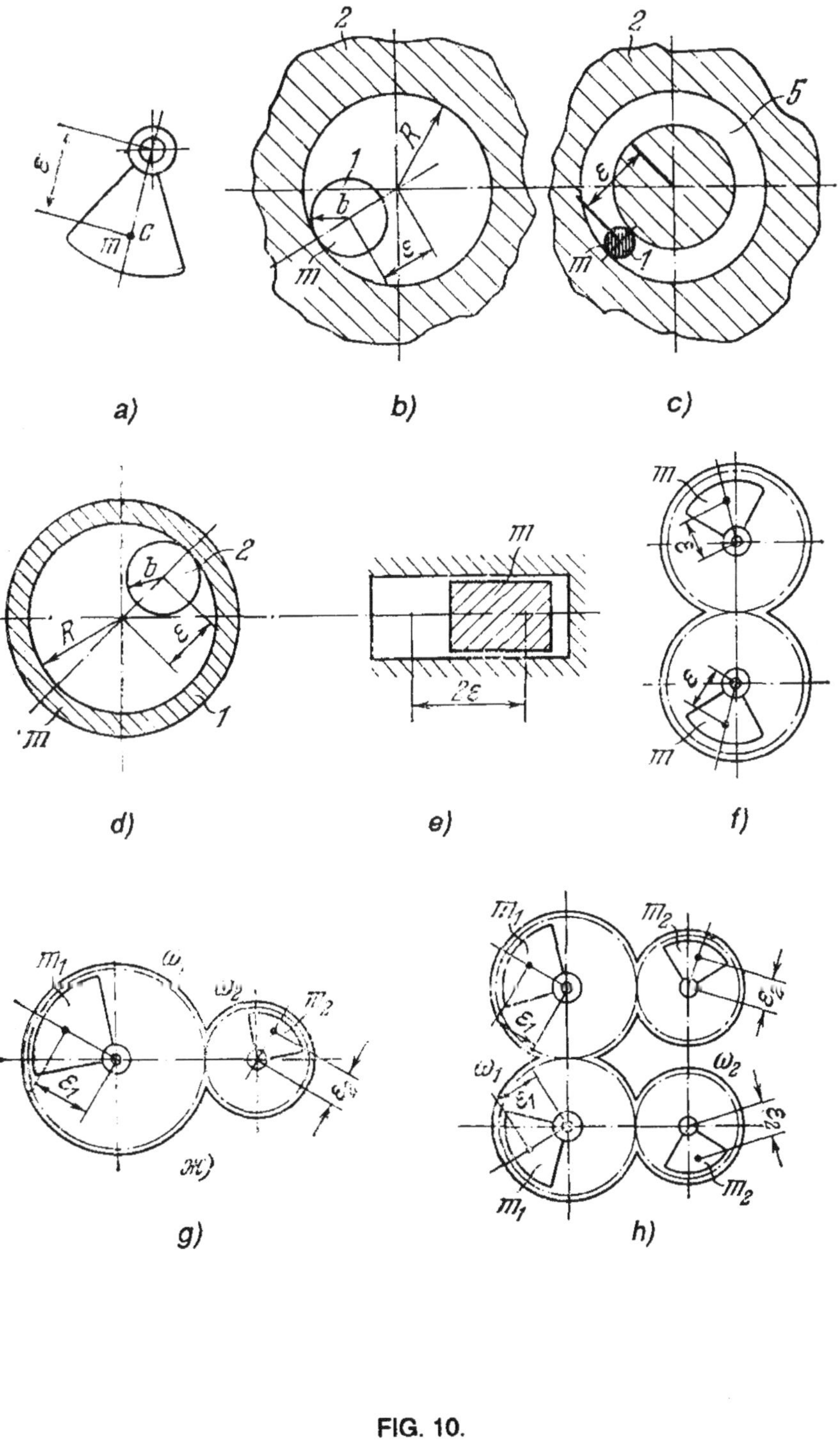

FIG. 10.

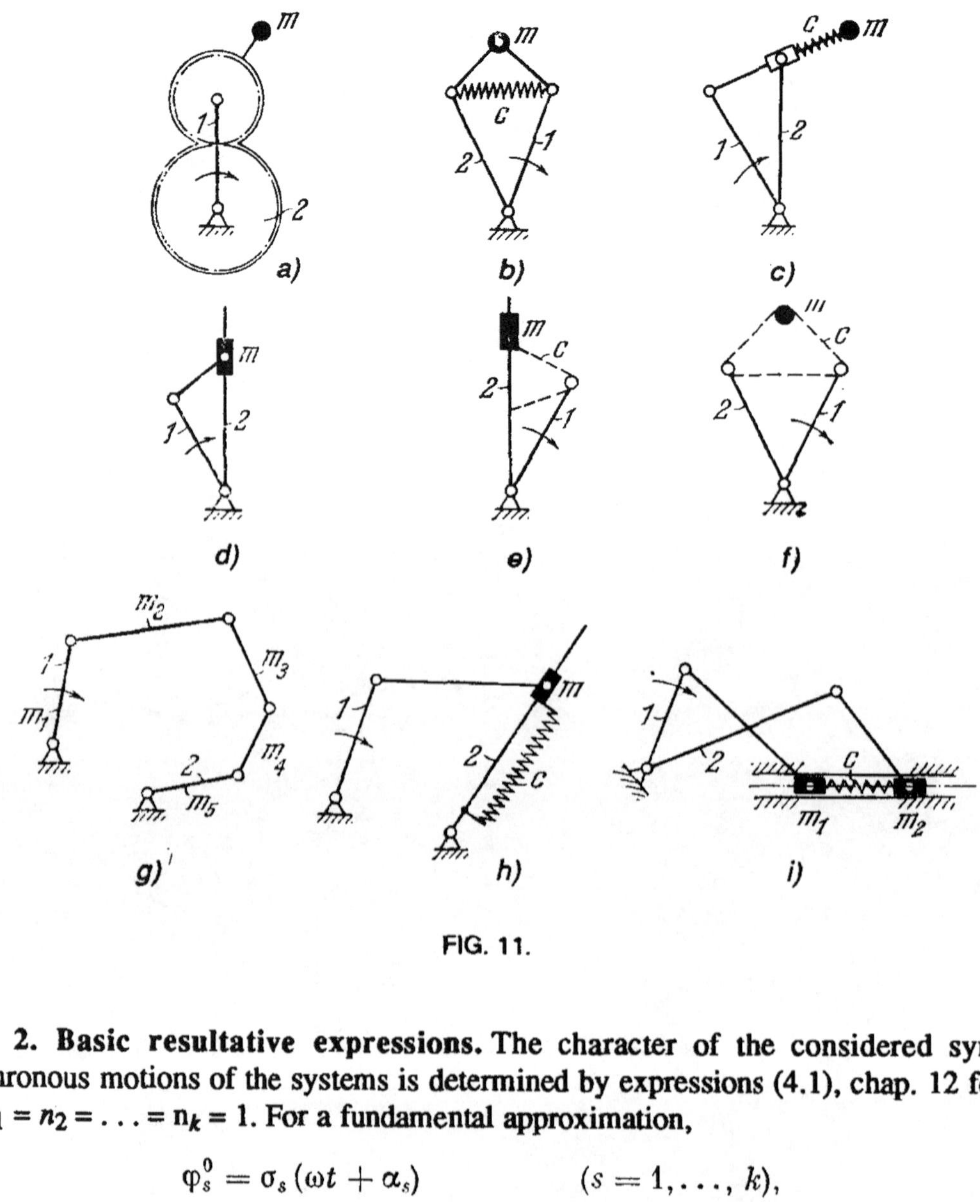

FIG. 11.

2. Basic resultative expressions. The character of the considered synchronous motions of the systems is determined by expressions (4.1), chap. 12 for $n_1 = n_2 = \ldots = n_k = 1$. For a fundamental approximation,

$$\varphi_s^0 = \sigma_s(\omega t + \alpha_s) \qquad (s = 1, \ldots, k),$$
$$u_r^0 = u_r^0(\omega t, \alpha_1, \ldots, \alpha_k) \qquad (r = 1, \ldots, \nu), \tag{8.2}$$

where u_r^0 are periodic in time (period $T = 2\pi/\omega$) solutions of equations

$$\left(\frac{d}{dt}\frac{\partial L}{\partial \dot{u}_r} - \frac{\partial L}{\partial u_r} = Q_r^{(0)}\right) \qquad (r = 1, \ldots, \nu). \tag{8.3}$$

Here and below, as before, additional redundant round brackets mean that the inserted expressions are calculated for the fundamental solution (8.2).

Firstly, an *external synchronization* problem is considered, when generalized forces $Q_s^{(1)}$, $Q_r^{(0)}$ are explicit functions of time t.

Initial phases $\alpha_1, \ldots, \alpha_k$ for the motions of inertial elements of vibro-exciters are determined from the following basic equations [see (4.6), (4.7) in chap. 12]:

$$P_s(\alpha_1, \ldots, \alpha_k) \equiv \frac{1}{k_s}(A_s - \mathscr{M}_s) \equiv \frac{1}{k_s}\left(\frac{\partial \Lambda}{\partial \alpha_s} + A_s\right) = 0 \qquad (s = 1, \ldots, k), \tag{8.4}$$

where

$$\Lambda = \langle (L) \rangle \tag{8.5}$$

is a Lagrange function averaged during period $T = 2\pi/\omega$ on the fundamental solution, and

$$A_s = Z_s - G_s,$$

$$Z_s = \sigma_s \langle (Q_s^{(1)}) \rangle, \quad G_s = -\sum_{r=1}^{\nu} \left\langle (Q_r^{(0)}) \frac{\partial u_r^0}{\partial \alpha_s} \right\rangle \tag{8.6}$$

is an *excess torque*, or average moment of nonconservative forces *referred to coordinate* φ_s. In the following, it will be called *torque reduced (referred) to the s-th exciter shaft*, or *just torque acting on s-th exciter*.

Equations (8.4), which are the same as (4.6) in chap. 12, can be considered as equilibrium equations for average torques acting on the s-th exciter. It is important to consider the structure and mechanical meaning of each term. Terms

$$\mathscr{M}_s(\alpha_1, \ldots, \alpha_k) = -\frac{\partial \Lambda}{\partial \alpha_s} \tag{8.7}$$

are *average potential torques acting on the exciters.* Since the Lagrange function of the system in this problem can be considered as not explicitly dependant on t, then expression (4.8) of chap. 12 takes place:

$$\sum_{s=1}^{k} \mathscr{M}_s = -\sum_{s=1}^{k} \frac{\partial \Lambda}{\partial \alpha_s} = 0. \tag{8.8}$$

This means that moments (torques) M_s are not directly connected with inflow or loss of energy in the system. They are just redistributing energy between the individual exciters.

To better understand the structure of the Lagrange function of the system, and thus of torques M_s, function L can be written as

$$L = L^* + L^{(\mathrm{I})} + L^{(\mathrm{II})}. \tag{8.9}$$

Here

$$L^* = \sum_{s=1}^{k} L_s + \Delta L^*, \quad \Delta L^* = \sum_{s=1}^{k} \Delta L_s^*, \tag{8.10}$$

where $L_s = L_s(\dot{\varphi}_s, \varphi_s)$ are "proper" or "natural" Lagrange functions of the exciters, i.e., their Lagrange functions when bodies of the supporting system are stationary and the supported connections are absent. Expressions

$$\Delta L_s^* = \Delta L_s^* (\dot{\varphi}_s, \varphi_s; \dot{u}_1, \ldots, \dot{u}_m; u_1, \ldots, u_m) \tag{8.11}$$

are incremental additions to these Lagrange functions due to the mobility of bodies constituting the supporting system. Here $u_1, \ldots, u_m$ are generalized coordinates locating bodies of the supporting system [$m \leq \nu$, see (8.3)].

The terms

$$L^{(I)} = L^{(I)}(\dot{u}_1, \ldots, \dot{u}_m; u_1, \ldots, u_m) \tag{8.12}$$

designate a "natural" Lagrange function of the supporting structure (connections of the first type), and

$$L^{(II)} = L^{(II)}(\dot{\varphi}_1, \ldots, \dot{\varphi}_k; \varphi_1, \ldots, \varphi_k; \dot{x}_1, \ldots, \dot{x}_\nu; x_1, \ldots, x_\nu) \tag{8.13}$$

designate a Lagrange function of the supported connections (connections of the second type). As was mentioned before, supported connections can be embodied by elastic shafts or couplings, connecting exciter rotors, electric shaftlike devices and also by centrifugal-inertial connections. These connections, as opposed to supporting ones, do not create an additional mobility of the exciter rotors. Usually they are introduced as the means for a forced synchronization if the self-synchronization effect would not provide stability of the required phase relations between the rotors (see § 11).

Correlation of functions $\Delta L_s{}^*$ and $L^{(II)}$, with $\varphi_1, \ldots, \varphi_k$, is assumed to be such that with a substitution like (8.1) they become (or remain) periodic functions of ωt with the period 2π. Therefore, all average values of Lagrange functions

$$\begin{gathered} \Lambda = \sum_{s=1}^{k} \Lambda_s + \Delta\Lambda^* + \Lambda^{(I)} + \Lambda^{(II)} = \Lambda^* + \Lambda^{(I)} + \Lambda^{(II)}, \\ \Lambda_s = \langle (L_s) \rangle, \quad \Lambda^* = \langle (L^*) \rangle, \quad \Delta\Lambda^* = \langle (\Delta L^*) \rangle, \\ \Lambda^{(I)} = \langle (L^{(I)}) \rangle, \ \Lambda^{(II)} = \langle (L^{(II)}) \rangle \end{gathered} \tag{8.14}$$

are 2π-periodic functions of each of the phases $\alpha_1, \ldots, \alpha_k$.

Since natural Lagrange functions of each exciter depend only on generalized coordinates and the velocities of this exciter, then

$$\frac{\partial \Lambda_s}{\partial \alpha_j} = 0 \qquad (j = 1, \ldots, k). \tag{8.15}$$

Accordingly,

$$\mathcal{M}_s = -\frac{\partial \Lambda}{\partial \alpha_s} = -\frac{\partial \Lambda_0}{\partial \alpha_s}, \quad \frac{\partial \Lambda^*}{\partial \alpha_s} = \frac{\partial}{\partial \alpha_s}(\Delta\Lambda^*), \tag{8.16}$$

where

$$\Lambda_0 = \langle (L_0) \rangle = \langle (\Delta L^* + L^{(I)} + L^{(II)}) \rangle = \Delta\Lambda^* + \Lambda^{(I)} + \Lambda^{(II)} \tag{8.17}$$

is an average during the period magnitude of the Lagrange function of the complete system of connections, calculated in the fundamental approximation.

In the considered problem, torque $\mathcal{M}_s$ is a sum of conservative parts $W_s^{(K)}$ and $S_s^{(K)}$ of the vibratory torque and the torque exerted by the forced synchronization system,* or the supported connections system,

*As a reminder, according to the definition in § 5, vibratory torques are additional terms in equilibrium equations for average torques (8.4), which are caused by vibrations of the bodies of supporting exciters.

$$\mathcal{M}_s = W_s^{(K)} + S_s^{(K)}. \tag{8.18}$$

In case of inertialess connections, when mobility of the supporting bodies is not reflected in kinetic energy of the supported connections, the following expression can be written:

$$W_s^{(K)} = -\frac{\partial\left(\Delta\Lambda^* + \Lambda^{(I)}\right)}{\partial\alpha_s}, \quad S_s^{(K)} = -\frac{\partial\Lambda^{(II)}}{\partial\alpha_s}. \tag{8.19}$$

The following expressions are true for torques $W_s^{(K)}$ and $S_s^{(K)}$, which are obtained in the same way as (8.8) and have the same meaning:

$$\sum_{s=1}^{k} W_s^{(K)} = 0, \quad \sum_{s=1}^{k} S_s^{(K)} = 0, \tag{8.20}$$

Terms A_s in (8.4) represent excess torques on the exciter shafts. Parameter $G_s = -\sum_{r=1}^{\nu}\left\langle \left(Q_r^{(0)}\right)\frac{\partial x_r^0}{\partial\alpha_s}\right\rangle$ in this problem can be considered as an average moment (torque) of nonconservative forces associated with vibratory coordinates and considered in the fundamental approximation. It can be written that

$$G_s = W_s^{(G)} + S_s^{(G)},$$

$$W_s^{(G)} = -\sum_{r=1}^{m}\left\langle \left(Q_r^{(0)}\right)\frac{\partial x_r}{\partial\alpha_s}\right\rangle, \quad S_s^{(G)} = -\sum_{r=m+1}^{\nu}\left\langle \left(Q_r^{(0)}\right)\frac{\partial x_r}{\partial\alpha_s}\right\rangle. \tag{8.21}$$

If a motion along the vibratory coordinates far from resonance is considered, as it was assumed, for example, in § 4, then nonconservative forces (usually viscous resistance forces) frequently could be neglected in the fundamental approximation. Then

$$Q_r^{(0)} = 0 \quad (r = 1, \ldots, \nu), \quad W_s^{(G)} = S_s^{(G)} = G_s = 0. \tag{8.22}$$

With this, basic equations (8.4) are significantly simplified not only due to disappearance of these moments (torques) but also due to simplification of expressions for u_r^0 $(\omega t, \alpha_1, \ldots, \alpha_k)$. The latter can now be determined from (8.3) for $Q_r^{(0)} = 0$.

Finally, parameters $Z_s = \sigma_s \langle (Q_s^{(1)}) \rangle$, which can be called *excess torques on exciter shafts*, represent in this problem the average values of differences between torques $(\sigma_s \langle L_s \rangle)$ transmitted from motors, and resistance torques $(\sigma_s \langle (R_s) \rangle)$ for motion of inertial elements of exciters,

$$Z_s = \sigma_s \langle (Q_s^{(1)}) \rangle = \sigma_s \langle (L_s) \rangle - \sigma_s \langle (R_s) \rangle. \tag{8.23}$$

In the simplest problem, considered in §§ 4–7, it was [see (4.20)]

$$Z_s = \sigma_s L_s(\sigma_s\omega) - \overset{\circ}{R}_s(\omega).$$

In more complicated cases, parameters $\sigma_s<(L_s)>$ and $\sigma_s<(R_s)>$ might depend on phases $\alpha_1, \ldots, \alpha_k$. For example, when exciters are driven by synchronous non-reactive motors

$$\sigma_s \langle (L_s) \rangle = \sigma_s L_s^* \left[\sigma_s (\alpha_s - \beta_s) \right]. \tag{8.24}$$

Here L_s^* is a synchronous torque transmitted from the motor, and B_s is the so-called *initial angle of the synchronous motor* selected to assure a required phase correlation of the exciter motions. Equation (8.24) is discussed in more detail in (57).

Similarly, in the case of planetary vibro-exciters, $\sigma_s<(R_s)>$ depends on all phases $\alpha_1, \ldots, \alpha_s$, since motion resistance for the inertial element of an exciter depends in a significant degree on the character of the supporting body vibrations. It can be written that

$$\sigma_s \langle (R_s) \rangle = \overset{\circ}{R}_s (\omega) + W_s^{(R)} (\alpha_1, \ldots, \alpha_k), \tag{8.25}$$

thus referring term $W_s^{(R)}$ to vibratory torque W_s. As a result, the latter is expressed as a sum of three terms,

$$W_s = W_s^{(K)} + W_s^{(R)} + W_s^{(G)}, \tag{8.26}$$

two of which are nonconservative and one, conservative. Torque S_s, due to existence of the supporting connections, can be analogously presented as

$$S_s = S_s^{(K)} + S_s^{(G)}. \tag{8.27}$$

Specific expressions for $W_s^{(R)}$ and $S_s^{(K)}$ can be found in (57); for some cases they are given in subsection 5.

If the introduced nomenclature and results of the above analysis are utilized, then basic equations (8.4) can be also presented in one of the following equivalent formats

$$P_s (\alpha_1, \ldots, \alpha_k) \equiv \frac{1}{k_s} \left[\sigma_s \langle (L_s) \rangle - \overset{\circ}{R}_s (\omega) - W_s - S_s \right] = 0 \tag{8.28}$$

$$(s = 1, \ldots, k),$$

$$P_s (\alpha_1, \ldots, \alpha_k) \equiv \frac{1}{k_s} \left(\frac{\partial \Lambda_0}{\partial \alpha_s} + Z_s - G_s \right) = 0 \tag{8.29}$$

$$(s = 1, \ldots, k).$$

These equations, as well as equations (4.19) derived in § 4 for the special case, can be rigorously derived by averaging the right-hand sides of the equations of motion for vibro-exciters [equations (4.2), chap. 12], in which instead of φ_s, u_r are substituted for their fundamental approximations (8.2).

The general conclusion of the analysis can now be formulated as follows:

For synchronous motions of the considered type to exist, it is necessary that basic equations (8.4) [or (8.28), (8.29)] would allow for real solutions in relation to *constants $\alpha_1, \ldots, \alpha_k$. Stable synchronous motions would correspond only to those solutions for which all roots of the k-th order algebraic equation*

$$\left|\frac{\partial P_s}{\partial \alpha_j} - \delta_{sj}\varkappa\right| = 0 \qquad (s, j = 1, \ldots, k) \tag{8.30}$$

have negative real parts. If there is even a single root with the positive real part, the respective motion is unstable. In case of zero or purely imaginary roots, a special analysis is required.

For an *internal synchronization* problem (e.g., self-synchronization), when equations of motion for the system are autonomous, phases $\alpha_1, \ldots, \alpha_k$ are determined from the basic equations (8.4) or (8.28), (8.29) only up to the constant (i.e., only differences $\alpha_s - \alpha_k$ are determined). However, the same equations, in general, allow us to find also a fundamental approximation for the synchronous angular velocity ω. In this case, (8.30) would certainly have a zero root which does not influence stability and is easily separated. After separation, a $(k-1)$-th order equation is obtained

$$\left|\frac{\partial (P_s - P_k)}{\partial \alpha_j} - \delta_{sj}\varkappa\right| = 0 \qquad (s, j = 1, \ldots, k-1), \tag{8.31}$$

This equation has the same format as the more detailed equation (4.24); a corresponding abbreviated notation is referred to in the footnote in § 2, chap. 10.

An equation for the energy breakdown in the system can be obtained as a summation of all equations (8.27) and considering (8.20), (8.25) and (8.26), and is written as follows

$$\omega \sum_{s=1}^{k} \sigma_s \langle (L_s) \rangle = \omega \sum_{s=1}^{k} (R_s^\circ + W_s^{(R)} + W_s^{(G)} + S_s^{(G)}),$$

In the case of debalanced vibro-exciters driven by induction motors with not too different partial angular velocities (see § 5), when resistance forces in the vibratory part of the system are not considered ($Q_r^{(0)} = 0$), the equation can be written

$$G_s = 0, \quad W_s^{(R)} = W_s^{(G)} = 0, \quad W_s = W_s^{(K)}, \quad Z_s = k_s(\omega_s - \omega), \tag{8.32}$$

In such a case, equations (8.28) are greatly simplified, assuming exactly the same format as (5.17),

$$P_s \equiv (\omega_s - \omega) - \frac{1}{k_s} W_s(\alpha_1, \ldots, \alpha_k) = 0 \qquad (s = 1, \ldots, k), \tag{8.33}$$

The difference is only in expressions $W_s = W_s^{(K)}$, which can now become more cumbersome than (4.20). By summing up equations (8.33), first multiplied by k_s, we arrive at the same expression (5.18) for the synchronous angular velocity ω. This has a meaning of energy breakdown also in this case.

3. A case of a quasi-linear vibratory system. Expressions of vibratory torques via harmonic influence coefficients. The general expressions derived above can be simplified if the supporting system is linear, at least in the fundamental approximation assuming the latter. It can also be assumed that the support connections (elements of the forced synchronization system) are massless and do not contribute additional degrees of freedom, so $m = \nu$ and expression (8.13) for the Lagrange function $L^{(II)}$ would not contain variables $u_1, \ldots, u_\nu$. The final assumption is that only motion along the vibratory coordinates far from the resonance is considered and non-conservative generalized forces corresponding to these coordinates in the fundamental approximation are neglected, that is, (8.22) are satisfied.

With these assumptions expressions (4.14), chap. 12, would apply, and considering (8.14) and (8.14) arrive at

$$\frac{\partial(\Lambda + \Lambda^{(I)} - \Lambda^{(II)})}{\partial\alpha_s} \equiv \frac{\partial(\Lambda^* + 2\Lambda^{(I)})}{\partial\alpha_s} \equiv \frac{\partial(\Delta\Lambda^* + 2\Lambda^{(I)})}{\partial\alpha_s} = 0. \quad (8.34)$$

Since in this case (8.19) is correct, then considering (8.16) and (8.4),

$$\mathcal{M}_s = W_s^{(K)} + S_s^{(K)} = -\frac{\partial\Lambda}{\partial\alpha_s} = -\frac{\partial\Lambda_0}{\partial\alpha_s} = \frac{\partial(\Lambda^{(I)} - \Lambda^{(II)})}{\partial\alpha_s},$$

$$W_s^{(K)} = -\frac{\partial(\Delta\Lambda^* + \Lambda^{(I)})}{\partial\alpha_s} = \frac{\partial\Lambda^{(I)}}{\partial\alpha_s} = -\frac{1}{2}\frac{\partial}{\partial\alpha_s}\Delta\Lambda^*, \quad (8.35)$$

$$S_s^{(K)} = -\frac{\partial\Lambda^{(II)}}{\partial\alpha_s}.$$

As a result, basic equations (8.4), (8.27) and (8.28) are also simplified and can be written in one of the following formats:

$$P_s(\alpha_1, \ldots, \alpha_k) \equiv \frac{1}{k_s}\left[\sigma_s\langle(L_s)\rangle - R_s^\circ - W_s^{(R)} - W_s^{(K)} - S_s^{(K)}\right] = 0$$

$$(s = 1, \ldots, k), \quad (8.36)$$

$$P_s(\alpha_1, \ldots, \alpha_k) \equiv \frac{1}{k}\left[\sigma_s\langle(L_s)\rangle - R_s^\circ - W_s^{(R)} - \frac{\partial(\Lambda^{(I)} - \Lambda^{(II)})}{\partial\alpha_s}\right] = 0. \quad (8.37)$$

In the case of a quasi-linear vibratory system, an effective method of deriving expressions for vibratory torques W_s can be applied. The method, independently proposed in a slightly different format in (283), and in (329, 330), expresses W_s via so-called *harmonic influence coefficients.*

Firstly, an assumption is made that the *vibro-exciters are regular debalanced ones*, see § 1 and Fig. 9a. On the other hand, assumptions about nonconservative forces $Q_r^{(0)}$ corresponding to vibratory coordinates being equal to zero, and about absence of resonance in these coordinates, are dropped. It will be assumed that $Q_r^{(0)}$ are viscous friction forces. In this case, $W_s^{(R)} = 0$, or resistance torque $\sigma_s\langle(R_s)\rangle$, does not depend on $\alpha_1, \ldots, \alpha_k$, and according to (8.25), (8.26)

$$\sigma_s\langle(R_s)\rangle = R^\circ, \quad W_s = W_s^{(K)} + W_s^{(G)}. \quad (8.38)$$

Let $o_s u_s v_s w_s$ ($s = 1, \ldots, k$) be a rectangular coordinate frame attached to the respective body of the supporting system, Fig. 12a. Axis $o_s w_s$ is directed along the rotation axis of the s-th exciter rotor, and origin o_s is selected in such a way that the center of gravity of the rotor C_s is always in the plane $o_s u_s v_s$.

Parameters (total number of k^2) are now introduced

$$k_{uu}^{(js)}(\omega), \quad k_{uv}^{(js)}(\omega), \quad k_{vu}^{(js)}(\omega), \quad k_{vv}^{(js)}(\omega),$$
$$\psi_{uu}^{(js)}(\omega), \quad \psi_{uv}^{(js)}(\omega), \quad \psi_{vu}^{(js)}(\omega), \quad \psi_{vv}^{(js)}(\omega),$$

which are defined as follows.

Let a harmonic force of the unitary amplitude and period $2\pi/\omega$ be applied to point o_j of the supporting body in the direction of axis $o_j u_j$. Besides this force, no other forces are acting on the vibratory system. Then the amplitude of the projection of displacement of point o_s to the axis $o_s u_s$ at steady forced vibrations, and the phase shift angle between said displacement and force, will be, respectively, equal to $k_{uu}^{(js)}(\omega)$ and $\psi_{uu}^{(js)}(\omega)$.

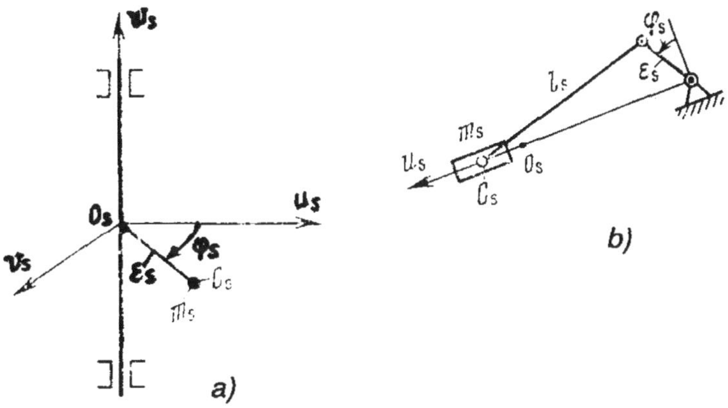

FIG. 12.

Thus, if only one given force $i_j \cos \omega t$ is acting on the vibratory system in point o_j, then projection of point o_s displacement to axis $o_s u_s$ in the steady vibration regime will be

$$s_{uu}^{(js)} = k_{uu}^{(js)}(\omega) \cos\left[\omega t - \psi_{uu}^{(js)}(\omega)\right]. \tag{8.39}$$

Analogously, parameters $k_{uv}^{(js)}(\omega)$ and $\psi_{uv}^{(js)}(\omega)$ are defined, respectively, as amplitude and phase shift of the projection of point o_s displacement to axis $o_s v_s$ under action of the unitary harmonic force of period $2\pi/\omega$ to point o_j in the direction of axis $o_j u_j$, etc.

The introduced coefficients can be determined as a result of solving corresponding linear problems or even experimentally (see below). In the case when friction in the vibratory part of the system is not considered, parameters $k_{uu}^{(js)}(\omega), \ldots, k_{vv}^{(js)}(\omega)$ become regular harmonic influence coefficients. This generic name will also be retained for systems with friction; the appropriate coefficients will be called *harmonic influence coefficients of the first kind.*

Harmonic influence coefficients comply with the reciprocal expressions (283)

$$k_{uu}^{(sj)}(\omega) = k_{uu}^{(js)}(\omega), \quad k_{vu}^{(sj)}(\omega) = k_{uv}^{(js)}(\omega), \quad k_{vv}^{(sj)}(\omega) = k_{vv}^{(js)}(\omega),$$
$$\psi_{uu}^{(sj)}(\omega) = \psi_{uu}^{(js)}(\omega), \quad \psi_{vu}^{(sj)}(\omega) = \psi_{uv}^{(js)}(\omega), \quad \psi_{vv}^{(sj)}(\omega) = \psi_{vv}^{(js)}(\omega). \tag{8.40}$$

For the full definition of phases $\psi_{uu}^{(sj)}, \ldots, \psi_{vv}^{(sj)}$, parameters $k_{uu}^{(sj)}, \ldots, k_{vv}^{(sj)}$ can be considered as definitely positive.

As shown in (283) and also in (57), vibratory torque W_s for debalanced vibro-exciters can be expressed (with accuracy up to insignificant small terms) via above introduced harmonic influence coefficients as follows:*

$$W_s = W_s^{(K)} + W_s^{(G)} =$$
$$= \frac{m_s \varepsilon_s \omega^4}{2} \sum_{j=1}^{k} m_j \varepsilon_j \left[k_{uu}^{(js)} \sin\left(\alpha_s - \alpha_j + \psi_{uu}^{(js)}\right) + \right.$$
$$+ \sigma_j k_{vu}^{(js)} \cos\left(\alpha_s - \alpha_j + \psi_{vu}^{(js)}\right) - \sigma_s k_{uv}^{(js)} \cos\left(\alpha_s - \alpha_j + \psi_{uv}^{(js)}\right) +$$
$$\left. + \sigma_j \sigma_s k_{vv}^{(js)} \sin\left(\alpha_s - \alpha_j + \psi_{vv}^{(js)}\right)\right]. \tag{8.41}$$

Here, as in § 4, m_s and ε_s designate, respectively, mass and eccentricity of the s-th exciter rotor; $\sigma_s = 1$ if a synchronous motion is considered in which the s-th exciter rotor rotates in the positive direction (see Fig. 12a); and $\sigma_s = -1$ if it rotates in the negative direction.

It is frequently more convenient to use, instead of introduced harmonic influence the coefficients of the first kind $k_{uu}^{(js)}, \ldots, k_{vv}^{(js)}, \psi_{uu}^{(js)}, \ldots, \psi_{vv}^{(js)}$ another set of parameters

$$K_{uu}^{(js)} = K_{uu}^{(sj)} = k_{uu}^{(js)} \cos \psi_{uu}^{(js)}, \quad K_{vu}^{(js)} = K_{uv}^{(sj)} = k_{vu}^{(js)} \cos \psi_{vu}^{(js)},$$
$$K_{vv}^{(js)} = K_{vv}^{(sj)} = k_{vv}^{(js)} \cos \psi_{vv}^{(js)};$$
$$G_{uu}^{(js)} = G_{uu}^{(sj)} = k_{uu}^{(js)} \sin \psi_{uu}^{(js)}, \quad G_{vu}^{(js)} = G_{uv}^{(sj)} = k_{vu}^{(js)} \sin \psi_{vu}^{(js)}, \tag{8.42}$$
$$G_{vv}^{(js)} = G_{vv}^{(sj)} = k_{vv}^{(js)} \sin \psi_{vv}^{(js)},$$

which were used in (329, 330); these parameters will be called *harmonic influence coefficients of the second kind*. It is obvious that by their physical meaning parameters $K_{uu}^{(js)}$ and $G_{uu}^{(js)}$, for example, are coefficients at $\cos \omega t$ and $\sin \omega t$, respectively, in the expression $s_{uu}^{(js)}$ for projection of the displacement of the point o_s to axis $o_s u_s$ under the action on point o_j of a unitary amplitude harmonic excitation directed along axis $o_j u$. In other words, under action of the said excitation, projection of displacement $s_{uu}^{(js)}$ of point o_s to axis $o_s u$ is presented not as (8.39), but as

$$s_{uu}^{(js)} = K_{uu}^{(js)} \cos \omega t + G_{uu}^{(js)} \sin \omega t.$$

It is easy to see that expressions (8.42) directly follow from a comparison of this equality with (8.39).

* The respective formula in (283) is different from (8.41) and the formula in (57) since in the former all $\sigma_s = 1$. Also there is a difference in the sign of W_s due to a different definition of vibratory torque.

Using harmonic influence coefficients of the second kind, vibratory torques expression (8.41) can be written as follows:

$$W_s = W_s^{(K)} + W_s^{(G)} =$$

$$= \frac{m_s\varepsilon_s\omega^4}{2}\sum_{j=1}^{k} m_j\varepsilon_j\Big[\big(K_{uu}^{(js)} - \sigma_j G_{vu}^{(js)} + \sigma_s G_{uv}^{(js)} + \sigma_j\sigma_s K_{vv}^{(js)}\big)\sin(\alpha_s - \alpha_j) +$$

$$+ \big(G_{uu}^{(js)} + \sigma_j K_{vu}^{(js)} - \sigma_s K_{uv}^{(js)} + \sigma_j\sigma_s G_{vv}^{(js)}\big)\cos(\alpha_s - \alpha_j)\Big]. \quad (8.43)$$

In the case when the considered motion is far from resonance in relation to vibratory coordinates and friction forces in the fundamental approximation are not considered, all parameters $\psi_{uu}^{(js)}, \ldots, \psi_{vv}^{(js)}$ are either zeros or π (a reminder: coefficients $k_{uu}^{(js)}, \ldots, k_{vv}^{(js)}$ are definitely positive). Thus, in this case

$$W_s^{(G)} = 0, \quad G_{uu}^{(js)} = G_{uv}^{(js)} = G_{vu}^{(js)} = G_{vv}^{(js)} = 0,$$

and expression (8.43) is simplified:

$$W_s = W_s^{(K)} = \frac{m_s\varepsilon_s\omega^4}{2}\sum_{s=1}^{k} m_j\varepsilon_j\Big[\big(K_{uu}^{(js)} + \sigma_j\sigma_s K_{vv}^{(js)}\big)\sin(\alpha_s - \alpha_j) +$$

$$+ \big(\sigma_j K_{vu}^{(js)} - \sigma_s K_{uv}^{(js)}\big)\cos(\alpha_s - \alpha_j)\Big]. \quad (8.44)$$

Now the case of directed action vibro-exciters is considered—so-called *piston exciters*, sketched in Fig. 10e and Fig. 12b. Inertial element m_s ("piston") in such exciters is driven, for example, by a central slider-crank mechanism. The ratio between the crank radius ε_s and conrod length l_s is assumed to be so small that motion of mass m_s at uniform crank rotation could be considered with an adequate accuracy as simple harmonic. The origin of the moving axis $o_s u_s$ is located at the median position of mass m_s in the process of its vibrations in the supporting body. In this case, use of only two harmonic influence coefficients of the first kind $k_{uu}^{(js)}$ and $\psi_{uu}^{(js)}$ is sufficient. The first is equal to the amplitude of projection of vibratory displacement of the point o_j of the supporting body to axis $o_j u$, said vibrations being caused by a unitary amplitude harmonic force having period $T = 2\pi/\omega$. The force is acting on the point o_s along the axis $o_s u_s$; $\psi_{uu}^{(js)}$ is the corresponding phase shift angle. Then the following expression for vibratory torque emerges:

$$W_s = W_s^{(K)} + W_s^{(G)} =$$

$$= \frac{\sigma_s m_s\varepsilon_s\omega^4}{2}\sum_{j=1}^{k} k_{uu}^{(js)}\sigma_j m_j\varepsilon_j \sin\big(\alpha_s - \alpha_j + \psi_{uu}^{(js)}\big). \quad (8.45)$$

Using harmonic influence coefficients of the second kind defined as

$$K_{uu}^{(js)} = K_{uu}^{(sj)} = k_{uu}^{(js)}\cos\psi_{uu}^{(js)},$$

$$G_{uu}^{(js)} = G_{uu}^{(sj)} = k_{uu}^{(js)}\sin\psi_{uu}^{(js)}, \quad (8.46)$$

expression (8.45) becomes

$$W_s = W_s^{(K)} + W_s^{(G)} = \frac{\sigma_s m_s \varepsilon_s \omega^4}{2} \sum_{j=1}^{k} m_j \varepsilon_j \sigma_j \left[K_{uu}^{(js)} \sin(\alpha_s - \alpha_j) + G_{uu}^{(sj)} \cos(\alpha_s - \alpha_j) \right]. \quad (8.47)$$

As before, when viscous friction forces in the vibratory part of the system are not considered, in a nonresonant case $W_s^{(G)} = 0$, $G_{uu}^{(js)} = 0$, and only the first term remains under the summation sign in (8.47).

As in § 5, all the derived expressions for vibratory torques can be expressed as sums of partial vibratory torques w_{sj} having, of course, the same physical meaning.

Expressions for vibratory torques via harmonic influence coefficients can also be derived for other, more complex exciter designs, such as for exciters generating excitation forces and torques of a very general character (214).

4. Integral stability criterion (extremal property) of synchronous motions and trend of vibro-exciters toward synchronization. What was said in § 7 about self-synchronization of vibro-exciters can be expanded to the much more general systems considered below.

For the integral stability criterion to be correct in a general case it is sufficient that the function B (potential of averaged nonconservative forces) exist, such that

$$\frac{\partial B}{\partial \alpha_s} = \Lambda_s = Z_s - G_s \qquad (s = 1, \ldots, k). \quad (8.48)$$

Then, according to § 8, chap. 10, and § 4, chap. 12, the potential function can be written as

$$D = -(\Lambda + B); \quad (8.49)$$

Stable synchronous motions of the vibro-exciters correspond to the rough minimum points of this function.

Since natural averaged Lagrange functions of exciters Λ do not depend on $\alpha_1, \ldots, \alpha_k$ (see subsection 2), the expression for function D can be written as follows

$$D = -(\Lambda_0 + B) = -(\Delta\Lambda^* + \Lambda^{(\mathrm{I})} + \Lambda^{(\mathrm{II})} + B), \quad (8.50)$$

Here $\Lambda_0 = \Delta\Lambda^* + \Lambda^{(1)} + \Lambda^{(\mathrm{II})}$ is an averaged value of the Lagrange function of the connection system.

In the case when the system is quasi-linear in relation to vibratory coordinates, and conservative forces by these coordinates are not considered (a motion far from resonance is analyzed), expressions (8.34) would hold. Thus, besides expressions (8.49), (8.50), one of the following expressions can be used as a potential function

$$D = \Lambda^{(\mathrm{I})} - \Lambda^{(\mathrm{II})} - B, \quad (8.51)$$

$$D = -\left(\frac{1}{2}\Delta\Lambda^* + \Lambda^{(\mathrm{II})} + B\right), \quad (8.52)$$

where $\partial B/\partial\alpha_s = \Lambda_s = Z_s$. In the case of the self-synchronization of debalanced and similar exciters, if supported connections are not present, $\Lambda^{(\mathrm{II})} = 0$; if partial velocities are identical, as in § 7, $B = \mathrm{const}$, and it can be taken that

$$D = \Lambda^{(I)} \quad \text{or} \quad D = -\Delta\Lambda^*. \tag{8.53}$$

In other words, in this typical case, either averaged during a period Lagrange function, or taken with an opposite sign addition (due to the mobility of the supporting bodies) to the Lagrange function of exciters, it can be taken as a potential function. For this, Lagrange functions are assumed to be calculated in the fundamental approximation.

An even more special, but important for applications, case is when bodies of the supporting systems are connected between themselves and also with the stationary foundation by very compliant elastic elements (however, geometric connections between the bodies can exist). In such a case, potential energy $\Pi^{(I)}$ of the supporting system can be neglected, and according to the first expression (8.53)

$$D = \langle\langle T^{(I)} \rangle\rangle. \tag{8.54}$$

Thus in this case *stable synchronous motions correspond to the minima of the kinetic energy of the bodies constituting the supporting system, averaged during the period.* Specifically, if such phase correlations between the exciters are possible, at which vibrations of the supporting system cease (in the fundamental approximation), then only this configuration would become stable. This statement, generalizing what was concluded in § 7 and called a paradox of nonworking connections, has important practical applications (see § 14).

A geometric format of the integral stability criterion proposed in (164, 165) [see also (57)] corresponds to the second expression in (8.53). It so happens that the term $-\Delta\Lambda^*$ in the considered systems represents a "loss" of kinetic energy of rotating rotors due to vibrations of their axes. A minimum of the expression $-\Delta\Lambda^* = -\Delta T^*$ is associated with a minimum of the sum of the amplitudes of the projection of the displacements of points o_s (Fig. 11a) to the direction of vector representing the eccentricities of rotor ε_s. A reminder: for an internal synchronization problem(with a special case of self-synchronization), from extremal conditions for function D

$$\frac{\partial D}{\partial \alpha_s} = -P_s(\alpha_1, \ldots, \alpha_k) = 0 \qquad (s = 1, \ldots, k) \tag{8.55}$$

phase shifts $\alpha_s - \alpha_k$, as well as the initial approximation for the synchronous angular velocity ω, can be determined.

A trend toward synchronization of vibro-exciters in the considered systems is predetermined by 2π-periodicity of functions Λ, Λ_0, $\Delta\Lambda^*$, $\Lambda^{(I)}$, $\Lambda^{(II)}$ by all phases $\alpha_1, \ldots, \alpha_k$. Namely (see § 9, chap. 10) if, for example, A_s are equal to zero or small relative to, respectively, $\partial\Lambda_0/\partial\alpha_s$, $\partial(\Lambda^{(I)} - \Lambda^{(II)})/\partial\alpha_s$ or $\partial(\frac{1}{2}\Delta\Lambda^* + \Lambda^{(II)})/\partial\alpha_s$, i.e., if a periodic potential or quasi-potential function exists, then locally stable synchronous motions would definitely exist. A specific case would be for debalanced or similar vibro-exciters with identical or quasi-identical partial angular velocities ω_s.

5. Synchronization of vibro-exciters attached to an elastically suspended rigid body experiencing planar vibrations. A broad class of vibratory devices can be idealized as the system sketched in Fig. 13. The supporting body B_0 (vibrating member of the machine) is assumed to be a rigid body which can perform small planar vibrations and is connected to the stationary foundation by a system of elastic and damping elements. The body carries k' debalanced and k'' planetary vibro-exciters. The total number of exciters is thus $k = k' + k''$.

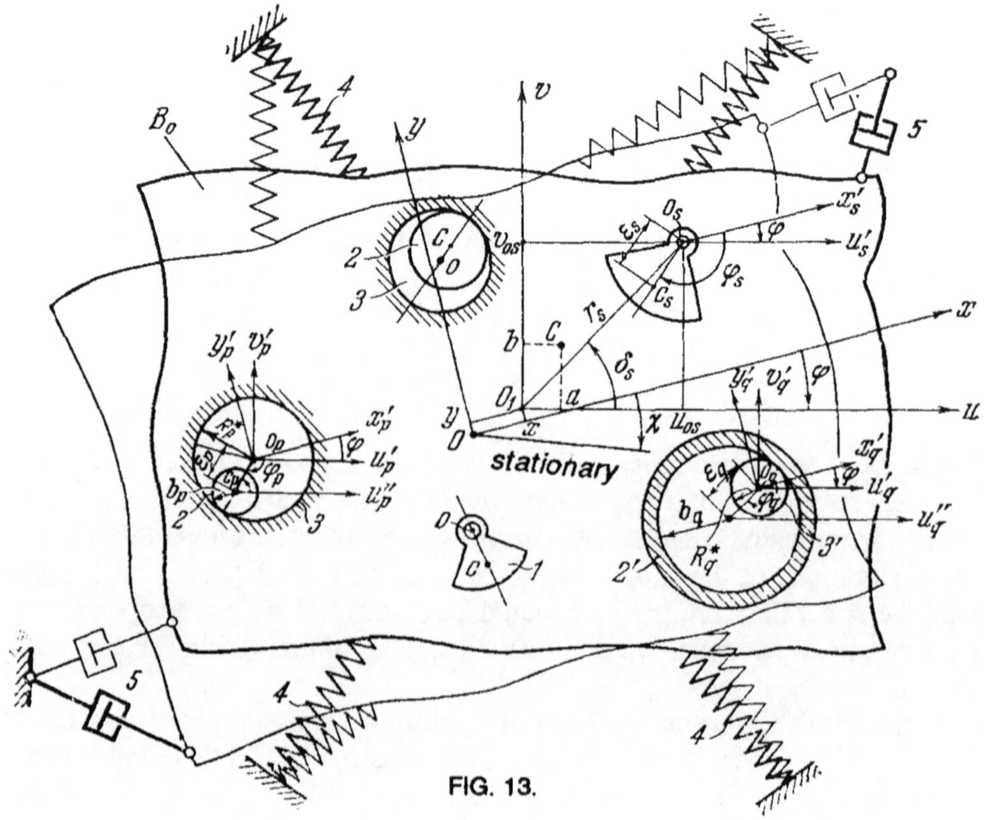

FIG. 13.

Debalanced exciters 1 are, as was described in § 1, unbalanced rotors driven by some motors. Planetary exciters could either have round cylindrical rollers 2 inserted into sleeves *3* fastened to the body B_0 (*internal planetary exciters*), or rings *2'* which can roll by their internal surfaces on shafts *3'* attached to the body B_0 (*external planetary exciters*). The exciters' axes are assumed to be perpendicular to the plane in which motion of the body B_0 occurs. It is also assumed that the rollers and sleeves, the rings and shafts of planetary exciters are always in contact, i.e., that non-positive constraints are not violated. Conditions for the latter are discussed in (57). Inertial elements of exciters (rotors, rollers, rings) can be interconnected by a system of linear elastic synchronizing elements (supported connections).

The synchronization problem for such a system is analyzed in (36, 57). The basic correlations are as follows.

Let uO_1v be a frame rigidly attached to the supporting body B_0, and whose origin O_1 is located in the center of gravity of a so-called *auxilliary body*, i.e., body consisting of B_0 with masses m_s of all inertial elements of the exciters attached to it, concentrated on the exciters' axes. Further, xOy is a stationary frame coinciding with uO_1v at the static equilibrium position of the system. Inclinations of the axes are selected in such a manner that quasi-elastic coefficient c_{xy} in the expression for potential energy $\Pi^{(I)}$ of the supporting body B_0 is zero (see below). Such axes selection allows us to somewhat simplify the following correlations. Plane

xOy is not specifically assumed to be a vertical one; potential energy of gravity of the inertial elements of the exciters in the case considered below cannot be considered (57).

Coordinates x, y of point O_1, in the frame xOy and angle φ of the body counted clockwise are assumed to be generalized coordinates of the supporting body B_0. Similarly, angles φ_s of the inertial elements of the exciters are counted from the stationary axis Ox. Expressions for kinetic and potential energy are as follows:

$$T = \sum_{s=1}^{k} T_s + T^{(\mathrm{I})} + \Delta T^*, \quad \Pi = \Pi^{(\mathrm{I})} + \Pi^{(\mathrm{II})}, \tag{8.56}$$

where

$$T_s = \frac{1}{2} I_s \dot{\varphi}_s^2, \quad T^{(\mathrm{I})} = \frac{1}{2} M (\dot{x}^2 + \dot{y}^2) + \frac{1}{2} I \dot{\varphi}^2$$

$$\Delta T^* = -\sum_{s=1}^{k} m_s \varepsilon_s \dot{\varphi}_s (\dot{x} \sin \varphi_s + \dot{y} \cos \varphi_s) +$$

$$+ \dot{\varphi} \sum_{s=1}^{k} \dot{\varphi}_s \left[m_s \varepsilon_s \cos (\varphi_s + \delta_s) - h_s \frac{\varepsilon_s \lambda_s^* (b_s + \varepsilon_s \lambda_s^*)}{b_s^2} \right]; \tag{8.57}$$

$$\Pi^{(\mathrm{I})} = \frac{1}{2} (c_x x^2 + c_y y^2 + c_\varphi \varphi^2 + 2c_{x\varphi} x\varphi + 2c_{y\varphi} y\varphi),$$

$$\Pi^{(\mathrm{II})} = \frac{1}{4} \sum_{s=1}^{k} \sum_{j=1}^{k} c_{sj} (\sigma_s \varphi_s - \sigma_j \varphi_j - \varkappa_s + \varkappa_j)^2,$$

Here m_s, I_s, ε_s are masses, effective (reduced) moments of inertia and eccentricities of exciters; r_s, δ_s are polar coordinates of the exciters' axes (see Fig. 12); M, I are mass and moment of inertia of the auxilliary body; $b_s = R_s^* - \varepsilon_s$ is the radius of the roller in case of an internal exciter, and $b_s = R_s^* + \varepsilon_s$ is the internal radius of the ring in the case of an external exciter (R_s^* is the radius of the sleeve or shaft);$\lambda_s^*=1$ if the roller or ring are rolling without sliding, and $\lambda_s^* = 0$ if they are sliding without rotation relative to the supporting body B_0; $h_s = 0$ for debalanced, and $h_s = 1$ for planetary exciters: $c_x, \ldots, c_{y\lambda}$ are quasi-elastic coefficients; c_{sj} are stiffnesses, and χ_s are "initial angles" for the system of elastic synchronizing elements which can be arbitrarily assigned, with their selection determined by the required phase correlations between exciters.

Motion far from resonance is considered, i.e., natural frequencies of body B_0 on elastic elements p_1, p_2, p_3 are sufficiently different from whole-number multiples of synchronous angular velocity ω of the exciters. Resistance forces along vibratory coordinates in the fundamental approximation are neglected ($Q_x^{(0)} = Q_y^{(0)} = Q_z^{(0)} = 0$).

Then the fundamental approximate solution would correspond to uniform rotations of inertial elements of exciters as $\varphi_s^0 = \sigma_s(\omega t + \alpha_s)$ since, as before, a simple synchronization is considered, and to small steady forced vibrations of the body B_0 at the mentioned motion of exciters. These vibrations are determined as $2\pi/\omega$-periodic solution of equations

$$
\begin{aligned}
M\ddot{x}^0 + c_x x^0 + c_{x\varphi}\varphi^0 &= \sum_{s=1}^{k} m_s\varepsilon_s\omega^2 \cos(\omega t + \alpha_s), \\
M\ddot{y}^0 + c_y y^0 + c_{y\varphi}\varphi^0 &= -\sum_{s=1}^{k} \sigma_s m_s\varepsilon_s\omega^2 \sin(\omega t + \alpha_s), \\
I\ddot{\varphi}^0 + c_\varphi\varphi^0 + c_{x\varphi}x^0 + c_{y\varphi}y^0 &= \sum_{s=1}^{k} \sigma_s m_s\varepsilon_s r_s\omega^2 \sin(\omega t + \alpha_s + \sigma_s\delta_s),
\end{aligned}
\tag{8.58}
$$

which can be written as follows

$$
\begin{aligned}
x^0 &= \sum_{s=1}^{k} \frac{m_s\varepsilon_s}{M}\left[A_s^\circ \cos(\omega t + \alpha_s) + \sigma_s B_s^\circ \sin(\omega t + \alpha_s)\right], \\
y^0 &= \sum_{s=1}^{k} \frac{m_s\varepsilon_s}{M}\left[C_s^\circ \cos(\omega t + \alpha_s) + \sigma_s D_s^\circ \sin(\omega t + \alpha_s)\right], \\
\varphi^0 &= \sum_{s=1}^{k} \frac{m_s\varepsilon_s}{M\rho}\left[M_s^\circ \cos(\omega t + \alpha_s) + \sigma_s N_s^\circ \sin(\omega t + \alpha_s)\right].
\end{aligned}
\tag{8.59}
$$

Here

$$
\begin{aligned}
A_s^\circ &= A_s^\circ(\omega^2) = b_{xx}(\omega^2) + b_{x\varphi}(\omega^2)\,\nu_s \sin\delta_s, \\
B_s^\circ &= B_s^\circ(\omega^2) = -b_{xy}(\omega^2) + b_{x\varphi}(\omega^2)\,\nu_s \cos\delta_s,
\end{aligned}
$$

$$
\begin{aligned}
C_s^\circ &= C_s^\circ(\omega^2) = b_{xy}(\omega^2) + b_{y\varphi}(\omega^2)\,\nu_s \sin\delta_s, \\
D_s^\circ &= D_s^\circ(\omega^2) = -b_{yy}(\omega^2) + b_{y\varphi}(\omega^2)\,\nu_s \cos\delta_s, \\
M_s^\circ &= M_s^\circ(\omega^2) = b_{x\varphi}(\omega^2) + b_{\varphi\varphi}(\omega^2)\,\nu_s \sin\delta_s, \\
N_s^\circ &= N_s^\circ(\omega^2) = -b_{y\varphi}(\omega^2) + b_{\varphi\varphi}(\omega^2)\,\nu_s \cos\delta_s; \\
\nu_s &= r_s/\rho, \quad \rho^2 = I/M;
\end{aligned}
\tag{8.60}
$$

$$
b_{xx}(\omega^2) = \frac{(p_y^2-\omega^2)(p_\varphi^2-\omega^2) - p_{y\varphi}^4}{D(\omega^2)}\,\omega^2, \quad b_{x\varphi}(\omega^2) = -\frac{p_{x\varphi}^2(p_y^2-\omega^2)\,\omega^2}{D(\omega^2)},
$$

$$
b_{xy}(\omega^2) = \frac{p_{x\varphi}^2 p_{y\varphi}^2 \omega^2}{D(\omega^2)}, \qquad b_{y\varphi}(\omega^2) = -\frac{p_{y\varphi}^2(p_x^2-\omega^2)\,\omega^2}{D(\omega^2)},
$$

$$
b_{yy}(\omega^2) = \frac{(p_x^2-\omega^2)(p_\varphi^2-\omega^2) - p_{x\varphi}^4}{D(\omega^2)}\,\omega^2, \qquad b_{\varphi\varphi} = \frac{(p_x^2-\omega^2)(p_y^2-\omega^2)}{D(\omega^2)}\,\omega^2;
$$

$$
D(p^2) = (p_x^2 - p^2)(p_y^2 - p^2)(p_\varphi^2 - p^2) - p_{y\varphi}^4(p_x^2 - p^2) - p_{x\varphi}^4(p_y^2 - p^2);
$$

$$
p_x^2 = \frac{c_x}{M}, \quad p_y^2 = \frac{c_y}{M}, \quad p_\varphi^2 = \frac{c_\varphi}{I}, \quad p_{x\varphi}^2 = \frac{c_{x\varphi}}{M\rho}, \quad p_{y\varphi}^2 = \frac{c_{y\varphi}}{M\rho},
$$

and, according to the assumption, frequency equation $D(p^2) = 0$ does not have roots $p = n\omega$ (where n is a whole number); thus, in any case, $D(\omega^2) \neq 0$.

Expressions for vibratory torques W_s, moments S_s of the forced synchronization system and resistance torques R_s° are as follows

$$W_s = W_s^{(K)} + W_s^{(R)},$$

$$S_s = S_s^{(K)} = -\frac{\partial \Lambda^{(II)}}{\partial \alpha_s} = \sum_{j=1}^{k} c_{sj}(\alpha_s - \alpha_j - \varkappa_s + \varkappa_j);$$

$$W_s^{(K)} = \frac{\partial \Lambda^{(I)}}{\partial \alpha_s} = \frac{1}{2}\frac{m_s\varepsilon_s\omega^2}{M}\sum_{j=1}^{k} m_j\varepsilon_j [P_{sj}\sin(\alpha_s - \alpha_j) + Q_{sj}\cos(\alpha_s - \alpha_j)],$$

$$W_s^{(R)} = \frac{1}{2} h_s f_s' \frac{m_s\varepsilon_s\omega^2}{M}\sum_{j=1}^{k} m_j\varepsilon_j [P_{sj}\cos(\alpha_s - \alpha_j) - Q_{sj}\sin(\alpha_s - \alpha_j)];$$

$$R_s^{\circ} = m_s\varepsilon_s\omega^2\left[\frac{1}{2} f_s d_s + h_s\left(f_s'\varepsilon_s - \frac{1}{2} f_s d_s\right)\right], \qquad (8.61)$$

where

$$\begin{aligned}
P_{sj} &= P_{js} = b_{xx} + \sigma_s\sigma_j b_{yy} + b_{x\varphi}(\nu_j \sin\delta_j + \nu_s \sin\delta_s) - \\
&- b_{y\varphi}\sigma_s\sigma_j(\nu_j\cos\delta_j + \nu_s\cos\delta_s) + b_{\varphi\varphi}\sigma_s\sigma_j\nu_s\nu_j\cos(\sigma_s\delta_s - \sigma_j\delta_j), \\
Q_{sj} &= -Q_{js} = (\sigma_s - \sigma_j) b_{xy} + b_{x\varphi}(\sigma_j\nu_j\cos\delta_j - \sigma_s\nu_s\cos\delta_s) + \\
&+ b_{y\varphi}(\sigma_s\nu_j\sin\delta_j - \sigma_j\nu_s\sin\delta_s) + b_{\varphi\varphi}\sigma_j\sigma_s\nu_j\nu_s\sin(\sigma_s\delta_s - \sigma_j\delta_j)
\end{aligned} \qquad (8.62)$$

Here f_s, f_s' designate the friction coefficient in the rolling bearings of the debalanced exciter, and the rolling resistance coefficient in the planetary exciter, respectively. It is taken, as usual, that in the former case resistance torque is $\frac{1}{2} f_s d_s N_s \operatorname{sgn} \dot{\varphi}_s$, and in the latter it is $f_s' \varepsilon_s N_s \operatorname{sgn} \dot{\varphi}_s$, where d_s is the diameter of the inner race of the rolling bearing friction, and N_s is the magnitude of the normal component of the reaction force at the bearing. A reminder: for debalanced exciters $h_s = 0$, $W_s^{(R)} = 0$.

Basic equations to determine initial phases α_s are written as (8.36) or (8.37), and stability depends on the signs of the real parts of the roots of equation (8.30) or, in the case of the internal synchronization problem, of equation (8.31).

In the case of planetary exciters the potential function does not exist (57). Here only a quasi-potential function can exist in the sense shown in § 9, chap. 10.

6. Regimes of developing synchronous motions of vibro-exciters. In some cases, not only is analysis of stable steady regimes of synchronous motions of vibro-exciters important but so is the process of developing such motions. Differential equations of said transient processes can be easily derived using the method of direct separation of motions, or the A.I. Lurie variational method (see §§ 3, 6, chap. 11).

Using the former method, it can be shown that equations (4.2), (4.3) of chap. 12, for the problem of simple synchronization ($n_1 = \ldots = n_k = 1$), can be written as

$$I_s\ddot{\varphi}_s + k_s(\dot{\varphi}_s - \sigma_s\omega) = \mu\Phi_s(\varphi_s, u, \omega t) \qquad (s = 1, \ldots, k),$$

$$Du = \sum_{s=1}^{k} F_s(\varphi_s) + \mu U(u, \varphi_1, \ldots, \varphi_k), \qquad (8.63)$$

where D is some differential operator, and functions Φ_s, F_s and U can depend not only on φ_s and u but also on their derivatives. Compare these equations with equations (4.12) of this chapter, corresponding to the simplest problem.

In accordance with the principal precondition of the method let us assume that the considered motions can be written as

$$\varphi_s = \sigma_s[\omega t + \alpha_s(t) + \psi_s(t, \omega t)], \quad u = u(t, \omega t), \tag{8.64}$$

where $\alpha_s(t)$ are slow-changing, and ψ_s, u are fast-changing variables. The possibility of such a representation of near-synchronous solutions of equations (8.63) is confirmed by experiments (see § 13).

Considering, in accordance with remark 2) of § 3, chap. 11, that torques $\mu\Phi$ are fast changing, equations (3.8) and (3.7) of chap. 11 can be written as follows

$$I_s\ddot{\alpha}_s + k_s\dot{\alpha}_s = \mu\sigma_s\langle\Phi_s\rangle \qquad (s = 1, \ldots, k), \tag{8.65}$$

$$I_s\ddot{\psi}_s + k_s\dot{\psi}_s = \mu\sigma_s(\Phi_s - \langle\Phi_s\rangle) \qquad (s = 1, \ldots, k), \tag{8.66}$$

$$Du = \sum_{s=1}^{k} F_s[\sigma_s(\omega t + \alpha_s + \psi_s)] +$$

$$+ \mu U[u, \sigma_1(\omega t + \alpha_1 + \psi_1), \ldots, \sigma_k(\omega t + \alpha_k + \psi_k)].$$

Searching first for the $2\pi/\omega$-periodic solution of the equation for fast motions (8.66) while slow variables α_s are "frozen," obtain $\psi_s^0 \equiv 0$, and for vector u^0 we obtain equation

$$Du^0 = \sum_{s=1}^{k} F_s[\sigma_s(\omega t + \alpha_s)], \tag{8.67}$$

corresponding to the $2\pi/\omega$-periodic vibrations of the supporting body system acted upon by exciters whose motion is described as $\varphi_s^0 = \sigma_s(\omega t + \alpha_s)$. Assuming that such a solution u^0 exists and is asymptotically stable, substitute it into the right-hand sides of the equations of slow motions (8.65) together with $\varphi_s = \varphi_s^0$. This results in equation

$$I_s\ddot{\alpha}_s + k_s\dot{\alpha}_s = \mu\sigma_s\langle(\Phi_s)\rangle \qquad (s = 1, \ldots, k),$$

Additionally, not playing a conventional role, round brackets embracing the function symbol mean that the function is calculated for the fundamental solution φ_s^0, u^0.

Or, in accordance with expressions (4.6) of chap. 12

$$I_s\ddot{\alpha}_s + k_s\dot{\alpha}_s = k_sP_s(\alpha_1, \ldots, \alpha_k) \equiv \frac{\partial\Lambda}{\partial\alpha_s} + A_s \quad (s = 1, \ldots, k). \tag{8.68}$$

Thus, differential equations for slow processes of developing of synchronous motions of vibro-exciters are easily constructed if expressions for fundamental functions P_s are known. These expressions are given by (8.68), also by formulas (8.28) and (8.29), and in the case of the quasi-linear vibratory system, by (8.36) and (8.37). In a steady case, α_k = const, (8.68) leads to corresponding basic equations. At least when $I_s\alpha_s$ is so small that it can be neglected, and also in the case of the existence of potential function D, (8.68) also leads to stability conditions, which are expressed via equations (8.30) and (8.31) (see subsection 2, § 6, chap. 12).

As is clear from subsection 2, § 6, chap. 12, equations for slow motions like (8.68) would be obtained also as a result of using the A.I. Lurie variational method. The subject nonfree orbital system can be considered as a corresponding system with vibro-exciters.

Equations (8.68) can be used for an a posteriori check of the assumption about slowness of the transient processes described by these equations. Using the expressions for vibratory torques given above, it is easy to conclude that for the problem of self-synchronization of vibro-exciters far from the resonance of the vibratory part of the system, natural frequencies of rotors near steady synchronous regimes, obtained from (8.68), are of the order of $\omega\sqrt{m/M}$ (m is mass of the exciter rotor, M is mass of the whole system). Since $M \gg m$, then $p/\omega \ll 1$; thus these vibrations are actually slow.

7. Other generalizations of the problem of synchronization of vibro-exciters (brief survey). Here publications on the theory of synchronization of vibro-exciters containing generalizations of the problem analyzed in §§ 4–7, but not used in this book, are surveyed.

Some studies deal with *synchronization of exciters in special practical and important vibratory systems and classes of systems with a finite number of degrees of freedom.* In (163) synchronization of exciters attached to a free ("soft suspended") rigid body is studied. The case when some exciters are of the so-called swinging or pendulum type is specially considered. Self-synchronization of vibro-exciters in a chainlike linear vibratory system is studied in (329).

Problems of self-synchronization of exciters in specific one-, two-, and three-mass vibratory systems were studied in (99a, 235, 309).

The basics of the theory of *synchronization of vibro-exciters generating practical arbitrary forces and moments* are developed in (214). This theory helped us to study the synchronization of gyroscopic exciters (5, 132, 224). A concept of generalized vibro-exciters is formulated in (57, 63); in the latter publication the problem of their synchronization as synchronization of constrained (nonfree) orbital systems is considered.

Synchronization of planetary exciters which serve as working organs in some machines, is discussed in (42, 57, 148).

Synchronization of vibro-exciters in distributed parameter systems was studied in (56, 108, 212, 215).

Synchronization of mechanical exciters in systems with impacts between the bodies of the supporting system was considered in (130, 131, 206, 215, 241).

Multiple synchronization of vibro-exciters was treated in (241) and later in (15, 16).

Papers (214, 215) develop *the theory of synchronization of vibro-exciters as quasi-conservative dynamic objects.* In basic important and practical cases, such as when the potential function exists, this theory, more refined and sophisticated than the "non-quasi-conservative" theory described above, leads to the same results (57). However, there are cases when results are different. Some problems are treated with the quasi-conservative model in (15, 283).

Self-synchronization of parametrically connected vibroexciters is discussed in (58).

Synchronization of vibro-exciters in systems with strong energy dissipation in the vibratory part was studied in (56, 91, 293). It is important for the design of vibratory machinery in which the mass of the process load is significant as compared with the mass of the working organ.

A systematic description of the synchronization theory and a survey of the publications before 1970 are given in (57).

§ 9) Stability of phasing of rotating rotors of self-synchronizing vibro-exciters. Relative intensity of vibratory coupling between exciters.

1. The concept of phasing stability at self-synchronization of vibro-exciters. The stability of the required phasing between the rotations of vibro-exciter rotors is not a sufficient condition to make the synchronization phenomenon for engineering applications. It is also necessary that the said phasing is not too sensitive to various imperfections, such as random scatter of the vibro-exciter and motor parameters in relation to their nominal values, caused by fabrication inaccuracies, and also to variations of the process load. If these imperfections cause a significant change in the required stable phasing, then the character of the vibrations of the working organ of the machine would noticeably change, which in turn can cause unacceptable distortion of the production process.

Phasing stability will be used to describe a property of a vibratory device with self-synchronizing vibro-exciters to maintain, in a prescribed allowable range, the phase difference between exciter rotors under the influence of various perturbing factors. Phasing stability is an important component of general stability of the working vibratory regime of a working organ of a vibratory device (49, 57). The latter is an important constituent of reliability of the vibratory machine, or its capacity to perform prescribed tasks.

Phasing stability can be analyzed by considering the basic equations determining the phase differential of rotating vibro-exciters. Knowing the possible deviations of the system parameters, these equations give deviations of phase differences $\Delta\,\alpha_s$ from values of these differences $\alpha_s^0 - \alpha_k^0$ in stable synchronous motions. In the case of small deviations in the parameters and angles $\Delta\,\alpha_s$, the basic equations can be linearized relative to these deviations; thus the search for $\Delta\alpha_s$ would be reduced to the solution of a system of $(k-1)$ linear algebraic equations (k is the number of vibro-exciters).

2. The case of the symmetrical design with two exciters. First, phasing stability is considered for a reasonably typical class of vibratory devices with two self-synchronizing debalanced exciters. Their basic equations can be written as

$$P_1 \equiv \frac{1}{k_1}\left[k_1(\omega_1 - \omega) \pm \widetilde{W}(\omega)\sin\alpha\right] = 0,$$
$$P_2 \equiv \frac{1}{k_2}\left[k_2(\omega_2 - \omega) \mp \widetilde{W}(\omega)\sin\alpha\right] = 0. \tag{9.1}$$

The same format would give basic equations in the case of vibro-exciters with positive and not too different partial velocity ω and ω_2 working in the simplest vibratory systems [see (6.3) and (5.16)], as well as in more complex systems that have a certain symmetry (see subsection 3, § 10). The magnitude of vibratory torque in (9.1) is designated $\widetilde{W}(\omega)$ (as before). Subtracting equations (9.1), we arrive at the following correlation for phase shift α:

$$\sin\alpha = \pm(\omega_2 - \omega_1)\frac{k_1 k_2}{k_1 + k_2}\frac{1}{\widetilde{W}(\omega)}. \tag{9.2}$$

It can be seen that with identical partial velocities of exciters ($\omega_1 = \omega_2$) in the considered systems only synphaseous ($\alpha = \alpha_0 = 0$) and opposed phase ($\alpha = \alpha_0 = \pi$) rotations of exciters are possible. Since according to (8.31),

$$\varkappa = \frac{\partial (P_1 - P_2)}{\partial \alpha} = \pm \frac{k_1 + k_2}{k_1 k_2} \widetilde{W}(\omega) \cos\alpha, \tag{9.3}$$

then one of these motions is always stable, and the other, unstable.

Phase discord angle $\Delta\alpha$ is a deviation of the phase shift angle from its values $\alpha_0 = 0$ or $\alpha_0 = \pi$ in a stable synchronous motion with identical partial velocities ω_1 and ω_2. Designating $\alpha = \alpha_0 + \Delta\alpha$ in (9.2), we obtain

$$|\sin\Delta\alpha| = |\omega_2 - \omega_1| \frac{k_1 k_2}{k_1 + k_2} \frac{1}{\widetilde{W}(\omega)}. \tag{9.4}$$

It follows from the (9.2) or (9.3) condition for self-synchronization that

$$|\omega_1 - \omega_2| < \frac{k_1 + k_2}{k_1 k_2} \widetilde{W}(\omega) \tag{9.5}$$

is thus reduced to a requirement that the difference between the partial velocities of the vibro-exciters would not be very large. Introducing

$$q_\omega = \frac{1}{2}\left|\frac{\omega_2 - \omega_1}{\omega}\right|_{\max} = \frac{1}{2}\left|\frac{(\omega_2 - \omega_0) - (\omega_1 - \omega_0)}{\omega}\right|_{\max},$$
$$\overset{\circ}{p}_\omega = \frac{\widetilde{W}(\omega)}{\omega k_0}, \quad k_0 = 2k_1 k_2/(k_1 + k_2), \tag{9.6}$$

the condition (9.5) gives

$$\varkappa_\omega = \overset{\circ}{p}_\omega / q_\omega > 1, \tag{9.7}$$

and from (9.4)

$$|\sin\Delta\alpha|_{\max} = 1/\varkappa_\omega. \tag{9.8}$$

Parameters q_ω and $\overset{\circ}{p}_\omega$ are, respectively, the maximum possible and maximum allowable parameters from self-synchronization condition deviations of partial velocities of exciters from their nominal value ω_0. The ratio $\varkappa_\omega = \overset{\circ}{p}_\omega / q_\omega$ can be named *self-synchronization margin*. Formulas (9.7) and (9.8) show that to achieve self-synchronization, it is necessary that the margin $\varkappa_\omega$ be more than 1, and to achieve the conditions for phasing stability

$$|\Delta\alpha|_{\max} < |\Delta\alpha|^*_{\max} \tag{9.9}$$

this margin should not be less than a certain minimal allowable value $\varkappa^*_\omega$:

$$\varkappa_\omega > \varkappa^*_\omega. \tag{9.10}$$

If scatter of the values of overall damping coefficients k_1 and k_2 is not too broad (it was assumed that exciters are nominally identical), then in formula (9.6) for k_0 it can be taken that $k_1 = k_2$; thus

$$k_0 = k_1 = k_2. \tag{9.11}$$

As follows from numerical analysis, the two major reasons for scatter of the nominal velocities of debalanced vibro-exciters follow.

a) Electric motor speed deviation at the nominal load due to manufacturing tolerances. According to the State Standard GOST 183-55, p. 30, the corresponding maximum relative deviation for general-purpose induction motors is

$$\overset{\circ}{q}_\omega = \frac{1}{8}\,\frac{n_c^{(e)} - n_0^{(e)}}{n_c^{(e)}}, \tag{9.12}$$

where $n_c^{(e)}$ and $n_0^{(e)}$ are synchronous and nominal rpm, respectively.

b) Scatter of friction coefficients in exciter bearings. The corresponding relative deviation of partial velocities is calculated as (57)

$$\overset{\circ}{q}_f = \frac{m\varepsilon d f\left[\frac{1}{2}\,\omega k^* + L_0(\omega)\right]}{k_0^2}\, q_f, \tag{9.13}$$

Here $m\varepsilon$ is the static moment of debalanced weights; d is the inner race diameter for the exciter bearings; f is the friction coefficient in bearings; $L_0(\omega)$ is the nominal torque at the exciter shaft; $q_f = 1/2\,|f_2 - f_1|/f$ is the maximum relative deviation of the friction coefficients in bearings $q_f \approx 0.3$. Damping coefficients k_0, k^* and k° and torque $L_0(\omega)$ can be determined from

$$k^* = i^2\eta\,\frac{30}{\pi}\,\frac{L_0^{(e)}}{n_c^{(e)} - n_0^{(e)}}, \quad k^\circ = \int m\varepsilon\, d\omega, \quad k_0 = k^* + k^\circ, \quad L_0 = i\eta L_0^{(e)}, \tag{9.14}$$

Here i and η are the transmission ratio, that is, the efficiency of transmission between motor and exciter shafts, respectively (if there is a transmission, otherwise $i = \eta = 1$); $L_0^{(e)}$ is the nominal torque at the motor shaft.

There are also other causes for scatter of partial angular velocities, such as the variation of static moments $m\varepsilon$ of the exciters; values i and η, etc., but they are much less important.

As a result, the total maximum deviation of the angular velocities of the exciters can be found as

$$|q_\omega| = \sqrt{(\overset{\circ}{q}_\omega)^2 + (\overset{\circ}{q}_f)^2}. \tag{9.15}$$

This formula considers the random character of the deviations, due to which simultaneous attainment by $\overset{\circ}{q}_\omega$ and $\overset{\circ}{q}_f$, their maximal values, is highly improbable.

The described technique of stability evaluation can frequently be simplified. It was suggested in (164, 166), and based on an obvious physical understanding, that phasing stability is determined by the competition of two factors. A stabilizing factor is vibratory coupling between the rotors, whose measure can be a magnitude of vibratory torque $\widetilde{W}(\omega)$, as was noted in § 5 and which also follows from (9.4) to (9.7). A destabilizing factor is composed of manufacturing inaccuracies and deviations of the process load from the nominal value.* These destabilizing factors, in the first approximation, can be assumed to be proportioned to the nominal motor torque reduced to the exciter shaft (if the motor is properly selected, that is, has an adequate but not excessive installed power). Thus, the following criterion can be introduced which, analogous to self-synchronization margin $\varkappa_\omega$, can serve as a measure of phasing stability for self-synchronizing vibro-exciters:

$$k_\omega = \frac{\widetilde{W}(\omega)}{L_0(\omega)}. \tag{9.16}$$

To achieve stability, the following condition has to be satisfied

$$k_\omega \geqslant k_\omega^*, \tag{9.17}$$

where k_ω^* is minimum allowable value of coefficient k_ω, which can be called vibratory coupling coefficient.

The coefficient k_ω as a stability measure is convenient since it depends on easily determined parameters: the torque L_0 is taken from motor catalogs; the magnitude of vibratory torque is also easy to calculate or estimate (see below).

There is a simple correlation between stability measures $\varkappa_\omega$ and k_ω if certain conditions are satisfied. Thus, if $(q_f^0)^2 \ll (q_\omega^0)^2$ and $k^0 \ll k^*$, then from the above formulas

$$\frac{1}{|\sin \Delta\alpha|} = \varkappa_\omega \approx 8k_\omega. \tag{9.18}$$

which can be written $\varkappa_\omega^* \approx 8k_\omega^*$. Computations at somewhat different conditions result in $\varkappa_\omega \approx 5k_\omega$ and $\varkappa_\omega^* = 5k_\omega^*$.

Coefficients $\varkappa_\omega$ and k_ω indicate the probability of the synchronization effect occurring. Thus, for an assumption that relative scatter of partial angular velocities is described by a normal distribution with a zero mathematical expectation and rms $\sigma_\omega = \frac{1}{3} q_\omega = \frac{1}{3} \cdot \frac{1}{2} \left| \frac{\omega_2 - \omega_1}{\omega} \right|_{max}$ ["three sigma rule", e.g., see (86)] then according to (9.6) and (9.7), the probability of developing self-synchronization is

$$P_\omega = P_\omega\left(\frac{1}{2}\left|\frac{\omega_2 - \omega_1}{\omega}\right| < p_\omega^0\right) =$$

$$= P_\omega\left(\frac{1}{2}\left|\frac{\omega_2 - \omega_1}{\omega}\right| < 3\sigma_\omega \varkappa_\omega\right) = \Phi(3\varkappa_\omega), \tag{9.19}$$

where $\Phi(z) = \frac{2}{\sqrt{2\pi}} \int_0^z e^{-t^2/2} dt$ — is the probability integral. Correlation between P_ω and $1/\varkappa_\omega$ is given in Fig. 14. When (9.18) is satisfied, the same plot

* Basic equations (9.1) do not consider some types of such a process load. Its consideration, however, is necessary only in cases when the weight of the material being processed is a significant part of the weight of the working part of the machine (see subsection 5).

gives P_ω vs. $1/(8k_\omega)$, and for $\varkappa_\omega = 5k_\omega$, P_ω vs. $1/(5k_\omega)$.

3. Assignment and enhancement methods for stability margins. Judicious assignment of minimum allowable k_ω^* and $\varkappa_\omega^*$ or angles $(\Delta\alpha)^*_{max}$ has a first-rate practical importance for various vibratory devices. It is obvious that larger values of these parameters should be assigned for such devices as classifiers, rather than for devices performing strictly material-handling functions. Unfortunately, this issue, which has to be addressed not only analytically but as to the application experience for the specific groups of machinery, is not yet fully resolved. It is recommended in (166) to assign $\varkappa^*$ not less than 3.5–5 for vibratory conveyers; 5–12 for vibratory feeders depending on their use; 12–20 for vibratory screens. Corresponding values of k_ω^* are for conveyers 0.5–1.0; for feeders 1–2.4; for screens 1.5–4. According to (9.8), the maximum values of $|\Delta\alpha|^*_{max}$ corresponding to the above values of the margins are 12–16° for conveyers; 5–12° for feeders; 3–5° for screens. However, there are now reasons to believe that these recommendations are too conservative. There are known cases of successful performance of vibratory screens with two self-synchronizing vibro-exciters, for which $|\Delta\alpha|_{max} \approx 30°$, i.e., $\varkappa_\omega \approx 2$ and $k_\omega \approx 0.3$.

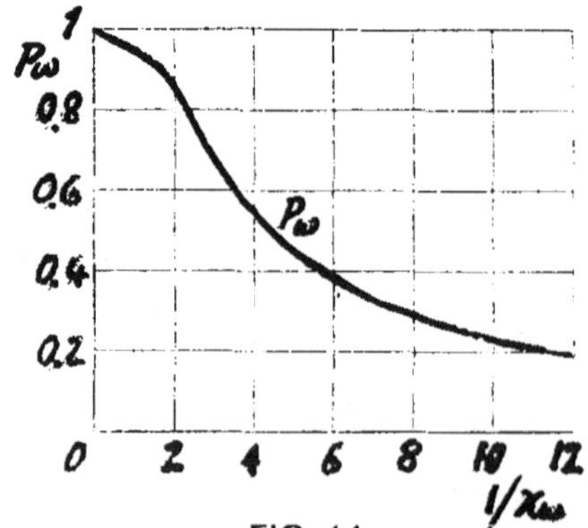

FIG. 14.

It can be seen from (9.6)–(9.10) and (9.16) that self-synchronization margin $\varkappa_\omega$ and vibratory coupling coefficient k_ω are directly proportional to the magnitude of vibratory torque $\tilde{W}(\omega)$. The latter parameter can be calculated using formulas from the special table in subsection 3, § 10 for some devices with two vibro-exciters; approximate estimates for $\tilde{W}(\omega)$ are given in § 12.

It follows from the aforesaid that stability can be enhanced by increasing torque $\tilde{W}(\omega)$. The latter can be done by changing parameters, such as increasing the frequency and amplitude of vibrations, and by changing the system structure, e.g., by increasing the number of degrees of freedom of the supporting bodies (see subsection 5).

Another method involves the reduction of scatter of the partial angular velocities q_ω of vibro-exciters. This is achieved, according to (9.15) and (9.16), by selection of motors with a minimum necessary rated power; by improving the machinery and assembly quality of bearing units (that is, reduction of q_f^γ); and also by selective matching of motors having a very small difference in partial angular verocities at nominal loading (reduction of q_ω^0).

4. Relative intensity of vibratory coupling between vibro-exciters. A possible conditional rating of the relative intensity of the vibratory coupling between exciters as a function of $\varkappa_\omega$ and k_ω is given in Table 1. This table is more detailed than suggested before in (64).

In the case of very strong coupling, self-synchronization and the accompanying effects can be expected with assurance. This cannot be said in the case of relatively strong coupling, but the possibility of such phenomena should certainly be considered. For a very weak coupling these phenomena cannot be considered, as a rule, and for a relatively weak coupling can be considered or not depending on the circumstances. Of course, these recommendations are somewhat conditional.

TABLE 1. Degrees of relative intensity of vibratory coupling.

Degrees of relative intensity of vibratory coupling	Very weak coupling	Relatively weak coupling	Relatively strong coupling	Very strong coupling
Coefficient $\varkappa_\omega$	$0 < \varkappa_\omega < 0.05$	$0.05 < \varkappa_\omega < 0.5$	$0.5 < \varkappa_\omega < 1$	$\varkappa_\omega > 1$
Coefficient k_ω (at $k_\omega = \varkappa_\omega /5$)	$0 < k_\omega < 0.01$	$0.01 < k_\omega < 0.1$	$0.1 < k_\omega < 0.2$	$k_\omega > 0.2$
Probability of developing self-synchronization P_ω	$P_\omega < 10\%$	$10\% < P_\omega < 50\%$	$50\% < P_\omega < 90\%$	$P_\omega > 90\%$

5. Stability analysis in more complex cases. In a general case of vibratory devices with numerous self-synchronizing vibro-exciters, the approach to stability analysis remains the same. Basically, it is as follows:

a) From experiments or theoretical analysis, the limits on history distortions of the vibration time for the working part of the vibratory machine (supporting body or system of bodies) are established from strength, process characteristics or other considerations.

b) From (8.50), maximum values of phase difference angles $|\Delta\alpha_s|^*_{max}$ of exciters are determined relative to the nominal phasing,* such that distortions of working organ vibrations are still acceptable.

c) Values of $|\Delta\alpha_s|^*_{max}$ are compared with the values of $|\Delta\alpha_s|_{max}$ determined from basic equations, such as (5.17), etc., which can occur at practically any possible manufacturing error of vibro-exciters, their motors, etc. As was noted in subsection 1, basic equations usually can be linearized relative to deviations of parameters and angles $\Delta\alpha_s$. If $|\Delta\alpha_s|_{max} < |\Delta\alpha_s|^*_{max}$, at least with a high probability, then the working regime of the machine vibrations and the corresponding phasing are sufficiently stable. In the opposite case, stability has to be considered as inadequate and efforts should be undertaken for its enhancement using the same methods as were discussed before for the case of two exciters.

Stability is discussed in more detail in (57). Additional recent studies include (91), which treats the influence of mechanical load on stability. This is important when the load is comparable with the overall machine weight. Paper (231) considers a method of stability enhancement by using pendulum suspensions attached to the supporting body by elastic pivots.

* That is, in relation to the stable phasing corresponding to a system not having manufacturing inaccurcies and not experiencing other distorting factors.

§ 10) Practical technique for analysis of devices with self-synchronizing vibro-exciters. Examples. Some results.

1. Analysis of devices with self-synchronizing debalanced vibro-exciters. Based on the aforesaid, the following sequence of analysis for vibratory devices employing self-synchronizing vibro-exciters can be formulated:*

1) Steady forced vibrations with $T = 2\pi/\omega$ are determined,

$$u = u^0(\omega t, \alpha_1, \ldots, \alpha_k) \tag{10.1}$$

from the supporting body or supporting system of bodies when vibro-exciter rotors are uniformly rotating with initial phases $\alpha_1, \ldots, \alpha_k$, i.e.,

$$\varphi_s = \varphi_s^0 = \sigma_s(\omega t + \alpha_s). \tag{10.2}$$

These vibrations are especially easy to find for a linear supporting system.

2) Basic equations, like (8.4), (8.28), (8.29), (8.36), (8.37), to determine constants α_s in possible synchronous motions are written. This can be done by one of three methods:

a) *Direct averaging of the equations of the motion for rotors.* Equations of the motion of rotors considering mobility of the supporting system are as follows:

$$I_s\ddot{\varphi}_s + k_s(\dot{\varphi}_s - \sigma_s\omega) = \Phi_s^*(\varphi_1, \ldots, \varphi_k, \ddot{u}, \dot{u}, u, \omega t) \tag{10.3}$$
$$(s = 1, \ldots, k),$$

Expressions for torques L_s from motors and resistance torques R_s near synchronous angular velocity are linearized according to (4.10), (4.11), (4.13). Then, the right-hand sides of equations (10.3) are averaged during one period $T = 2\pi/\omega$, assuming that φ_s and u vary as in (10.1), (10.2). This leads to a basic equation in the format

$$P_s(\alpha_1, \ldots, \alpha_k) \equiv \frac{\sigma_s}{k_s}\langle(\Phi_s^*)\rangle \equiv$$

$$\equiv \frac{\sigma_s}{k_s}\langle\Phi_s(\varphi_1^0, \ldots, \varphi_k^0, \ddot{u}^0, \dot{u}^0, u^0, \omega t)\rangle \qquad (s = 1, \ldots, k). \tag{10.4}$$

This method is the most universal one. It is also applicable in the case when there are some planetary vibro-exciters. However, it differs from the other two methods by the need to construct not only equations for forced vibrations of bodies of the supporting system, but also equations of motion for vibro-exciters that consider the mobility of the supporting bodies.

b) *Use of expressions (8.7) for potential moments $\mathcal{M}_s$ expressed via an averaged Lagrange function for the system, or the use of the integral stability criterion (extremal property) of synchronous motion.* In this technique, expressions for potential Π and kinetic T energies of the system are constructed (or energies Π_0 and T_0 for the connection system). After this, Lagrange function $L = T - \Pi$ or $L_0 = T_0 - \Pi_0$ is

*The most frequent case of simple (nonmultiple) synchronization is considered. Cases where this assumption is not required are specially identified. Before analysis, it is useful to get assurance that the system was not considered before. It is advisable to look at the table in subsection 3, § 10, or more the comprehensive Table 5 at the end of (57). Some classes of systems were analyzed in a general formulation (see subsections 5 and 7, § 8).

written. This function is averaged during one period at $\varphi_s = \varphi_s^0$ and $u = u^0$, and is determined from (10.1), (10.2). The obtained function Λ (or Λ_0) is used to determine moments $\mathcal{M}_s = -\partial\Lambda/\partial\alpha_s = -\partial\Lambda_0/\partial\alpha_s$ in basic equations (8.4). In the case of the existence of potential of averaged nonconservative forces B (see subsection 4, § 8) the function is used for constructing a potential function $D = -(\Lambda + B)$ (or $D = -(\Lambda_0 + B)$, see (8.50), and subsequent application of the integral stability criterion. Basic equations are derived in this case from stationary conditions $\partial D/\partial\alpha_s = 0$ of function D. If friction in the supporting system is neglected in the first approximation, then the potential of averaged nonconservative forces B would certainly exist, and

$$B = \sum_{s=1}^{k} \left[\sigma_s L_s(\sigma_s\omega) - \overset{\circ}{R}_s(\omega)\right]\alpha_s. \tag{10.5}$$

If, in addition, the supporting system is linear in the first approximation, then instead of a Lagrange function for the complete system, only a Lagrange function for the supporting system $L^{(I)} = T^{(I)} - \Pi^{(I)}$ and parameter $\Lambda^{(I)} = \langle(L^{(I)})\rangle$ could be calculated. With this, according to (8.51), $D = \Lambda^{(I)} - B$, and vibratory torques $W_s^{(K)} = \partial\Lambda^{(I)}/\partial\alpha_s$ [see (8.35)].

c) *Use of expressions for vibratory torques via harmonic influence coefficients.* This method, usable in the case of quasi-linear supporting systems, is described in subsection 3, § 8. Its use requires only the construction and solution of linear differential equations, describing in the first approximation vibrations of the supporting system under harmonic forces applied to the axes of the vibro-exciters. This method is applicable, in principle, in cases of any vibro-exciters and when resistance forces are considered during vibrations.

3) From basic equations $P_s = 0$, phase differences $\alpha_s - \alpha_k$ are determined in the possible synchronous motions of the exciters, together with the initial approximation to synchronous angular velocity ω. The latter can also be found directly from (5.18). Conditions for existence of real solutions of basic equations relative to $\alpha_s - \alpha_k$ are the conditions for the possibility of self-synchronization of vibro-exciters.

4) Those phase combinations are selected from the obtained solutions of basic equations which correspond to the stable synchronous motions. The selection is performed either on the basis of the algebraic equation (8.31), whose root should have negative real parts for a stable phasing, or on the basis of the integral stability criterion, according to which stable phase combinations correspond to a rough minimum of potential function D.

5) Selected $\alpha_s - \alpha_k$ are substituted into the expression for u^0, and thus the first approximation for vibrations of the supporting systems in stable synchronous motions is obtained.

6) If the phasing of the exciters at which the required vibrations of the supporting system is obtained is stable, then the system stability is checked according to § 9. If the required phasing is unstable, achieving its stability is attempted either by one of the methods from § 11, or by using forced synchronization means (also see § 11). If the required phasing is stable, but varies in time, then vibration parameters or the supporting system structure has to be changed.

2. Examples of study self-synchronization of vibro-exciters. Some simple examples are described; more sophisticated examples can be found in (57).

Example 1. *Self-synchronization of debalanced exciters in a simple single-degree-of-freedom vibratory system (see Fig. 6).* Since methods a) and b) for deriving basic equations are essentially illustrated in §§ 4, 7, an expression for vibratory torque $W_s^{(K)}$ via harmonic influence coefficients will be constructed. It is obvious that in the simplistic case being considered, all coefficients but $K_{uu}^{(js)}$ are zeros, and coefficients $K_{uu}^{(js)}$ are identical for all s and j; thus $K_{uu}^{(js)} = K_{uu}$. The term K_{uu} is determined as a solution of

$$M\ddot{x} + c_x x = \sin \omega t, \tag{10.6}$$

describing vibrations of the platform under the action of a harmonic force of unitary amplitude and directed along axis $O_1 u$. From this equation $x = \frac{1}{M\left(p_x^2 - \omega^2\right)}$ $\sin \omega t$ from which $K_{uu} = 1/\left[M\left(p_x^2 - \omega^2\right)\right]$, where, as before, $p_x = \sqrt{c_x/M}$.

Substituting K_{uu} into (8.43), we arrive at formula (4.20) for vibratory torque $W_s^{(K)}$.

Example 2. *Self-synchronization of two debalanced vibro-exciters on a soft-suspended solid body performing planar vibrations (see Fig. 15).* Supporting body B_0 can move parallel to a plane which is perpendicular to rotational axes of exciter rotors which are assumed identical. The center of gravity of the body lies in the plane passing through these axes at a distance r from both of them. Elastic supports are assumed to be very soft; thus, the frequencies of free vibrations of body B_0 on the supports is much lower than the synchronous angular velocity ω of the rotors. In such a case, potential energy $\Pi^{(1)}$ and corresponding members in the equations of motion for B_0 can be neglected. Thus, if rotors rotate as described by (10.2), vibrations of B_0 would be described by equations

$$\begin{aligned} M\ddot{x}^0 &= F[\cos(\omega t + \alpha_1) + \cos(\omega t + \alpha_2)], \\ M\ddot{y}^0 &= -F[\sigma_1 \sin(\omega t + \alpha_1) + \\ &\quad + \sigma_2 \sin(\omega t + \alpha_2)], \\ I\ddot{\varphi}^0 &= Fr[\sigma_1 \sin(\omega t + \alpha_1) - \\ &\quad - \sigma_2 \sin(\omega t + \alpha_2)], \end{aligned} \tag{10.7}$$

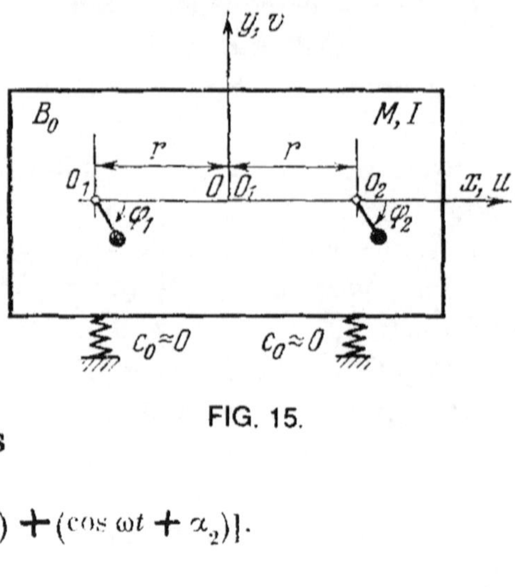

FIG. 15.

where M and I are the mass and moments of inertia of the auxilliary body, and $F = m\varepsilon\omega^2$ is the amplitude of force developed by each exciter. Steady vibratory solution of (10.7) is

$$\begin{aligned} x^0 &= -\frac{F}{M\omega^2}[\cos(\omega t + \alpha_1) + (\cos \omega t + \alpha_2)], \\ y^0 &= \frac{F}{M\omega^2}[\sigma_1 \sin(\omega t + \alpha_1) + \sigma_2 \sin(\omega t + \alpha_2)], \\ \varphi^0 &= -\frac{Fr}{I\omega^2}[\sigma_1 \sin(\omega t + \alpha_1) - \sigma_2 \sin(\omega t + \alpha_2)]. \end{aligned} \tag{10.8}$$

To derive basic equations, the integral stability criterion will be used. Since partial velocities of exciters are assumed to be identical and positive, the role of the potential function is played in this case by an average value of the Lagrange function of the auxiliary body. Due to the assumed softness of the elastic supports, the latter is simply equal to the average value of kinetic energy of the body. Performing the averaging, and considering (10.8) and (4.17), we arrive at

$$D = \Lambda^{(I)} = \langle\langle T^{(I)}\rangle\rangle = \frac{1}{2}\langle M[(\dot{x}^0)^2 + (\dot{y}^0)^2] + I(\dot{\varphi}^0)^2\rangle$$
$$\frac{F^2}{2M\omega^2}\left(1 + \sigma_1\sigma_2 - \sigma_1\sigma_2\frac{Mr^2}{I}\right)\cos(\alpha_2 - \alpha_1) + C_1. \tag{10.9}$$

Here C_1 is a parameter not dependent on α_1, α_2. As could be expected, function D depends in this case only on phase difference $\alpha = \alpha_2 - \alpha_1$. Equating $\partial D/\partial\alpha$ to zero, we arrive at the following basic equation for phase difference α in possible synchronous motions.

$$\sin\alpha = 0. \tag{10.10}$$

This equation has two essentially different solutions:

$$(\alpha)_1 = (\alpha_2 - \alpha_1)_1 = 0, \quad (\alpha)_2 = (\alpha_2 - \alpha_1)_2 = \pi, \tag{10.11}$$

The first one will be called synphaseous, and the second opposed-phaseous. If the following condition stands,

$$1 + \sigma_1\sigma_2 - \sigma_1\sigma_2\frac{Mr^2}{I} < 0 \tag{10.12}$$

then function D has a minimum at $\alpha = (\alpha)_1 = 0$ and a maximum at $\alpha = (\alpha)_2 = \pi$. Thus, according to the integral stability criterion, synphaseous motion is stable and opposed-phaseous is unstable.

Let exciter rotors first rotate in the same directions, $\sigma_1\sigma_2 = 1$. Then, if

$$Mr^2/I > 2 \tag{10.13}$$

synphaseous rotation will be stable and, as can be seen from (10.8), body B_0 would perform circular translational vibrations with amplitude $r_0 = 2m\varepsilon/M$, (Fig. 16a). If the opposite inequality $Mr^2/I < 2$ occurs, the opposed-phaseous rotation would be stable, and body B_0 would perform angular vibrations with amplitude $\varphi_0 = 2m\varepsilon r/I$ (Fig. 16b).

When the rotors are rotating in opposite directions, $\sigma_1\sigma_2 = -1$, condition (10.12) is never complied with. Thus, the opposed-phaseous rotation is always stable. This leads, as can be seen from (10.8), to linear translational harmonic vibrations of B_0 perpendicular to the plane containing the rotor axes (Fig. 16c), with amplitude $A_0 = 2m\varepsilon/M$.

The expression for vibratory torques, whose magnitudes are required for a stability check (see § 9), is

$$W_1^{(K)} = -W_2^{(K)} = \frac{\partial\Lambda^{(I)}}{\partial\alpha_1} = \frac{F^2}{2M\omega^2}\left(1 + \sigma_1\sigma_2 - \sigma_1\sigma_2\frac{Mr^2}{I}\right)\sin\alpha. \tag{10.14}$$

The same results, as is easy to see, can be obtained using harmonic influence coefficients.

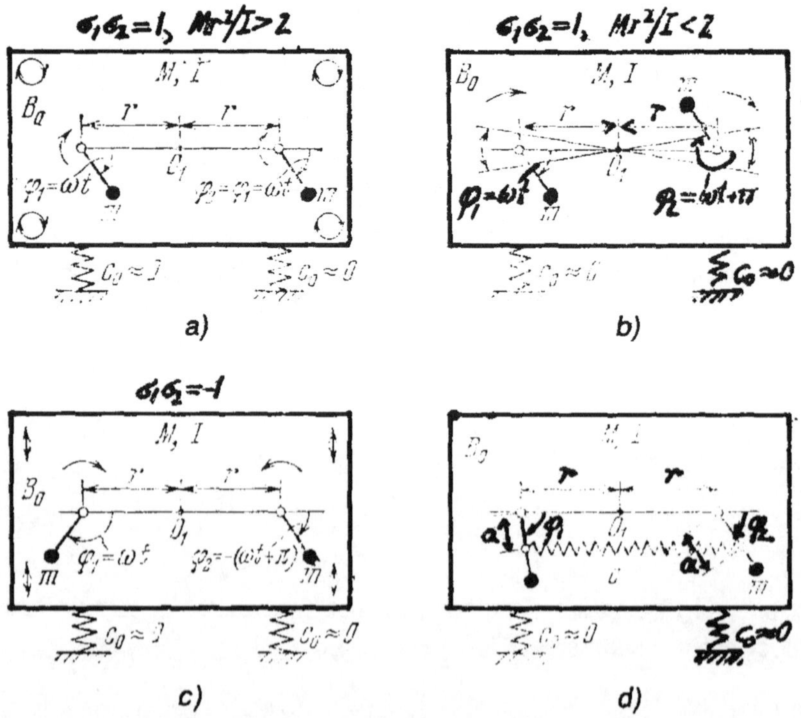

FIG. 16.

The features of self-synchronization of the two vibro-exciters in the simple system established above are widely used in vibratory machines and devices.

3. Stability conditions for the most important synchronous motions of vibro-exciters in some vibratory systems. Table 2 shows stability conditions of the most important regimes of synchronous rotation of vibro-exciter rotors in some practical and important vibratory systems. Expressions for magnitudes of vibratory torques are given; vibrations of supporting bodies in the corresponding regimes are described; and references on sources where the analysis is performed are given. A more complete table is given in (57).

Some general remarks to Table 2 are as follows:

1) If there is no special comment, induction motor drive is assumed with partial velocities of all exciters identical and positive.

2) "Soft suspended solid body" means that natural frequencies of the suspended body are much lower than the lowest angular velocity ω of the exciter rotors (the corresponding elastic supports are marked $c_0 \approx 0$).

3) In the table M, I_x, I_y, I_z are mass and the principal central moment of inertia of auxiliary bodies about the respective axes (auxiliary bodies are the ones obtained from the physical bodies of the masses of all debalanced and planetary exciters and are concentrated in rotational centers of their vectors—eccentricities and phases of all exciters of directed action—in their midpositions); O is the center of gravity of an auxiliary body; m, ε are mass and eccentricity of the exciter; oo is the rotational axis of an exciter rotor.

4) If there are double signs in formulae of column 4 of the table, either upper or lower signs have to be considered.

5) In a correlation between the actual system and one of the presented sketches, special attention should be paid to correctness in consideration of the essential degrees of freedom of the supporting bodies, since with a change of these degrees of freedom the result usually significantly changes (see subsection 2, § 11, and subsection 8, § 14).

6) The systems presented in Table 2, parts 2-5, 7, 8, 14, 16-18, 20, and 21 are special cases of the system studied in subsection 5, § 8 (Fig. 12), and systems in parts 11-13, and 15 are special cases of the systems considered in (163).

7) For a practical application of the majority of the design, compliance with phasing stability (see § 9) is necessary besides general stability conditions.

§ 11) Synthesis of vibratory machinery having several mechanical vibro-exciters.

1. The problem of creating a specified vibratory field and a synthesis of excitation systems. The problem of designing vibratory devices, mentioned in § 2, leads, besides other things, to consideration of the synchronization of vibro-exciters from a broader but somewhat more formal viewpoint.

Useful application of mechanical vibrations as exemplified in the development of vibratory technology makes the following problem a time one:

Accurate assurance is required that certain points of elastic bodies are vibrating with prescribed (and not in opposition with constraints) time histories.*

For example, for designing vibration-stimulated conveying machinery, it is necessary to force an elastic beam or shell to vibrate in a manner only slightly different from its rectilinear translational vibrations in a rigid body mode.

The problem can be called the problem of creating a specified vibratory field. Its features and solution difficulties are mostly due to the following. First, contemporary vibration exciters develop forces distributed along a small segment of the surface of the elastic bodies comprising the vibratory system. These forces can be considered concentrated. Second, the number of exciters is practically always limited and should be minimized due to economic and maintenance reasons. Finally, action of real-life exciters on a vibratory system cannot always be reduced to the action of the prescribed exciting forces, as is done in the forced vibration theory. These forces significantly depend on vibrations of areas of the elastic system to which exciters are attached. In other words, exciters together with the elastic system comprise a unified vibratory system with larger number of DOF (degrees of freedom) than the original system has due to additional DOF from exciters. Equations of motion of such a combined system are usually nonlinear.

This is illustrated by system (4.12), which would become linear if equations of motion for exciter rotors were removed and their motions considered uniform as given. However, such a simplification would make the solution of the synchronization problem impossible, and would not allow us to consider many important effects.

The problem can still be simplified considerably due to the fact that the reactive effect of the vibrations of the elastic system on exciters can usually be considered

* Determining optimal or just rational time histories and parameters of vibrations is a special problem in the theory of vibratory processes.

TABLE 2. Stability conditions for the most important

	Brief description of systems. Sections of this book or publications where the system is considered.	Dynamic model; nomenclature
1	Arbitrary number of identical debalanced vibro-exciters installed on a rigid body with a single degree of freedom. See § 4–7, chap. 3, and (31, 57).	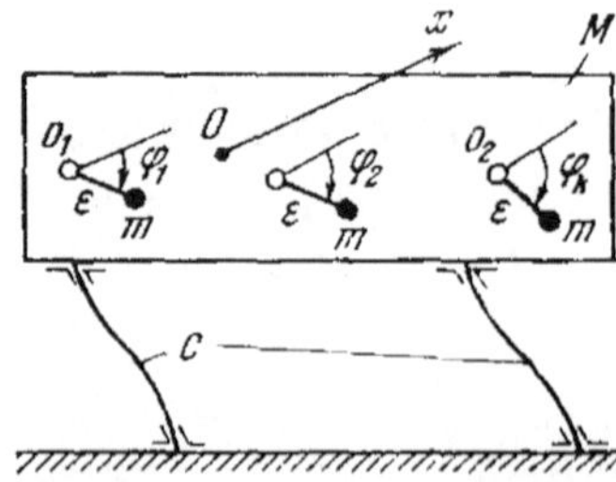

Note: 1. Results are also true for directional action exciters. 2. If number of

2	Two identical debalanced exciters symmetrically installed on a soft-suspended rigid body performing planar vibrations; exciter axes are parallel and coplanar with center of gravity O, and have the same distance r from the center of gravity. See subsection 2, § 10, chap. 3, and (36, 57).	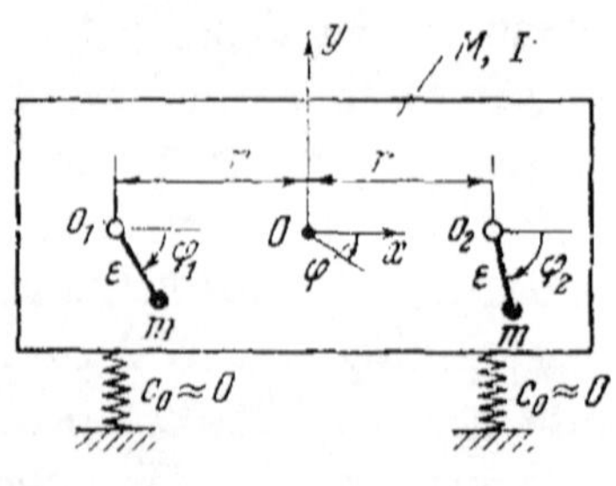
3	Same as in part 2, but the rotors are connected by a spring (stiffness c) attached at distance a from the rotor axes. See §11, chap. 3, and (57).	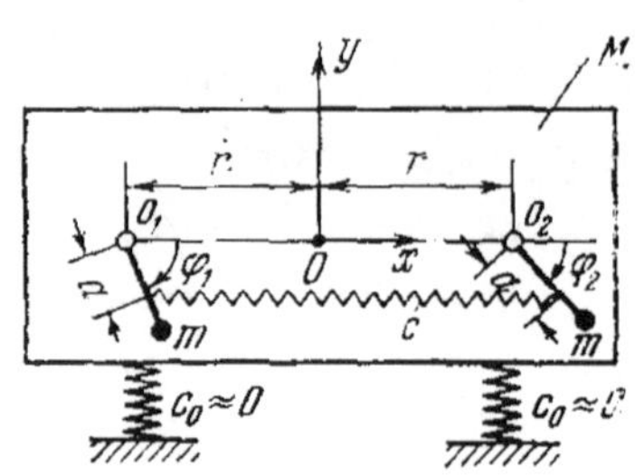

synchronous motions of exciters in some vibratory systems.

	Character of synchronous motion (first approximation) of		Conditions of stable synchronous motion	Expressions for magnitudes of vibratory torques $\tilde{W}(\omega)$
	exciters	supporting bodies		
	a) $\lvert\varphi_1\rvert=\lvert\varphi_2\rvert=\dots$ $\dots=\lvert\varphi_k\rvert=\omega t$	Harmonic translational vibrations.	$\omega < p = \sqrt{c/M}$ (*p* is the frequency of free vibrations)	For $k = 2$
	b) $\lvert\varphi_1\rvert=\lvert\varphi_2\rvert=\dots$ $\dots=\lvert\varphi_{k/2}\rvert=$ $=\lvert\varphi_{k/2+1}+\pi\rvert=\dots$ $\dots=\lvert\varphi_k+\pi\rvert=\omega t$ (*k* is even)	No vibrations.	$\omega > p = \sqrt{c/M}$	$\tilde{W}(\omega)=\frac{1}{2}\times$ $\times\frac{m^2\varepsilon^2\omega^4}{M}\left\lvert\frac{1}{p^2-\omega^2}\right\rvert$

vibro-exciters $k > 2$, there may also exist other synchronous motions.

	a) $\varphi_1=\pm\omega t$, $\varphi_2=\mp(\omega t+\pi)$ (rotors of exciters are rotating in opposite directions)	Linear translational harmonic vibrations along axis Oy with amplitude $A_0=2m\varepsilon/M$	Regime is stable at any combination of parameters if $r \neq 0$.	$\tilde{W}(\omega)=$ $=\frac{1}{2}\frac{m^2\varepsilon^2\omega^2}{M}\frac{Mr^2}{I}$
	b) $\varphi_1=\pm\omega t$, $\varphi_2=\pm\omega t$ (exciter rotors are rotating in the same direction)	Circular translational vibrations with amplitude $r_0=2m\varepsilon/M$	$\frac{Mr^2}{I}>2$	$\tilde{W}(\omega)=$ $=\frac{1}{2}\frac{m^2\varepsilon^2\omega^2}{M}\times$ $\times\left\lvert\frac{Mr^2}{I}-2\right\rvert$
	c) $\varphi_1=\pm\omega t$, $\varphi_2=\pm(\omega t+\pi)$ (exciter rotors are rotating in the same direction)	Angular harmonic vibrations with amplitude $\varphi_0=2m\varepsilon r/I$	$\frac{Mr^2}{I}<2$	
	a) Same as in part 2a.		$\frac{Mr^2}{I}>\frac{Mca^2}{m^2\varepsilon^2\omega^2}$	Same as in part 2a.
	b) Same as in part 2b.		$\frac{Mr^2}{I}>2-\frac{Mca^2}{m^2\varepsilon^2\omega^2}$	Same as in parts 2b and 2c.
	c) Same as in part 2c.		$\frac{Mr^2}{I}<2-\frac{Mca^2}{m^2\varepsilon^2\omega^2}$	

	Brief description of systems. Sections of this book or publications where the system is considered.	Dynamic model; nomenclature
4	Same system as in part 2, but the rotors are connected by elastic or electric link (stiffness c), or are driven by identical and identically installed synchronous two-pole motors (stiffness of load-velocity characteristic in the work point $dL^*/d\theta = -c$. See § 11, chap. 3, and (57).	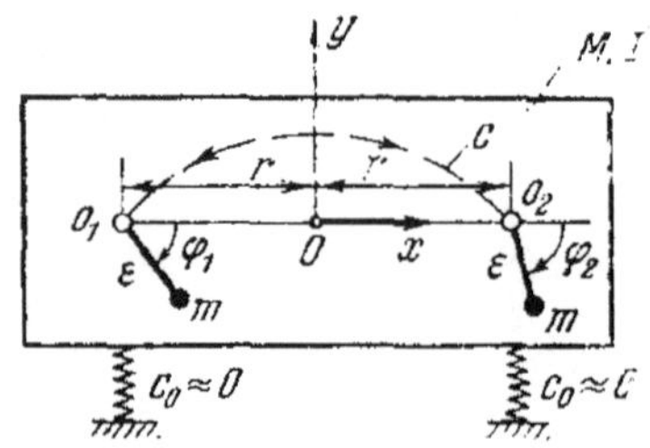
5	Same as in part 2, but the solid body is connected with the stationary foundation by a symmetric set of elastic supports (stiffness c_x, c_y, $c_\varphi (c_{x\varphi} = c_{y\varphi} = 0)$. See (57, 79).	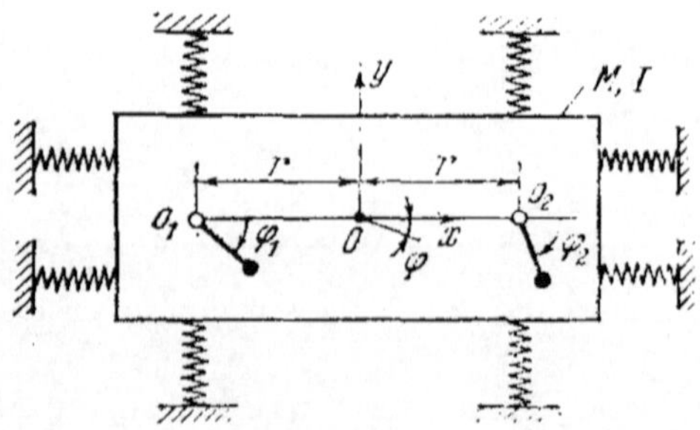 $p_x^2 = \frac{c_x}{M}$, $p_y^2 = \frac{c_y}{M}$, $p_\varphi^2 = \frac{c_\varphi}{M}$, $\lambda_x = \frac{p_x}{\omega}$, $\lambda_y = \frac{p_y}{\omega}$, $\lambda_\varphi = \frac{p_\varphi}{\omega}$
6	Same as in part 2, but there is an additional symmetrically located mass M_2 connected with the solid body via	

Table 2 *(continued).*

Character of synchronous motion (first approximation) of: exciters	supporting bodies	Conditions of stable synchronous motion	Expressions for magnitudes of vibratory torques $\widetilde{W}(\omega)$
a) Same as in part 2b.		$c > -\frac{1}{2}\frac{m^2\varepsilon^2\omega^2}{M}\times \times\left(\frac{Mr^2}{I}-2\right)$	Same as in parts 2b and 2c.
b) Same as in part 2c.		$c < -\frac{1}{2}\frac{m^2\varepsilon^2\omega^2}{M}\times \times\left(\frac{Mr^2}{I}-2\right)$	
a) Same as in part 2a, but the amplitude of translational vibrations along Oy $A_0=\frac{2m\varepsilon}{M}\frac{1}{\mid\lambda_y^2-1\mid}$		$\frac{1}{\lambda_x^2-1}-\frac{1}{\lambda_y^2-1}+\frac{Mr^2}{I}\times\frac{1}{\lambda_\varphi^2-1}<0$	$\widetilde{W}(\omega)=\frac{1}{2}\frac{m^2\varepsilon^2\omega^2}{M}\times\left\|\frac{1}{\lambda_x^2-1}-\frac{1}{\lambda_y^2-1}+\frac{Mr^2}{I}\frac{1}{\lambda_\varphi^2-1}\right\|$
b) Same as in part 2b, but the trajectory of translational vibrations is an ellipse with semi-axes $A_x=\frac{2m\varepsilon}{M}\frac{1}{\mid\lambda_x^2-1\mid}$, $A_y=\frac{2m\varepsilon}{M}\frac{1}{\mid\lambda_y^2-1\mid}$		$\frac{1}{\lambda_x^2-1}+\frac{1}{\lambda_y^2-1}-\frac{Mr^2}{I}\times\frac{1}{\lambda_\varphi^2-1}>0$	$\widetilde{W}(\omega)=\frac{1}{2}\frac{m^2\varepsilon^2\omega^2}{M}\times\left\|\frac{1}{\lambda_x^2-1}+\frac{1}{\lambda_y^2-1}-\frac{Mr^2}{I}\frac{1}{\lambda_\varphi^2-1}\right\|$
c) Same as in part 2c, but the amplitude of angular vibrations $\varphi_0=\frac{2m\varepsilon r}{I}\frac{1}{\mid\lambda_\varphi^2-1\mid}$		$\frac{1}{\lambda_x^2-1}+\frac{1}{\lambda_y^2-1}-\frac{Mr^2}{I}\times\frac{1}{\lambda_\varphi^2-1}<0$	
a) Same as in part 2a, (except formulae for amplitudes)		Same as in part 3a.	$\widetilde{W}(\omega)=\frac{1}{2}\frac{m^2\varepsilon^2\omega^2}{M}\frac{Mr^2}{I}$

	Brief description of systems. Sections of this book or publications where the system is considered.	Dynamic model; nomenclature
	elastic elements (stiffness c) and damping elements (resistance coefficient ß). See (56, 57).	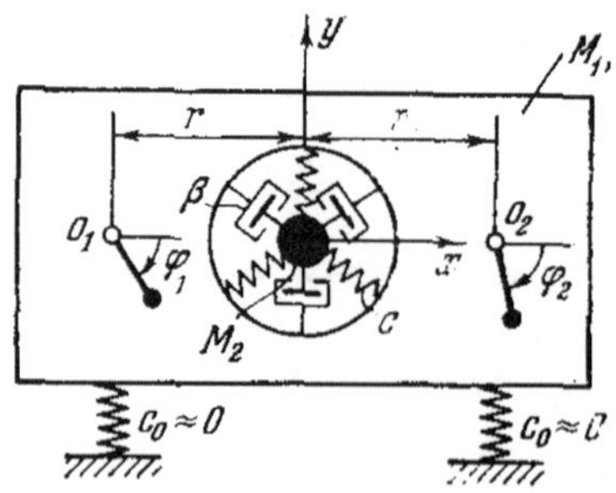 $p^2 = \frac{c}{M^*}$, $M^* = \frac{M_1 M_2}{M_1 + M_2}$, $2n = \frac{\beta}{M^*}$

Note: Systems in which more complicated elements are attached

7	Same as in part 2, but the exciters' rotors are not coplanar with center of gravity O. See (38, 42, 57).	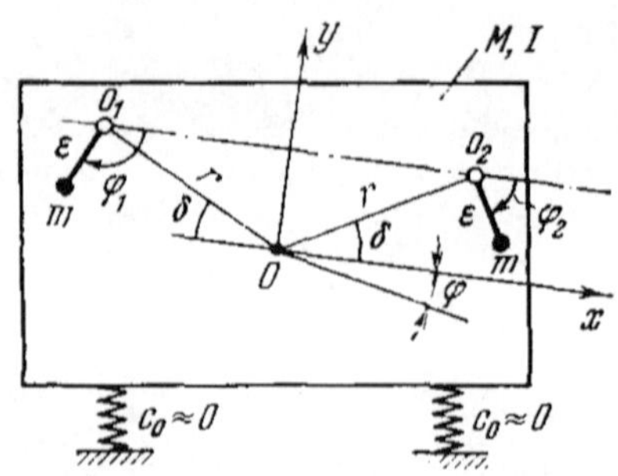
8	Same as in part 2, but the exciters are not debalanced but monoharmonic or directional action; lines of action or exciting forces are coplanar with O and equidistant from it. See (42, 57).	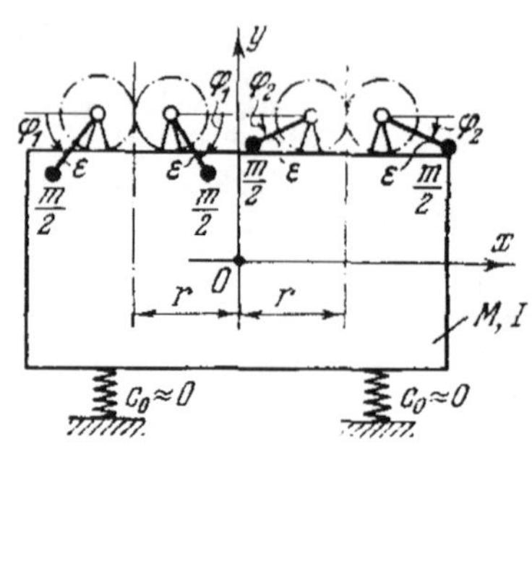

Table 2 *(continued)*.

	Character of synchronous motion (first approximation) of: exciters	Character of synchronous motion (first approximation) of: supporting bodies	Conditions of stable synchronous motion	Expressions for magnitudes of vibratory torques $\tilde{W}(\omega)$
	b) Same as in part 2b (except expressions for amplitudes).		$\frac{Mr^2}{I} > 2 - \frac{2(\lambda^2-1)}{(\lambda^2-1)^2+4n_1^2}\times\frac{M_2}{M_1}$ ($\lambda = p/\omega$, $n_1 = n/\omega$)	$\tilde{W}(\omega) = \frac{1}{2}\frac{m^2\varepsilon^2\omega^2}{M}\left\lvert\frac{Mr^2}{I} + 2\left(\frac{1}{\lambda^2-1}\times\frac{M_2}{M_1} - 1\right)\right\rvert$
	c) Same as in part 2c (except expressions for amplitudes).		$\frac{Mr^2}{I} < 2 - \frac{2(\lambda^2-1)}{(\lambda^2-1)^2+4n_1^2}\times\frac{M_2}{M_1}$	
	to body $M1$, see (56, 57).			
	$\varphi_1 = \pm\omega t$, $\varphi_2 = \mp(\omega t + \pi)$ (exciter rotors are rotating in opposite directions)	Rectilinear translational harmonic vibrations with amplitude $A_0 = 2m\varepsilon/M$, directed along Oy (perpendicular to the plane of rotor axes o_1o_2).	The regime is stable at any combination of parameters if $r \neq 0$.	$\tilde{W}(\omega) = \frac{1}{2}\frac{m^2\varepsilon^2\omega^2}{M}\frac{Mr^2}{I}$
	a) $\varphi_1 = \pm\omega t$, $\varphi_2 = \pm\omega t$	Rectilinear translational harmonic vibrations with amplitude $A_0 = 2m\varepsilon/M$, directed along Oy (perpendicular to the plane of exciter rotor axes).	$\frac{Mr^2}{I} > 1$	$\tilde{W}(\omega) = \frac{1}{2}\frac{m^2\varepsilon^2\omega^2}{M}\times\left\lvert 1 - \frac{Mr^2}{I}\right\rvert$
	b) $\varphi_1 = \pm\omega t$. $\varphi_2 = \pm(\omega t + \pi)$	Angular harmonic vibrations with amplitude $\varphi_0 = 2m\varepsilon r/I$	$\frac{Mr^2}{I} < 1$	

	Brief description of systems. Sections of this book or publications where the system is considered.	Dynamic model; nomenclature
9	Two identical debalanced exciters attached to one of the masses of a two-mass chain-type vibratory system. See (57, 329).	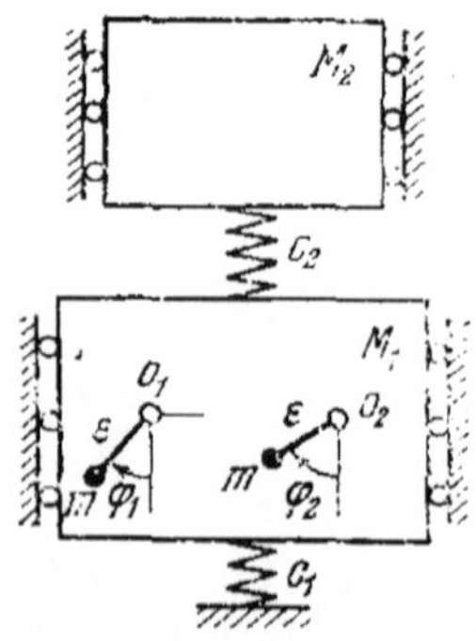

Note: For more general chain systems with an arbitrary number of masses

10	Same as in part 9, but masses can move in orthogonal directions and exciters are installed on the "external" mass. See (42, 57, 246, 330).	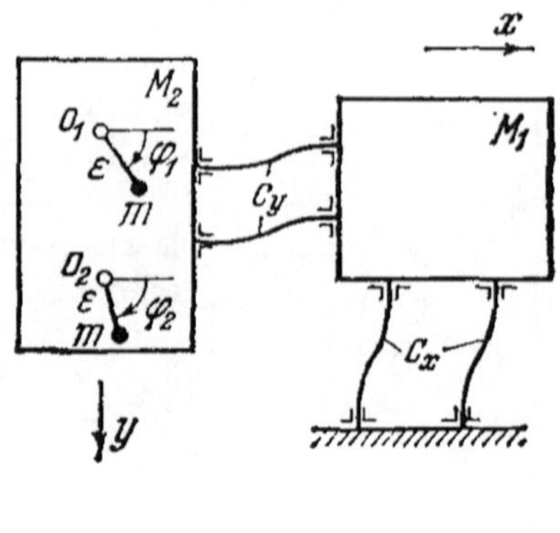 $p_x^2 = \frac{c_x}{M}$, $p_y^2 = \frac{c_y}{M_2}$, $M = M_1 + M_2$, $\lambda_x = \frac{p_x}{\omega}$, $\lambda_y = \frac{p_y}{\omega}$

Table 2 *(continued)*.

Character of synchronous motion (first approximation) of: exciters	supporting bodies	Conditions of stable synchronous motion	Expressions for magnitudes of vibratory torques $\tilde{W}(\omega)$
a) $\lvert\varphi_1\rvert=\lvert\varphi_2\rvert=\omega t$	Harmonic vibrations of both masses.	$(\tilde{p}^2-\omega^2)/(p_1^2-\omega^2)(p_2^2-\omega^2)>0$, where $\tilde{p}=c_2/M_2$, and p_1 and p_2 are the roots of equation $p^4-p^2\left(\frac{c_1}{M_1}+\frac{c_2}{M_1}+\frac{c_2}{M_2}\right)+\frac{c_1c_2}{M_1M_2}=0$	$\tilde{W}(\omega)=\frac{1}{2}\frac{m^2\varepsilon^2\omega^4}{M_1}\times\left\lvert\frac{\tilde{p}^2-\omega^2}{(p^2-\omega^2)}\times\frac{1}{(p_2^2-\omega^2)}\right\rvert$
b) $\lvert\varphi_1\rvert=\lvert\varphi_2+\pi\rvert=\omega t$	No vibrations.	$(\tilde{p}^2-\omega^2)/(p_1^2-\omega^2)(p_2^2-\omega^2)<0$	

and an arbitrary number and location of exciters, see (57, 249, 329).

a) $\varphi_1=\pm\omega t$, $\varphi_2=\pm\omega t$ (exciter rotors are rotating in the same direction)	Harmonic vibrations of M_2 in x,y directions, and M_1 in x direction.	$\frac{1}{\lambda_x^2-1}+\frac{M}{M_2}\times\frac{1}{\lambda_y^2-1}>0$	$\tilde{W}(\omega)=\frac{1}{2}\frac{m^2\varepsilon^2\omega^2}{M}\times\left\lvert\frac{1}{\lambda_x^2-1}+\frac{M}{M_2}\frac{1}{\lambda_y^2-1}\right\rvert$
b) $\varphi_1=\pm\omega t$, $\varphi_2=\pm(\omega t+\pi)$ (exciter rotors are rotating in the same direction)	No vibrations.	$\frac{1}{\lambda_x^2-1}+\frac{M}{M_2}\frac{1}{\lambda_y^2-1}<0$	
c) $\varphi_1=\pm\omega t$, $\varphi_2=\mp\omega t$ (exciter rotors are rotating in opposite directions)	Harmonic vibrations of both masses only in x direction.	$\frac{1}{\lambda_x^2-1}-\frac{M}{M_2}\frac{1}{\lambda_y^2-1}>0$	$\tilde{W}(\omega)=\frac{1}{2}\frac{m^2\varepsilon^2\omega^2}{M}\times\left\lvert\frac{1}{\lambda_x^2-1}-\frac{M}{M_2}\frac{1}{\lambda_y^2-1}\right\rvert$

	Brief description of systems. Sections of this book or publications where the system is considered.	Dynamic model; nomenclature
11	Two identical debalanced exciters installed on soft-suspended body with six degrees of freedom; the rotor axes coincide with one of the principal central inertia axes of the body; the planes of rotation of rotor centers of gravity are equidistant from the center of gravity of the body. See (42, 57).	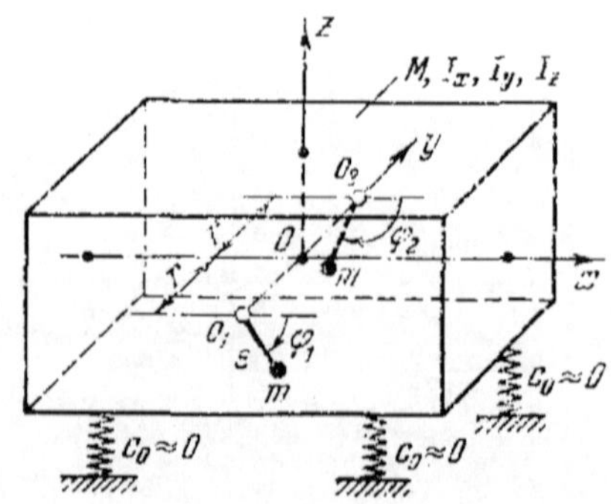
12	Same as in part 11, but exciter axes are parallel to plane xOz, equidistant from it, and inclined at angle $\pi/2-\beta$ to plane xOy (o_1Oo_2 is the plane of rotation of rotor centers of gravity; axes $Oxyz$ are the principal central inertia axes). See (57, 164, 166).	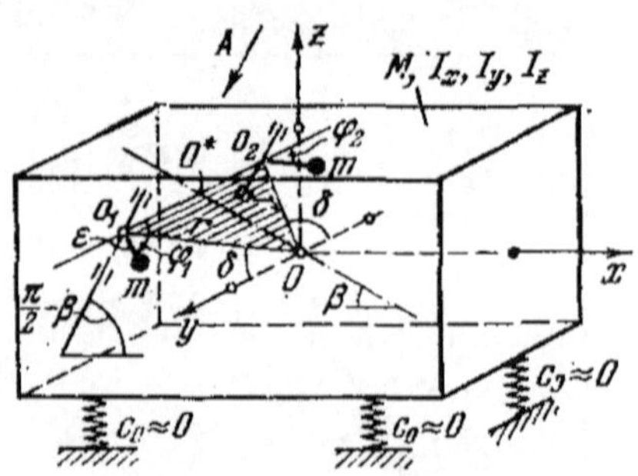 View A (rotated) 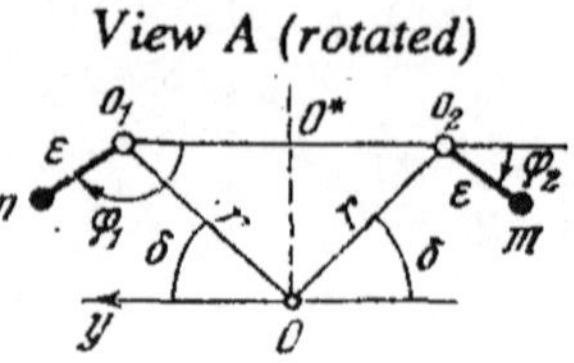

Note: The system is a three-dimensional version

13	Same as in part 11, but the exciters are not debalanced ones, but monoharmonic in directional action; the lines of action of exciting forces o_1o_1 (cont'd p. 92)	

Table 2 *(continued)*.

Character of synchronous motion (first approximation) of		Conditions of stable synchronous motion	Expressions for magnitudes of vibratory torques $\widetilde{W}(\omega)$
exciters	supporting bodies		
d) $\varphi_1 = \pm\omega t$, $\varphi_2 = \mp(\omega t + \pi)$ (exciter rotors are rotating in opposite directions)	M_1 is stationary, M_2 is vibrating only in y-direction.	$\frac{1}{\lambda_x^2 - 1} - \frac{M}{M_2}\frac{1}{\lambda_y^2 - 1} < 0$	
$\varphi_1 = \varphi_2 = \pm\omega t$	Circular translational vibrations parallel to plane xOz with amplitude $r_0 = 2m\varepsilon/M$	$\frac{Mr^2}{I''} > 2$ $\left(I'' = \frac{I_x I_z}{I_x + I_z}\right)$	$\widetilde{W}(\omega) = \frac{1}{2}\frac{m^2\varepsilon^2\omega^2}{M} \times$ $\times \left\vert 2 - \frac{Mr^2}{I''}\right\vert$
$\varphi_1 = \pm\omega t$ $\varphi_2 = \mp(\omega t + \pi)$ (exciter rotors are rotating in opposite directions)	Rectilinear translational harmonic vibrations with amplitude $A_0 = 2m\varepsilon/M$, parallel to line OO^*.	Regime is stable at any combination of parameters (if $r \neq 0$).	$\widetilde{W}(\omega) =$ $= \frac{1}{2}\frac{m^2\varepsilon^2\omega^2}{M}\frac{Mr^2}{I'}$, $I' = \frac{I_x I_z}{I_{xz}}$, $I_{xz} = I_z \sin^2\beta +$ $+ I_x \cos^2\beta$

of the system in part 7.

a) $\vert\varphi_1\vert = \vert\varphi_2\vert = \omega t$	Rectilinear translational vibrations with amplitude $A_0 = 2m\varepsilon/M$, parallel to the line (cont'd p. 93)	$\frac{Mr^2}{I'} > 1$	$\widetilde{W}(\omega) = \frac{1}{2}\frac{m^2\varepsilon^2\omega^2}{M} \times$ $\times \left\vert 1 - \frac{Mr^2}{I'}\right\vert$

	Brief description of systems. Sections of this book or publications where the system is considered.	Dynamic model; nomenclature
	and o_2o_2 are parallel to and equidistant from plane xOz and inclined at angle ß to plane xOy. See (42, 57).	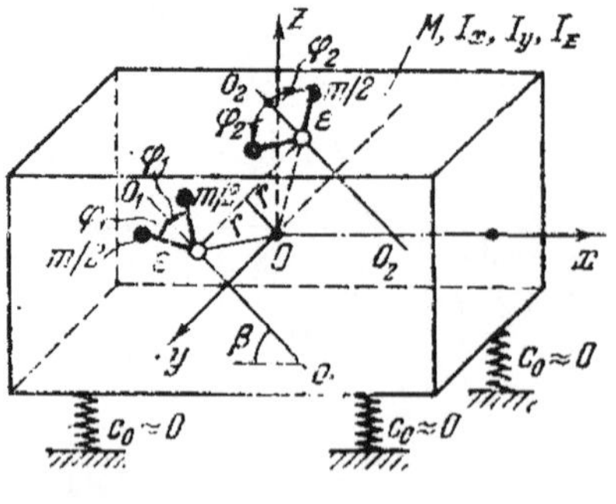

Note: The system is a three-dimensional version of the system

14	Two identical monoharmonic exciters of directional action attached to the body suspended on a symmetrical system of elastic elements and performing planar vibrations; the lines of action of the exciting forces coincide and have a distance h from the center of gravity. See (42, 51, 57).	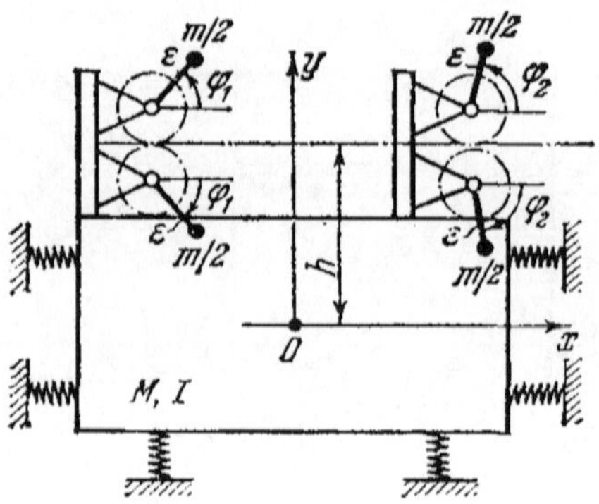 $p_x^2 = \frac{c_x}{M}$, $p_\varphi^2 = \frac{c_\varphi}{M}$, $\lambda_x = \frac{p_x}{M}$, $\lambda_\varphi = \frac{p_\varphi}{\omega}$
15	Two identical debalanced exciters attached to a soft-suspended body with six degrees of freedom; $Oxyz$ are principal central axes of inertia; exciter axes are inclined at angle γ to plane xOy. See (57, 118, 163, 164, 330).	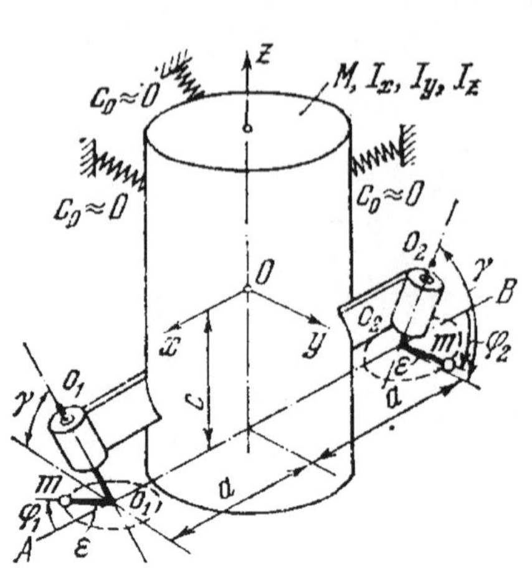

Table 2 *(continued)*.

	Character of synchronous motion (first approximation) of		Conditions of stable synchronous motion	Expressions for magnitudes of vibratory torques $\widetilde{W}(\omega)$
	exciters	supporting bodies		
		of action of exciter forces o_1o_1 and o_2o_2.	$I' = \dfrac{I_x I_y}{I_{xy}}$, $I_{xy}=I_x\cos^2\beta+ +I_y\sin^2\beta$	
	b) $\lvert\varphi_1\rvert= =\lvert\varphi_2+\pi\rvert=\omega t$	Angular harmonic vibrations about some axis situated in plane xOz.	$\dfrac{Mr^2}{I'}<1$	
in part 8.				
	a) $\lvert\varphi_1\rvert= =\lvert\varphi_2+\pi\rvert=\omega t$	No vibrations.	$\dfrac{1}{\lambda_x^2-1}+\dfrac{Mh^2}{I}\times \times\dfrac{1}{\lambda_\varphi^2-1}<0$	$W(\omega)=\dfrac{1}{2}\dfrac{m^2\varepsilon^2\omega^2}{M}\times \times\left\lvert\dfrac{1}{\lambda_x^2-1}+ +\dfrac{Mh^2}{I}\dfrac{1}{\lambda_\varphi^2-1}\right\rvert$
	b) $\lvert\varphi_1\rvert=\lvert\varphi_2\rvert= =\omega t$	Planar nontranslational vibrations.	$\dfrac{1}{\lambda_x^2-1}+\dfrac{Mh^2}{I}\times \times\dfrac{1}{\lambda_\varphi^3-1}>0$	
	$\varphi_1=\pm\omega t$, $\varphi_2=\pm\omega t$	Screw-like harmonic vibrations about axis z; amplitude of vertical component $A_0=2m\varepsilon\cos\gamma/M$ angular amplitude $\varphi_0=2m\varepsilon a\sin\gamma/I_z$	$\dfrac{Ma^2}{I_z}<2+\dfrac{Mc^2}{I_x}+ +\dfrac{M(c^2+a^2\cos^2\gamma)}{I_y\sin^2\gamma}$	$\widetilde{W}(\omega)=\dfrac{1}{2}\dfrac{m^2\varepsilon^2\omega^2}{M}\times \times\sin^2\gamma\left\lvert\dfrac{Ma^2}{I_z}- -2-\dfrac{Mc^2}{I_x}- -\dfrac{M(c^2+a^2\cos^2\gamma)}{I_y\sin^2\gamma}\right\rvert$

	Brief description of systems. Sections of this book or publications where the system is considered.	Dynamic model; nomenclature
16	Two identical biharmonic debalanced exciters symmetrically installed on a soft-suspended body that can perform planar vibrations. Planes *1–1′* and *2–2′* of axes of each exciter are parallel and equidistant from center of gravity *O*. See (42, 57).	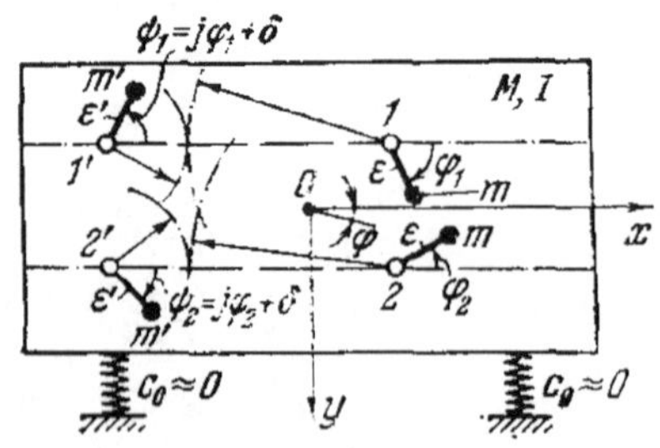 j is the transmission ratio between exciter shafts.
17	Three debalanced exciters, symmetrically installed on a soft-suspended body that can perform planar vibrations; the plane containing all rotor axes also contains the center of gravity; external exciters are identical. See (42, 57).	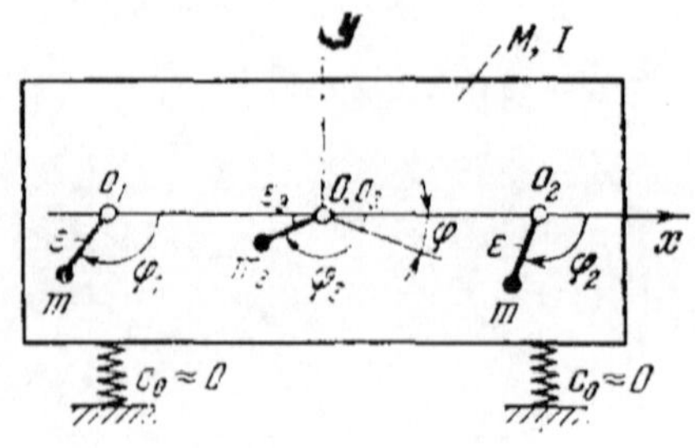
18	Four identical exciters symmetrically installed on a soft-suspended body that can perform planar vibrations. See (42, 57).	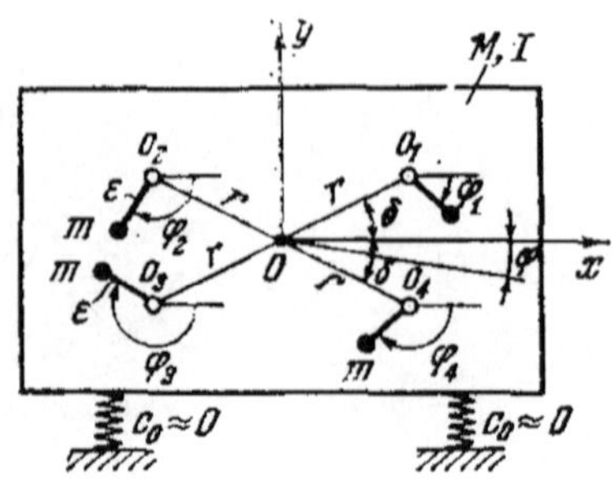

Note: Stability of regimes in which there are no vibrations of the supporting body for more

Table 2 *(continued)*.

	Character of synchronous motion (first approximation) of		Conditions of stable synchronous motion	Expressions for magnitudes of vibratory torques $\tilde{W}(\omega)$
	exciters	supporting bodies		
	$\varphi_1 = \pm\omega t$ $\varphi_2 = \pm\omega t$	Rectilinear translational biharmonic vibrations parallel to Ox with time history $x = \frac{2m\varepsilon}{M}\cos\omega t + \frac{2m'\varepsilon'}{M}\cos j\omega t$	Regime is stable at any combination of parameters, but other synchronous regimes can also be stable. At $j = 2$ such additional regimes are absent if $4m'\varepsilon' < m\varepsilon$	
	$\varphi_1 = \varphi_2 = \pm\omega t$, $\varphi_3 = \pm(\omega t + \pi)$ (rotors of all exciters rotate in the same direction)	Circular translational vibrations with amplitude $r_0 = \frac{2m\varepsilon - m_3\varepsilon_3}{M}$	$\frac{Mr^2}{I} > 2 - \frac{m_3\varepsilon_3}{m\varepsilon}$	
	a) $\varphi_1 = \pm\omega t$, $\varphi_2 = \pm(\omega t + \pi)$, $\varphi_3 = \mp(\omega t + \pi)$, $\varphi_4 = \mp\omega t$ (rotors of third and fourth exciters rotate in the opposite direction relative to first and second rotors) b) $\varphi_1 = \pm\omega t$, $\varphi_2 = \pm(\omega t + \pi)$, $\varphi_3 = \pm\omega t$, $\varphi_4 = \pm(\omega t + \pi)$ (rotors of all exciters rotate in the same direction)	No vibrations.	Regimes stable at any combination of parameters.	

general systems is addressed in subsection 4, § 8, and subsections 6 and 7 of § 14, chap. 3.

	Brief description of systems. Sections of this book or publications where the system is considered.	Dynamic model; nomenclature
19	Two identical debalanced or directional action exciters installed on a simply supported uniform beam with constant cross section. See (57, 330).	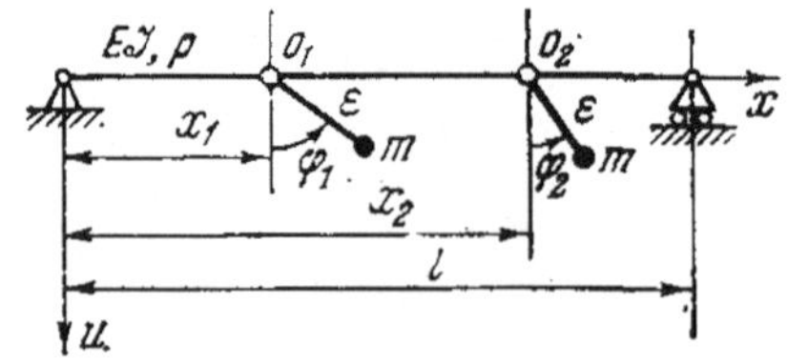 $E\mathcal{J}$ — bending stiffness ρ — linear density of beam

Note: The case of a soft suspended beam with an arbitrary number of identical

20	One debalanced (*1*) and one passive (without motor) external planetary exciter (*2*), coaxially installed on a free body (*3*) which can perform planar vibrations. See (251, 252, 42, 57).	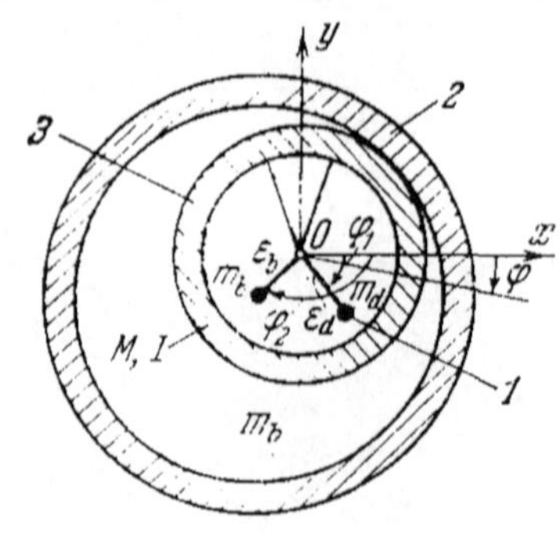

Note: The result is also true when the planetary exciter

Table 2 *(continued)*.

	Character of synchronous motion (first approximation) of		Conditions of stable synchronous motion	Expressions for magnitudes of vibratory torques $\widetilde{W}(\omega)$
	exciters	supporting bodies		
	a) $\lvert\varphi_1\rvert=\omega t$, $\lvert\varphi_2\rvert=\omega t$	Elastic vibrations under synphaseous harmonic exciting forces $F=m\varepsilon\omega^2$, applied at points o_1 and o_2.	$\widetilde{k}(\omega^2)>0$, $\widetilde{k}(\omega^2)=\frac{2}{\pi}\frac{l^3}{E\mathcal{J}}\times \times\sum_{j=1}^{\infty}\frac{1}{j^4-q^4}\times \times\sin j\pi\xi_1\sin j\pi\xi_2$, $\xi_1=x_1/l$, $\xi_2=x_2/l$, $q^4=\rho l^4\omega^2/\pi E\mathcal{J}$	$\widetilde{W}(\omega)=\frac{m^2\varepsilon^2\omega^4}{2}\times \times\lvert\widetilde{k}(\omega^2)\rvert$
	b) $\lvert\varphi_1\rvert=\omega t$, $\lvert\varphi_2\rvert=\omega t+\pi$	Elastic vibrations under antiphaseous exciting forces $F=m\varepsilon\omega^2$, applied at points o_1 and o_2.	$\widetilde{k}(\omega^2)<0$	

unidirectional exciters. See (212) and also (57).

	$\varphi_1=\pm(\omega t+\pi+ +\rho'+\arcsin q)$, $q=\frac{\varepsilon_b}{\widetilde{\mathcal{A}}_0}\sin\rho'$, $\varphi_2=\pm\omega t$, $\widetilde{\mathcal{A}}_0=m_d\varepsilon_d/m'_a$, $\rho'=\mathrm{arctg}\, f'$ (f' is the rolling friction coefficient),	Circular translational vibrations.	Regime is stable in all the domains of its existence determined by $\frac{\varepsilon_b}{\widetilde{\mathcal{A}}_0}\leqslant\frac{\sqrt{1+(f')^2}}{f'}$, if $k_2/k_1\ll 1$ (k_s is the overall damping coefficient).	

is internal.

	Brief description of systems. Sections of this book or publications where the system is considered.	Dynamic model; nomenclature
21	Two identical debalanced exciters and one passive internal planetary exciter symmetrically installed on a soft-suspended body that can perform planar vibrations. See (42, 57).	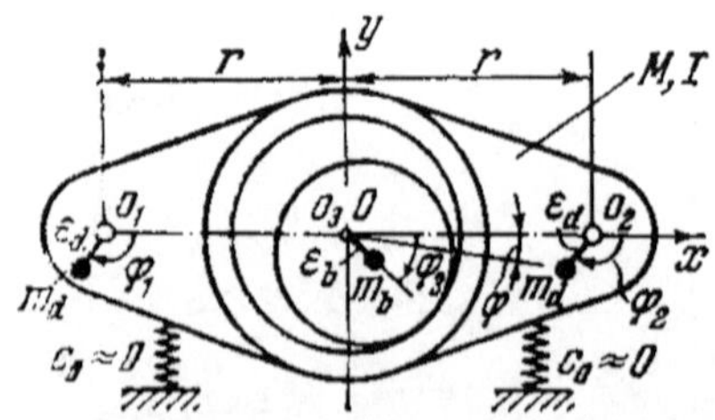

Note: 1. Regime exists at $\frac{e_b}{\tilde{\mathscr{A}}_0} \leqslant \frac{\sqrt{1+(f')^2}}{f'}$.

Table 2 *(continued)*.

Character of synchronous motion (first approximation) of		Conditions of stable synchronous motion	Expressions for magnitudes of vibratory torques $\widetilde{W}(\omega)$
exciters	supporting bodies		
$\varphi_1=\pm\omega t$, $\varphi_2=\pm\omega t$, $\varphi_3=\pm\left(\omega t+\rho'+\pi+\arcsin q\right)$, $q=\frac{\varepsilon_b}{\widetilde{\mathscr{A}}_0}\sin\rho'$, $\widetilde{\mathscr{A}}_0=2m_d\varepsilon_d/m_a$, $\rho'=\operatorname{arctg} f'$ (f' is the rolling friction coefficient), $m_a=M-m_b$	Circular translational vibrations.	$\frac{Mr^2}{I}>2\left\{1-\frac{m_b\varepsilon_b}{m_a\widetilde{\mathscr{A}}_0}\frac{1}{1+(f')^2}\times\left[\left((1+(f')^2\times\left(1-\frac{\varepsilon_b^2}{\widetilde{\mathscr{A}}_0^2}\right)\right)^{1/2}-(f')^2\frac{\varepsilon_b}{\widetilde{\mathscr{A}}_0}\right]\right\}$, if $k_b/k_d\ll 1$ (k_b and k_d are respective overall damping coefficients).	

2. Result is also true when the planetary exciter is external.

main mass M_1, via spring (stiffness c_2, with c_2, M_1, M_2 satisfying a simple correlation) or, paradoxically enough, to remove the guideways (see Fig. 17b, c, respectively). In both cases the desired result is due to the same factor—a change in the number of DOFs of the vibratory part of the system.*

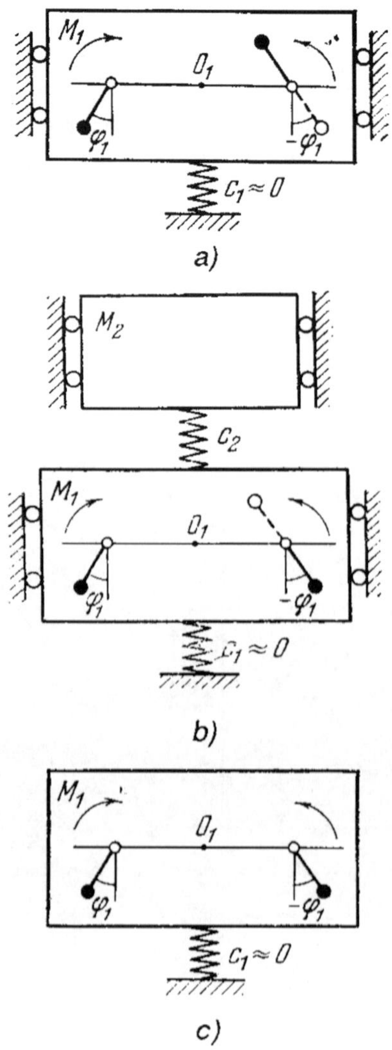

FIG. 17.

In the other example, circular translational vibrations of a soft-suspended rigid body are required. This can be achieved by using two identical debalanced exciters driven by induction motors with the same direction of rotation. Rotor axes are coplanar with and equidistant from the center of gravity O of the body (distance r). As shown in subsection 2, § 10 (also part 2, Table 2), in this case if

$$Mr^2/I > 2 \qquad (11.1)$$

then the synchronous and synphaseous rotation of the rotors is stable, which develops the required vibrations (Fig. 18a). If (11.1) is not satisfied, stable rotor rotation is in antiphase (Fig. 18b), and the body would perform angular vibrations about O_1.

Frequently, compliance with (11.1) is difficult. Thus, it is desirable to achieve stable synphaseous rotation of the two rotors with $Mr^2/I < 2$. As shown in part 17, Table 2, the system should be equipped with an additional centrally located debalanced exciter with the same partial velocity and the same rotation direction. If

$$\frac{Mr^2}{I} > 2 - \frac{m_3\varepsilon_3}{m\varepsilon} \qquad (11.2)$$

where $m\varepsilon$, $m_3\varepsilon_3$ are the static moments of rotors of the main and additional exciters, respectively, then the rotors of the main exciters are synphaseous in the stable synchronous regime, while the additional exciter would be antiphaseous to them (Fig. 18c). Vibrations would be circular translational, although (11.1) is not satisfied, but their amplitude would be somewhat reduced.

*The body vibrations in Fig. 17a could be, of course, maintained by just one debalanced exciter. However, in this case an unbalanced force in the perpendicular direction would develop, which is usually undesirable.

The same problem can be solved by the first method, by the elastodamping attachment of some additional mass M_2 to M_1 [see part 6, Table 2, and (57)].

Based on these synthesis techniques formulated from theoretical studies, several vibratory devices with self-synchronizing exciters were developed [see (57)], and also (108, 114–117, 167, 246).

Some methods would enhance the stability of a synchronous regime, since at certain conditions they lead to an increasing magnitude of vibratory torque. This concept is used, for example, in (231).

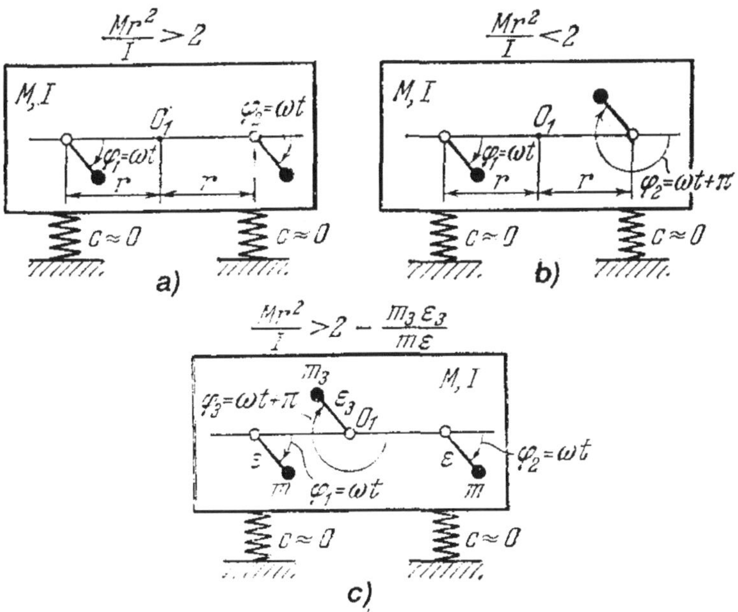

FIG. 18.

3. Methods of forced synchronization and synthesis of corresponding devices. The methods described for assuring stability of the required synchronous regime broaden substantially the potential for using the self-synchronization effect. However, in some cases, such designs are not optimal or are even impossible. Then, some forced synchronization technique could be used, such as:

1) Use of elastic elements (shafts, springs, couplings, etc.) to connect exciter shafts.

2) Use of induction nonreactive motors to drive exciters.

3) Use of servosystems like electric shafts or systems similar to systems of phaseous self-adjustment of frequency (see §§ 1 and 2, chap. 6).

Most systems of forced synchronization, both mechanical and electrical, are characterized by terms like $c_{sj}(\alpha_s - \alpha_j - \varkappa_s + \varkappa_j)$ in basic equations, where $\varkappa_s$, $\varkappa_j$ are so-called reference angles, and c_{sj} are quasi-elastic coefficients defining coupling between the s-th and j-th rotors, e.g., see (8.61). In the case of electrical synchronization, c_{sj} are represented by slopes ("stiffnesses") of respective torque characteristics.

By assigning reference angles $\varkappa_s$ and large enough c_{sj}, the existence and stability of any specified phasing of the rotors can always be obtained.

It is desirable, however, that the forced synchronization systems satisfy some optimality conditions. It is natural to request, for example, that the number of connections between the shafts of the exciters be the smallest, and the "stiffnesses" of the connections, the minimal allowable.

As shown in (39, 57), sometimes such an optimal selection of a forced synchronization system can be easily done. The problem of assuring the stability of a required phasing of the exciter rotors by applying a minimal number of rigid (kinematic) connections between the rotors can also be solved in a general formulation, e.g., see (57).

Here a simple example of the analysis of the forced synchronization of exciters in the system of Fig. 16 is given.

It is assumed that in the conditions of the example from subsection 2, § 10, exciter rotors are connected with springs (stiffness c) whose ends are attached to rotors at distances a from the rotation axes (Fig. 16d). The spring is not deformed when the rotor rotations φ_1 and φ_2 are the same. In such a case, according to (8.51), the potential function is expressed as

$$D = \Lambda^{(I)} - \Lambda^{(II)} = \langle (T^{(I)}) \rangle + \langle (\Pi^{(II)}) \rangle \tag{11.3}$$

and differs from (10.9) by the term

$$\langle (\Pi^{(II)}) \rangle = \left\langle \left(\frac{1}{2} c\Delta^2 \right) \right\rangle \tag{11.4}$$

The latter represents an average for a period of potential energy of the connecting spring (potential energy of supported connections), as calculated in the fundamental approximation (10.2). Elongation of the spring, assuming $a/r \ll 1$ and neglecting the powers of a/r higher than 1,

$$\Delta = l - l_0 = l - 2r \approx a(\cos\varphi_1 - \cos\varphi_2) \tag{11.5}$$

As a result,

$$D = -\frac{F^2}{2M\omega^2}\left(1 + \sigma_1\sigma_2 - \sigma_1\sigma_2\frac{Mr^2}{I} - \frac{Mca^2}{m^2\varepsilon^2\omega^2}\right)\cos\alpha + C_2 \tag{11.6}$$

where $\alpha = \alpha_2 - \alpha_1$, and C_2 is a constant that doesn't depend on α_1, α_2. As in subsection 2, § 10 (with the absence of spring) from $\partial D/\partial\alpha = 0$ it follows that both synphaseous $\alpha = (\alpha)_1 = (\alpha_2 - \alpha_1)_1 = 0$ and anti-phaseous motions $\alpha = (\alpha)_2 = (\alpha_2 - \alpha_1)_2 = \pi$ are possible. If

$$1 + \sigma_1\sigma_2 - \sigma_1\sigma_2\frac{Mr^2}{I} - \frac{Mca^2}{m^2\varepsilon^2\omega^2} < 0, \tag{11.7}$$

then function D has a minimum of $\alpha = (\alpha)_1 = 0$ and a maximum of $\alpha = (\alpha)_2 = \pi$. Thus, in accordance with the integral criterion, synphaseous motion of the rotors is stable, and antiphaseous is nonstable.

When rotors rotate in the same directions, or $\sigma_1\sigma_2 = 1$, condition (11.7) becomes

$$\frac{Mr^2}{I} > 2 - \frac{Mc^2a^2}{m^2\varepsilon^2\omega^2} = 2 - \frac{4ca^2}{MA_0^2\omega^2}. \tag{11.8}$$

If (11.8) is not satisfied, the antiphase motion is stable. At $c = 0$ this inequality becomes (10.13). Thus, the presence of a spring connecting the rotors alleviates compliance with the condition of stability of synphase rotation of rotors in the same directions. Moreover, by selection of a large enough stiffness, c stability of this motion can always be assured.

It is curious that for an antiphase rotation the spring is not needed and even counterproductive, since it can cause instability of this regime. The appropriate inequality is given in part 3a, Table 2; it coincides with (11.8).

A somewhat different condition would be obtained if the exciter rotors were connected with an elastic shaft of stiffness c which, for example, is not twisted at the synphaseous rotation of rotors in same direction; the corresponding condition is given in part 4a, Table 2. For $Mr^2/I > 2$, this condition is automatically satisfied at any stiffness c of the shaft; thus a synchronizing connection is not needed. By an appropriate selection of c, the stability of the synphaseous rotation in the same direction can also be assured at $Mr^2/I < 2$, i.e., in the case when such synchronous motion is unstable "by itself." An analogous result occurs when exciter rotors are connected either with an electric shaft with the same equivalent stiffness c, or are driven by synchronous nonreactive motors with the same stiffness of synchronous torque characteristic in the working point (57).

§ 12) Vibratory maintenance and entrainment of rotation of unbalanced rotor. Expressions for magnitude of vibratory torque and effective amplitude of rotor axis vibrations.

Vibration with some frequency ω of rotational axis of an unbalanced rotor may lead to a steady rotation of the rotor with the same or a fractional frequency, notwithstanding the fact that without the vibration the rotor, driven from some motor, was rotating with another frequency ω_0. In other words, there occurs an entrainment of rotor rotation by the vibration of its axis. Moreover, the vibration can cause and maintain rotation of the rotor with frequency ω or ω/n (n is a whole number), even if there was no rotation before the vibrations started. The latter effect is called *vibratory maintenance of rotation of an unbalanced rotor*. It is utilized, for example, in the "Hula-hoop" exercise, and in some vibratory machines and devices (see § 15).

In § 6, the vibratory maintenance of the rotation phenomenon was analyzed using the theory of synchronization of two vibro-exciters in a simple vibratory system. Here both phenomena will be analyzed in a more general case of vibrations of the rotor axis, but in an idealized formulation when these vibrations are given. This formulation corresponds to the problems of external synchronization (§ 1, chap. 1). Equations of motion of an unbalanced rotor (in Fig. 19), whose horizontal axis performs harmonic vibrations in two mutually orthogonal directions (with an elliptical trajectory)

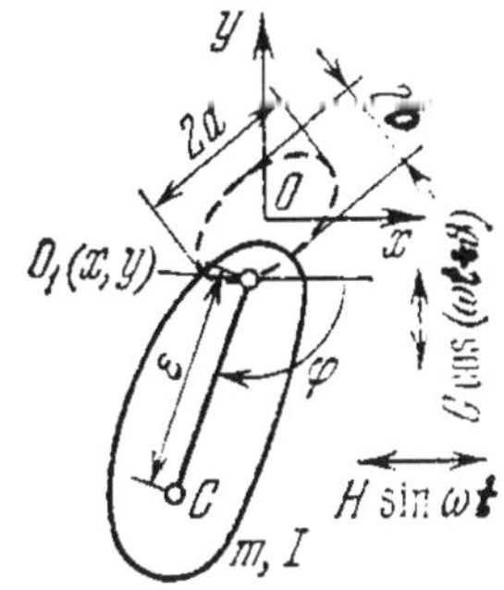

FIG. 19.

$$x = H \sin \omega t, \quad y = G \cos (\omega t + \vartheta) \tag{12.1}$$

(H, G are amplitudes, ω is frequency, and ϑ is phase shift angle), are as follows:

$$I\ddot{\varphi} = mg\varepsilon \cos\varphi - \\ - m\varepsilon\omega^2 [H \sin\omega t \sin\varphi + G\cos(\omega t + \vartheta)\cos\varphi] + L(\dot{\varphi}) - R(\dot{\varphi}) \tag{12.2}$$

Here φ is the angle of rotor rotation being counted clockwise; m, I, ε are the mass, moment of inertia, and eccentricity of rotor, respectively; g is the acceleration of gravity. As in § 4, $L(\dot{\varphi})$ and $R(\dot{\varphi})$ are, respectively, torque from induction motor and resistance torque.

Considering rotor rotation with an angular velocity close to $\dot{\varphi}^0 = \sigma\omega$, where $\sigma = \pm 1$, torques $L(\dot{\varphi})$ and $R(\dot{\varphi})$ are replaced by their linearized expressions like (4.10). After this, equation (12.2) can be written analogously to (4.12),

$$I\ddot{\varphi} + k(\dot{\varphi} - \sigma\omega) = \mu\Phi(\varphi, \omega t). \tag{12.3}$$

Here [see also formulas (4.13)]

$$k = k^* + k^\circ,$$

$$\mu\Phi(\varphi, \omega t) = mg\varepsilon \cos\varphi - \\ - m\varepsilon\omega^2[H \sin\omega t \cos\varphi + G\cos(\omega t + \vartheta)\cos\varphi] + \\ + L(\sigma\omega) - \sigma R^\circ(\omega). \tag{12.4}$$

The problem can be considered, for example, as a case of the problem analyzed in § 8 which corresponds to external synchronization of a single vibro-exciter. Then the basic equation (8.4), i.e., the equation for determining parameter α of the fundamental solution

$$\varphi^0 = \sigma(\omega t + \alpha), \tag{12.5}$$

will be written as

$$P(\alpha) \equiv \mu\sigma\langle(\Phi)\rangle \equiv \sigma L(\sigma\omega) - R^\circ(\omega) - W(\alpha) = 0. \tag{12.6}$$

The stability condition [see (8.30)] in such a simplistic case is reduced to inequality

$$\varkappa = \left.\frac{dP}{d\alpha}\right|_{\alpha=\alpha^*} = -\left.\frac{dW}{d\alpha}\right|_{\alpha=\alpha^*} < 0, \tag{12.7}$$

where α^* are solutions of equation (12.6).

Parameter W in equation (12.6) represents a vibratory torque which, as in the problems analyzed before, describes the specifics of the behavior of a rotor with a vibrating axis. This torque is described in this case as

$$W = \widetilde{W}(\omega)\cos(\alpha - \chi), \tag{12.8}$$

where

$$\widetilde{W}(\omega) = FA, \quad F = m\varepsilon\omega^2;$$

$$A = \frac{1}{2}\sqrt{G^2 + H^2 + 2\sigma GH\cos\vartheta} = \frac{1}{2}(a + \sigma^* b),$$

$$a = \frac{\sqrt{2}GH\,|\cos\vartheta|}{\sqrt{G^2 + H^2 - \sqrt{(G^2 + H^2)^2 - 4G^2H^2\cos^2\vartheta}}},$$

$$b = \frac{\sqrt{2}GH\,|\cos\vartheta|}{\sqrt{G^2 + H^2 + \sqrt{(G^2 + H^2)^2 - 4G^2H^2\cos^2\vartheta}}}; \tag{12.9}$$

$$\sigma^* = \sigma\,\mathrm{sgn}\,(\cos\vartheta) = \pm 1,$$

$$A\cos\chi = \frac{1}{2}(H + \sigma G\cos\vartheta),$$

$$A\sin\chi = \frac{1}{2}\sigma G\sin\vartheta.$$

Here, as in § 5, $\widetilde{W}(\omega)$ represents the magnitude of the vibratory torque W, and F is the excitation force developed by the rotor when its axis is stationary. Parameters a and b are the major and minor axes of the elliptical trajectory of vibrations as described by (12.1). Values $\cos\vartheta > 0$ correspond to a clockwise motion of the rotor axis around the ellipse, and $\cos\vartheta < 0$ correspond to a counterclockwise motion. Thus, $\sigma^* = \sigma\,\mathrm{sgn}\,(\cos\vartheta)$ is equal to 1 if rotor rotation in the direction coinciding with the direction of the rotor axis motion along the elliptical trajectory (12.1) is considered, and is equal to –1 if these directions do not coincide. Parameter A, corresponding to one introduced in § 5, would be called an *effective amplitude of rotor axis vibrations*. According to (12.9), effective amplitude is equal to one half of the sum of the ellipse half-axes if the motions coincide ($\sigma^* = 1$), and is equal to one half of the difference between the ellipse half-axes when the motion directions do not coincide ($\sigma^* = -1$).

In the case of rectilinear vibrations of the axis the $a = 0$, $b = H$, and as in § 5, effective amplitude is equal to one half of the rotor axis vibration amplitude. When the rotor axis is vibrating around a circle with radius $r_0 = a = b$, then the effective amplitude is $A = r_0$ if the direction of rotor rotation coincides with the direction of the axis motion around the circle. When these directions are opposite, effective amplitude and magnitude of vibratory torque $\widetilde{W}(\omega)$ are zeros.

The first three expressions (12.9) can be used to estimate the magnitude of vibratory torque $\widetilde{W}(\omega)$, if vibration amplitudes of the rotor axis in two mutually perpendicular directions (both perpendicular to the axis) are at least approximately known. Such an estimate is important, for example, for analyzing the vibration stability in devices with self-synchronizing vibro-exciters, see § 9.

First, *the case when a motor is absent or shut off* ($L = 0$), *i.e., the problem of vibratory maintenance of rotation*, is considered. In this case, (12.6) considering (12.8) can be written as

$$P(\alpha) \equiv -R^\circ(\omega) - \widetilde{W}(\omega)\cos(\alpha - \chi) = 0, \tag{12.10}$$

and stability condition as

$$\sin(\alpha - \chi) < 0. \tag{12.11}$$

If the following inequality is satisfied

$$R^\circ(\omega) < \widetilde{W}(\omega) = FA = m\varepsilon A\omega^2 \tag{12.12}$$

then equation (12.10) tolerates two essentially different solutions

$$\alpha_1 = \chi + \delta, \quad \alpha_2 = \chi - \delta,$$
$$\left(\frac{1}{2}\pi \leqslant \delta = \arccos\left[-\frac{R^\circ(\omega)}{m\varepsilon A\omega^2}\right] < \pi\right), \tag{12.13}$$

The first corresponds to an unstable rotation, and the second, to a stable rotation of the rotor with an average angular velocity ω.

The condition (12.12) of existence of the basic regime of vibratory maintenance of rotation of an unbalanced rotor coincides with condition (6.12). The latter is derived in a different way, but has the same physical meaning; resistance torque for the rotor rotation should not exceed some limiting value $R^\circ_{max}(\omega)$ equal to the magnitude $\widetilde{W}(\omega)$ of vibratory torque W or, which is the same thing, power $N(\omega) = R^\circ(\omega)\omega$ necessary to overcome resistance torque $R^\circ(\omega)$ should not exceed some limiting value $N_{max} = \widetilde{W}(\omega)\omega$:

$$N(\omega) = R^\circ(\omega)\omega < N_{max}(\omega) = \widetilde{W}(\omega)\omega = m\varepsilon A\omega^3. \tag{12.14}$$

As was noted in § 6, this power in real applications can be quite high (see the numerical example there). This is of utmost importance for the practical utilization of self-synchronization and vibratory maintenance of rotation phenomena.

These conditions do not guarantee development of such a steady regime of rotor rotation at arbitrary initial conditions; this regime can be rigidly excited. The issue is analyzed in (57, 148, 159, 196); experimental results are given in § 13.

It follows from (12.14) that the condition for maintaining rotor rotation in the direction of its axis motion around the elliptical trajectory is more "soft" than for the opposed direction; in the former case, effective amplitude A is, according to (12.9), larger than in the latter one. In the case of axis vibrations with a circular trajectory, rotor rotation in the opposite direction is totally impossible, ($A = 0$). Rectilinear vibrations, when both directions of rotation are equivalent, are an exception.

For *the case of the energized motor* ($L \neq 0$), *or the entrainment problem*, an excess torque, analogous to § 5, is

$$Z(\sigma, \omega) = \sigma L(\sigma\omega) - R^\circ(\omega). \tag{12.15}$$

According to (12.8), if

$$|Z(\sigma, \omega)| < \widetilde{W}(\omega) \tag{12.16}$$

then the basic equation (12.6) would tolerate two essentially different solutions—one stable and one unstable. The meaning of (12.16) is simple: for feasibility of the considered regime, the excess torque should not exceed the magnitude of the vibratory torque.

Partial velocity ω^* of the rotor (i.e., the angular velocity of the rotor on a stationary foundation counted, contrary to φ, in the direction of rotor rotation, see § 5) satisfies, according to (5.7),

$$\sigma L(\sigma\omega^*) = R^\circ(\omega^*)\,\mathrm{sgn}\,\omega^*. \tag{12.17}$$

Vibrations of the rotor axis can cause, in principle, rotation in the direction opposite to the direction of its driving motor. In such a case (representing a regime in which the driving motor performs as a generator) it may happen that $\sigma\dot{\varphi}^0 = \omega^* < 0$, or partial velocity is negative. As was noted in § 6, the concept of partial velocity can be considered a generalization of the self-excited vibrations concept. However, let $\omega^* > 0$, i.e., motor working in a normal regime. Then it follows from (12.15) and (12.16) that if the vibration frequency of axis ω coincides with ω^*, then $Z(\sigma, \omega) = Z(\sigma, \omega^*) = 0$, and condition (12.16) is always satisfied. In other words, if without axis vibrations the rotor rotates in a steady regime with angular velocity $\dot{\varphi}^0 = \sigma\omega^*$, then with vibrations of frequency $\omega = \omega^*$ it would also be capable of rotating with the same velocity. The considered regime can, however, also exist in the case when ω is slightly different from ω^*, such as when the vibratory torque counterbalances the excess torque $Z(\sigma,\omega)$, or the difference between the driving and resistance torques. Thus, there might exist a range of frequency variation

$$-\Delta_1 < \omega - \omega^* < \Delta_2, \tag{12.18}$$

in which rotor rotation is entrained by the frequency of external excitation. The width $\Delta = \Delta_1 + \Delta_2$ of this range is called the entrainment band.

The entrainment band for a given system can be reasonably wide. For example, ω^* can be zero (no driving motor or it is shut off), and still the rotation can be entrained by vibrations.

Besides the range (12.18), containing frequency ω^*, there could be other ranges of ω variation, in which such a rotation regime would exist. These ranges can be revealed by plotting $\sigma L(\sigma\omega)$, $R^\circ(\omega)$, $Z(\sigma, \omega)$, and $\widetilde{W}(\omega)$, see (57). Assuming, as in (5.14), near $\omega = \omega^*$

$$\begin{aligned} Z(\sigma, \omega) &= Z(\sigma, \omega^*) - k_Z(\omega - \omega^*), \\ \widetilde{W}(\omega) &= \widetilde{W}(\omega^*) + k_W(\omega - \omega^*), \end{aligned} \tag{12.19}$$

where

$$\begin{aligned} k_Z &= -\left.\frac{d[\sigma L(\sigma\omega) - R^\circ(\omega)]}{d\omega}\right|_{\omega=\omega^*} > 0, \\ k_W &= \left.\frac{d\widetilde{W}}{d\omega}\right|_{\omega=\omega^*} = 2m\varepsilon A\omega^* > 0, \end{aligned} \tag{12.20}$$

we arrive at

$$\Delta_1 = \omega^* - \omega_1 = \frac{\widetilde{W}(\omega^*)}{k_Z + k_W}, \quad \Delta_2 = \omega_2 - \omega^* = \frac{\widetilde{W}(\omega^*)}{k_Z - k_W}. \tag{12.21}$$

Usually $k_Z > k_W$, and parameter k_Z is quite similar to the overall damping coefficients k_s introduced in § 5.

It can be seen from (12.21) that there is no entrainment threshold. This is a value of effective vibration amplitude at which the entrainment band disappears. For a primitive self-generating vibratory system this fact was established in (6).

Vibrations of the rotor axis with frequency ω can also sustain its steady rotation

$$\varphi \approx \varphi^0 = \sigma\left(\frac{\omega}{n} t + \alpha\right) \qquad (n = 2, 3, \ldots), \tag{12.22}$$

i.e., with an average angular velocity, which is a whole fraction of ω (multiple forced synchronization), analysis at $n = 2$ resulted in the following basic equation for angle α (57, 148, 241):

$$P(\alpha) \equiv \sigma L\left(\frac{\sigma\omega}{2}\right) - R^\circ\left(\frac{\omega}{2}\right) - W_2(\alpha) = 0, \tag{12.23}$$

where the vibratory torque

$$W_2(\alpha) = \sigma \frac{4(m\varepsilon)^2 Ag}{I} \sin(2\alpha - \chi) \tag{12.24}$$

depends on acceleration of gravity g and does not depend on frequency ω. As before, from the condition of existence of real solutions $\alpha = \alpha^*$ for equation (12.23), the conditions of existence of the considered regimes would follow. The stability condition still has the format of (12.7). In this case, there are two stable regimes of rotation with different phases α.

The condition for vibratory maintenance of rotation, which follows from (12.23) and (12.24), and corresponds to inequality (12.14), now looks like

$$N\left(\frac{\omega}{2}\right) = R^o\left(\frac{\omega}{2}\right)\frac{\omega}{2} < N_{(2)\max}\left(\frac{\omega}{2}\right) = \frac{2(m\varepsilon)^2 A\omega g}{I}. \tag{12.25}$$

It is important that power $N_{(2)\max}(\omega/2)$ is usually much less than power $N_{\max}(\omega)$ for the principal rotational regime $\dot\varphi^0 = \sigma\omega$. Thus, in the example from § 6 at $I = 1$ kgm^2, $N_{(2)\max} = 2 \times 10^2 \times 0.25 \times 10^{-2} \times 314 \times 9.81 \approx 1600$ N · m/s ≈ 1.6 kW, while in § 6 800 kW was obtained for the principal regime.

Respectively, the entrainment band for the regime $\dot\varphi^0 = \sigma\omega/2$ is usually much narrower than for the regime $\dot\varphi^0 = \sigma\omega$. To enhance $N_{(2)\max}$ and widen the entrainment band (as well as the area of existence of multiple-synchronous motions of the vibro-exciters in self-synchronization regimes), the exciter could be equipped with a spring (Fig. 20), one end of which is pivotally attached to the rotor at some distance from its axis, and the other to the foundation (16, 59). For such a device, g in (12.24) is replaced by $(mg + T_0)/m$, where T_0 is an average spring force (or T_0/m, if the rotor rotates in a horizontal plane). As a result, the magnitude of the vibratory torque can be significantly increased, which is desirable. There are other methods of enhancement of the trend toward multiple synchronization also based, in effect, on enhancement vibrations of the rotor with frequency 2ω. The role of these vibrations follows from the analyses in (16, 57).

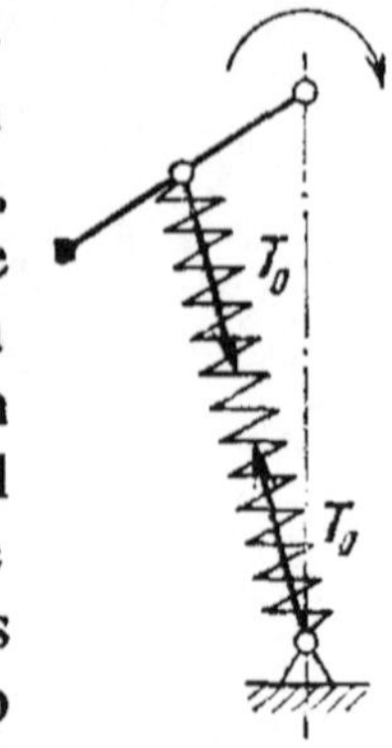

FIG. 20.

The problem of vibratory maintenance of rotation of an unbalanced rotor was discussed in (68), and then a generalized formulation was given in (31, 32, 42). The latter results were briefly described above. In (57) the case of the planetary vibro-exciter is also considered, and results of analyzing these problems as self-synchronization problems and forced synchronization problems (entrainment) are also compared. It is shown that the results coincide if the time history of the vibrations of the rotor axis for the solution of the entrainment problem is assumed to be the same as the time history of vibrations excited by all other exciters during their synchronous motion. The same conclusion can be obtained from analysis of the structure of the appropriate expression for vibratory torque, see (4.20), (8.45), (8.61), etc.

After (31, 32, 69) there appeared, both in the USSR and abroad, some other publications on the effect of vibratory maintenance of the rotation of an unbalanced rotor. Some of them are surveyed in (57); one of more the recent publications (80) studies vibratory maintenance of the rotation of an unbalanced rotor with a rotor with a pendulum attached off-center, i.e., a double physical pendulum.

§ 13) Experimental validation and study of self-synchronization and vibratory maintenance of rotation.

As noted in § 3, the self-synchronization and vibratory maintenance of rotation effects were initially discovered experimentally by chance (232). This led to the mathematical analysis (31, 32) which established new features and discovered the potential for practical applications. Thus a need for a careful experimental test of theoretical conclusions arose, as well as a need for the experimental study of some issues important for applications but not fully resolved theoretically.

These experiments were performed on special test rigs, in the "Mekhanobr" Institute and other places, including on commercial vibratory machines.

One such rig (from "Mekhanobr") is shown in Fig. 21. Its main part is a rigid pipelike vibrating part *1* (supporting body). In the pipe are located three debalanced vibro-exciters with horizontal axes *2* (Fig. 21a). The middle debalanced exciter can be easily replaced with a planetary one. Fig. 21b represents a case when such a planetary exciter is formed by two symmetrical sleeves *9*, in which cylindrical rollers could be easily inserted. The axis of the middle exciter extends through the center of gravity of the vibrating body, and the axes of the outside exciters are coplanar with and equidistant from the axis of the middle one.

Debalanced exciters are driven by induction motors *3* via flexible rubber shafts *10*. The unbalanced weights of the debalanced exciters are made of round washers *11* with threaded holes in their central parts. By changing washer positions on the radial rods/studs attached to the exciter rotors, their static moments can be varied in a broad range. Each of the debalanced exciters is equipped with special brakes *4* which allow variations in the resistance torque of the rotor and, thus, its partial angular velocity.

The vibratory part (supporting body) is suspended from the frame *8* by two springs *5* whose stiffness is selected to be so low that natural frequencies of the vibrating part are much lower than the angular velocity of the rotors. In addition, the vibrating part can be attached to the frame by a system of round elastic rods *6*. Their active length can be easily changed by shifting bracket *6* along the frame. This allows us to vary the stiffness of the elastic system and, thus, the natural frequencies of the vibrating part in a sufficiently broad range.

At a horizontal location of the axes of the elastic rods, as shown in Fig. 21a, the vibratory part has three essential degrees of freedom, since it can arbitrarily move in the vertical plane. The test rig allows, however, a vertical placement of the rod axes. In such a case, due to the horizontal location of the exciter axes, the vibrating part has only a single essential degree of freedom corresponding to displacement along the pipe, i.e., perpendicular to the rod axes.

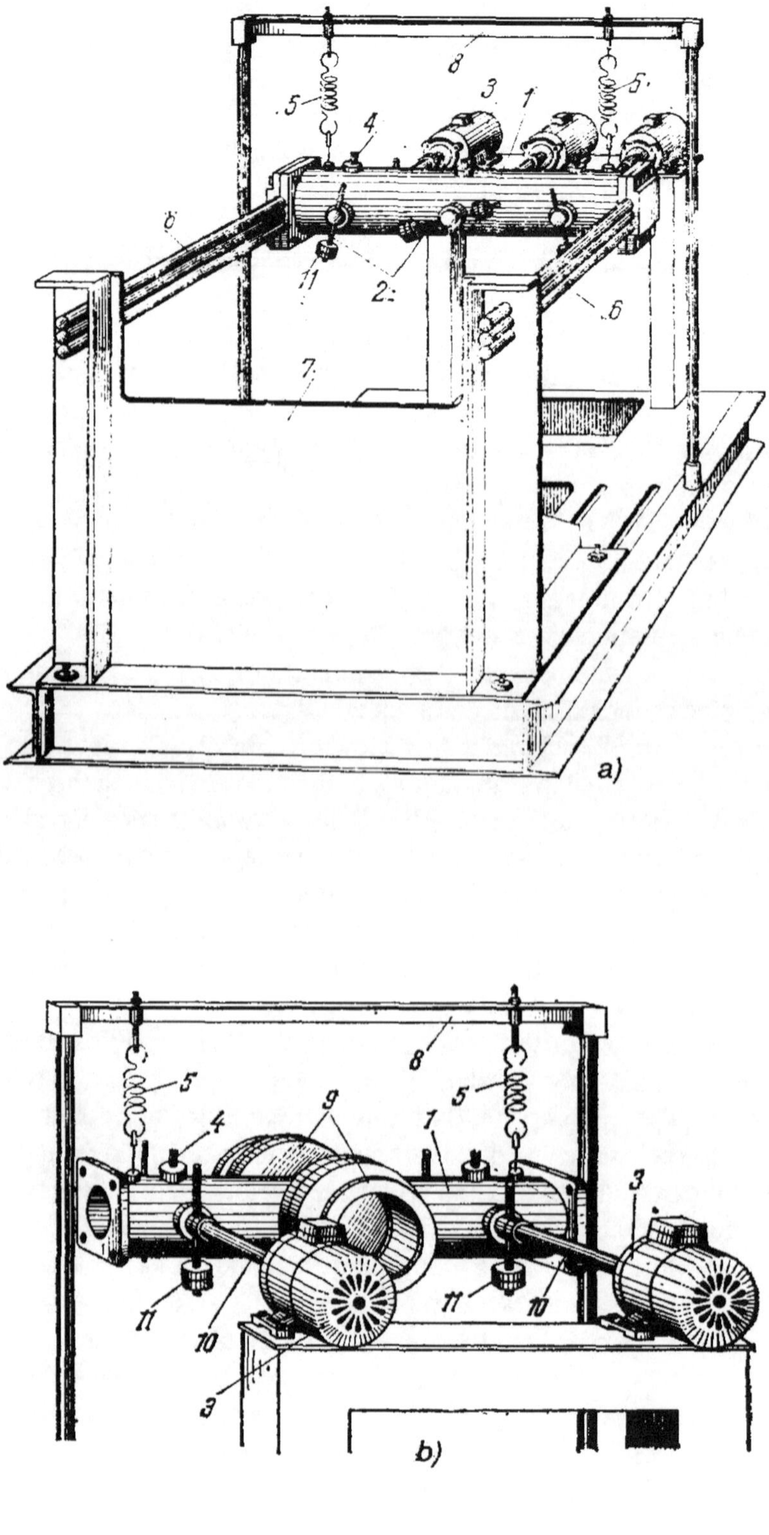

FIG. 21.

During the experiments, rotating rotors were observed in stroboscopic light with adjustable frequency of flashing. When the latter coincides with the rpm, the shaft looks stationary. This method not only allows us to monitor the identity of the average rotational velocities of the rotors (i.e., establish the presence or absence of self-synchronization) but also to observe the phases of rotor rotation. Besides, to get more accurate and objective data, the running rig was photographed with very short exposures, to obtain sharp images of the unbalanced weights.

One of the tests was for the case of the self-synchronization of two identical debalanced exciters that were symmetrically attached to a soft suspended vibrating body. As it was established theoretically, see subsection 2, § 10, and subsection 2, Table 2, when rotors rotate in the same direction and

$$Mr^2/I > 2 \tag{13.1}$$

then the synphaseous rotation is stable, causing circular translational vibrations of the vibrating portion. When the opposite inequality is true, then antiphaseous rotation is stable, corresponding to purely angular vibrations. For the opposite rotation of the rotors, the antiphase rotation is always stable, which results in rectilinear translational vibrations of the vibrating portion in a direction orthogonal to the plane containing the rotor axes.

All these results were experimentally confirmed.

Figs. 22 and 23 show photographs of the running rig* at exposures of 1/2500 s (at 1500 rpm). Figs. 22a–c show rotors rotating in the same direction ($\sigma_1\sigma_2 = 1$). For Figs. 22a–b (13.1) is satisfied, while for Fig. 22c it is not. Fig. 22a shows the case when both motors are energized; Fig. 22b when only one motor is energized. In the latter case, the stable rotation of the rotor with deenergized motor is sustained due to the vibrations of the vibrating portion, thus vibratory maintenance of rotation takes place.

In Fig. 23, the rotors are rotating in opposite directions ($\sigma_1\sigma_2 = -1$), with Fig. 23a corresponding to when both motors are energized, and Figure 23b to when one motor is energized.

As can be seen, phasing of the rotor rotations always fully complies with theoretical findings. Unplugging a motor of one of two running exciters during the tests led to only a small reduction of angular velocity of the synchronous rotation of the rotors, and to a quite small distortion of phasing (syn- or anti-) of rotation. Synchronous rotation of the rotor with the deenergized motor could continue for an indefinite time.

With the brakes disengaged and with very close partial angular velocities of exciters, an abrupt transition from stable synphaseous to stable antiphaseous rotation in violation of (13.1) was observed. Thus, if with $Mr^2/(2I) = 1.03$, the rotation is synphaseous, then an antiphaseous rotation already took place at $Mr^2/(2I) = 0.97$.

Other tests aimed at checking theoretical results for *vibratory maintenance of the rotation of an unbalanced rotor* were obtained in §§ 6 and 12. Initially, both exciter motors were energized. Then, after the stationary regime of synchronous rotation was established, one of the motors was deenergized. In some combinations of the parameters, the exciter with the deenergized motor continued rotation synchronously with another, driven, exciter, while in other combinations it came to a stop. By continuously changing the static moments of the exciters, the boundary of the do-

* In Figures 22 and 23, the spring suspension 5 of the vibrating portion shown in Figure 21 is replaced by soft rubber cables, which does not make a significant difference.

main of existence for the effect of the vibratory maintenance of rotation was established. It coincided with minor errors with the domain described by (6.12) and (12.12).

The same rig was also used to *study self-synchronization of three debalanced exciters*, specifically the stability condition given in part 17, Table 2. A high-speed movie camera was used to record the positions of the unbalanced weights. This allowed us to record the changes in the phases of the rotor rotation in the steady regime, which occur with frequencies equal to two and three times the synchronous rotation velocity. This complies with theoretical predictions (57) and explains the intensive noise and fast wear of gears, chains, and other means for forced kinematic synchronization in cases where they are used. Self-synchronization systems do not have such problems.

FIG. 22.

FIG. 23.

The same method was also used to study *regimes of the establishment of synchronous rotation of vibro-exciters*, as well as *the effect of vibratory excitation and maintenance of the roller rotation in planetary exciters*. The principal result of experiments conducted both on test rigs and on vibratory machinery with debalanced exciters driven from commercial, nominally identical, induction motors is as follows: self-synchronization is always established a very short time after starting the motors, if only the (local) theoretical conditions of existence and stability of the steady synchronous motion are satisfied.

Thus, the conditions of self-excitation of synchronous motions (entrainment in the synchronous regime) are in this case no more "rigid" than appropriate conditions of existence and stability of the steady regimes. This is very important for applications.

The other important experimental fact is that in the case of debalanced exciters, vibratory maintenance of the rotation of the unbalanced rotor is not, as a rule, self-excited. In other words, if motors of some exciters in a given machine or installation are started, then other (not energized) exciters usually are not entrained into a synchronous regime, although it exists and is stable (locally). After establishment of synchronous rotation of all rotors, e.g., as a result of energizing their motors for some time or an appropriate "push" of the rotor, one of several motors can be deenergized and still would not lose the synchronism. Thus, in the case of debalanced exciters, compliance with the conditions of existence and (local) stability of vibratory maintenance of rotation does not, as a rule, assure the development of rotation of arbitrary initial conditions. Exceptions are the cases when the amplitude of the steady vibrations of the rotor axis has about the same order of magnitude as an effective length of the physical pendulum corresponding to the unbalanced rotor. Such cases occur very seldom; usually the amplitudes of the vibrating organ are at least 5 to 10 times smaller than rotor eccentricity.

However, for planetary exciters it was found that the principal regime of steady motion of its roller is self-excited, for almost all purposes, in the whole domain of its existence.

Besides the described tests, which are more fully described in (57), there were many other experimental studies, e.g., in (116, 164, 172, 223, 299, 307, 320, 323). Some of these studies were performed on the universal vibratory rig described in subsection 1, § 15.

Considering all these publications, one can conclude that the theory of synchronization and vibratory maintenance of rotation, as presented in this chapter, is in good correlation with experiments and can be reliably used for the design of vibratory machinery.

§ 14) Principal features and paradoxes of self-synchronization of vibroexciters.*

1. Tendency of debalanced exciters for synchronization; vibratory maintenance of rotation. Debalanced exciters with sufficiently equal positive partial velocities ω_s, after being attached to a lightly damped vibratory system, show a tendency for self-synchronization. In this case, there exists at least one locally stable synchronous motion. In other words, if the exciter shafts, driven by inductionlike motors, are rotating with quasi-identical rpm while installed on a stationary foundation, they would certainly self-synchronize after being attached to an appropriate system of vibrating bodies.

Under certain conditions, self-synchronization is possible with strong damping and with not very close ω_s, such as with some $\omega_s = 0$ (disconnected driving motors of respective exciters). In such a case, the effect of vibratory maintenance of the rotation of unbalanced rotors takes place. This effect would also occur with appropriate conditions for planetary exciters.

Moreover, it is possible the self-synchronization of exciters with negative partial velocities ω_s, when some motors are working in generating regimes, have been induced to rotate in the direction opposite to their rotation while on a stationary foundation.

---

* Here a brief synopsis is given of the preceding paragraphs.

2. Effect of averaging partial velocities. At self-synchronization under the conditions mentioned in subsection 1, angular velocity of synchronous rotation ω is not higher than the highest, and not lower than the lowest partial velocity ω_s, and can accurately enough be calculated as

$$\omega = \frac{k_1\omega_1 + \ldots + k_k\omega_k}{k_1 + \ldots + k_k}, \tag{14.1}$$

where $k_s > 0$ are the overall damping coefficients of the exciters.

It is extended for all mechanical exciters shown in Fig. 10 except planetary ones.

3. Effect of transmission of large powers. The equalization of the partial angular velocities of vibro-exciters can be treated as a result of transferring power through the vibratory system from "faster" to "slower" exciters. In the special case of a deenergized (or missing) motor of an exciter, the energy required to drive its rotor is transferred from the exciters with effective motors via the vibratory system. It is very important for applications of self-synchronization and vibratory maintenance of rotation effects that this power be very significant. It has an order of magnitude of the product of the exciting force F developed by an exciter by the so-called effective amplitude A of the exciter axis and by synchronous velocity ω. For circular vibrations of the exciter axis, and for the same direction of the rotor rotation and axis movement along the vibration trajectory, the effective amplitude is equal to the radius of the trajectory of the axis vibrations, and for the rectangular harmonic vibrations, it is one half the amplitude of these vibrations. For example, for $F = m\varepsilon\omega^2 = 10^6\, N$, $A = 0.25 \times 10^{-2}$ m, and $\omega = 314\ \mathrm{s}^{-1}$ (quite realistic for the state of the art vibratory devices), maximum potential transmitted power is about 800 kW (see the example in subsection 1, § 6).

4. Correlation between phases of motion of inertial elements of exciters. During the stable synchronous motion of exciters, some specific values of the initial phases α_s of the inertial elements of exciters develop. For internal synchronization, e.g., for self-synchronization, this will be true for differences of initial phases $\alpha_s - \alpha_k$. In some cases, especially for a large number of exciters, several such (locally) stable sets of phases are possible.

5. Integral stability criterion (extremal property) of synchronous motions. In the case of self-synchronization of exciters (except planetary) with identical partial velocities, and of light damping, locally stable phasings correspond to points of rough minima of the Lagrange function (averaged for the period) relative to differences of the initial phases $\alpha_s - \alpha_k$. The Lagrange function period is computed for the fundamental approximation of the vibratory part of the system.* In many other cases, a stable phasing corresponds to the points of rough minima of a somewhat more structurally complex function of phases α_s—so-called potential function D—or certain adjacent points.

6. Effect of mutual balancing (principle of minimum of average kinetic energy) for exciters installed on a soft-suspended rigid body† (generalized Laval Principle). When several debalanced exciters with identical positive partial angular velocities are installed on a self-suspended rigid body, if dissipation is light, then there always exists a (locally) stable synchronous

* Rough minimum concept, see in § 8, chap. 10.

† See note 2) in p. 3, § 10.

motion at which unbalanced forces and torques generated by the exciters are mutually compensated. This means that averaging for a kinetic energy period is minimized. Specifically, if in principle such a phasing of the exciters is possible when total compensation of unbalaned forces and torques occurs, then this phasing, which is accompanied by zero kinetic energy, is stable. In such a case, the body is practically vibrationless. Such phasing will be called *compensating*.

This feature which follows one formulated in subsection 3, is extended also to many other types of mechanical exciters and, approximately, occurs in case of exciters with slightly different positive velocities, and also for planetary exciters t isiportant fr designig group foundationsfor unbalanced machines and devices (see subsection 2, § 15), and also devices for automatic balancing of unbalanced rotors, see chap. 4.

This feature can be considered as a peculiar generalization of the Laval Principle, i.e., self-balancing of a disc on a flexible rotating shaft in the aftercritical area (e.g., [30]). In comparison with the considered system with vibroexciters, though, case of shaft with disc is very special and degenerate.*

7. Paradox of nonworking connections. During the synchronization of mechanical exciters with identical positive partial velocities, there may exist, and be stable, synchronous motions at which connection system "does not work." The effect described in the previous paragraphs—when the supporting body in synchronous motion is stationary—is one of the examples of such a feature.

One can conclude erroneously that in such cases connections do not have any role and are not needed for synchronization. Actually, at a random perturbation of the motion, e.g., at a change of some phase α_s of the stable compensating phasing of the exciters, the connection system "turns on" and "works" till the perturbation ceases. Thus, in the case described in subsection 6, a perturbation of motion would cause vibrations of the rigid body which would not stop until the exciters phasing becomes compensating again.

This feature is analogous to the one described and explained by Huygens in the case of the synchronization of the pendulum clock (see the Introduction and chap. 5).

8. Character of stable phasing of exciters vs. number of DOF of connections. The character of stable phasing can change substantially by changing the number of essential degrees of freedom of the system of supporting bodies or the system of supported connections. For example, attachment of an additional mass on the spring, pendulum, etc., to the supporting body carrying exciters, can cause the stable synphaseous motion to become unstable, and vice versa (see subsection 2, § 11). Similarly, stable phasing changes after shifting the angular velocity ω from one range between natural frequencies of the supporting system to another range. It is characteristic, for example, of the system in part 19, Table 2 [more details in (57, 329)].

9. Paradox of forcing. This is well illustrated by the example in subsection 2, §11, (see Fig. 17). When, in this example, vertical guideways are used to assure vertical vibrations of the soft-suspended rigid body, then the compensating phasing of two exciters with oppositely rotating rotors is stable. Without the guideways (forcing of sorts), such phasing would be stable, which specifically assures the required vibrations of the body.

---

*Another interesting generalization of the Laval Principle—for a rigid body rotating on a string suspension—is discussed in (192).

This paradox can be considered as a special case of the feature described in the preceding point.

10. Character of stable phasing vs. number of exciters. It is clear from the example in subsection 2, § 11 that the character of exciters phasing in the stable synchronous motion can significantly change if one more exciter is added. This feature, as well as one noted in subsection 8, can be used to perform synthesis of devices with self-synchronizing exciters (see subsection 2, § 11).

11. Deterioration of self-synchronization and vibratory maintenance of motion conditions for multiple-synchronous regimes. The domains of existence and stability, as well as domains of attraction in cases of multiple self-synchronization and vibratory maintenance of rotation, are usually (without special devices) much narrower than in cases of simple self-synchronization/vibratory maintenance of rotation, e.g., see § 12.

§ 15) Applications of self-synchronization and vibratory maintenance of rotation effects.

1. Development of a new class of vibratory machinery. The discovery of the self-synchronization effect and development of its theoretical explanation led to the implementation of a new class of vibratory machines and devices with superior performance characteristics and economics.

Currently, the most popular are vibratory machines with two self-synchronizing debalanced exciters using the schematics in parts 2–15, Table 2.

FIG. 24.

Fig. 24 shows vibratory screens—machines for separating granular materials into two or more parts depending on the size of the granules. These machines are widely used in ore improvement plants, iron reduction furnaces, etc. The main working part of the screen is a grid fixed in a special box or grill, on which material is screened and transported simultaneously under vibrations. Screen boxes shown in the figure are induced to perform rectilinear translational harmonic vibrations at some angle to the screen plane. Such vibrations are generated by two self-synchronizing debalanced exciters (35). In the screen in Fig. 24a, the exciters are laid out according to the planar schematic of subsection 7, Table 2, while in the screen in Fig. 24b they are arranged by so-called 3-D schematic, as in subsection 12, Table 2.

The latter schematic is also used in the vibratory conveyer in Fig. 25. Conveyance of granular material in the tube of this conveyer also occurs under the influence of rectilinear translational harmonic vibrations of the pipe, occurring at a sharp angle relative to its axis.

Vibratory conveyers—elevators with two self-synchronizing exciters are shown in Fig. 26; they are designed according to the schematic of part 15, Table 2. Screw-like vibrations cause conveyance of granular material up along the threaded surface.

The use of self-synchronizing exciters in these machines instead of traditional kinematically synchronized exciters allows us not only to eliminate gear and chain transmissions, undesirable for vibratory devices, but also to create some other advantages. They include the possibility of using machines like those shown in Figs. 24b and 25 for hot materials; locating exciters, if necessary, at significant distances from one another; creating devices with working part vibrations that are easily adjustable, etc.

FIG. 25.

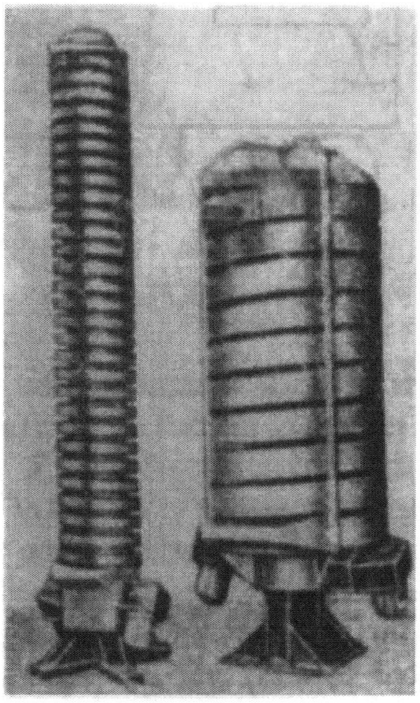

FIG. 26.

Successful examples of utilizing these possibilities are the unusual screens in Figs. 27 and 28, as well as the vibratory shakers and the vibratory ball mill shown in Figs. 29 and 30.

The first screen (104) contains a soft-suspended frame *1* on which two debalanced vibro-exciters *2* are symmetrically located. A certain number of grids *4* (there are four in Fig. 27) are attached to the frame by inclined flat springs *3*. The springs are soft in the transverse direction, but very rigid in the longitudinal direction; thus they provide a specific direction of translational vibrations while the frame is vibrating in a vertical direction due to the exciters. Studies have shown that this synchronous motion is stable in such a system, as it is stable in the simpler system in part 2a, Table 2.

Another screen, the so-called granulometer shown in Fig. 28, is designed for automatic monitoring of chunk sizes in crushed ore (267). In its housing *1* two grids (meshes) *2* and *3* are installed which are made as half-coils of a helical surface with a vertical axis. On the grid edges there are slots for the passage of material. On the granulometer housing, which is suspended on soft springs, are two

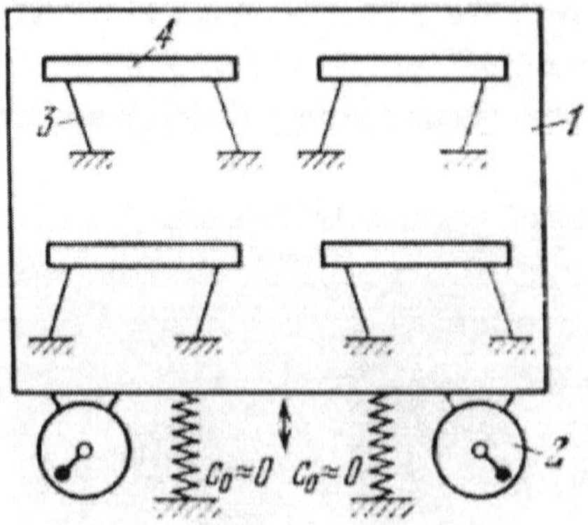

FIG. 27.

FIG. 28.

parts of the debalanced exciters *4* driven by independent induction motors *5*. Only the motors of the two exciters on opposite sides of the box are energized at a given time (schematic of part 15, Table 2). The exciters are self-synchronizing and generate screwlike vibrations of the box, inducing conveyance of material upward along the grids (meshes) if one pair of exciters is energized, and downward if the other pair or exciters is energized (the different direction of conveyance is achieved by an appropriate inclination of crossed axes of the exciter pairs). The energizing of one pair and deenergizing of another pair of exciters is performed by a special automatic device.

The device works as follows. After a sample of ore is loaded, the pair of exciters moving the material upwards along the helical screen surfaces is energized. With this, when the material leaves one screen, it enters another, etc. After a specific time interval has passed, which is needed for sufficiently complete screening of the fine fraction, this fraction, fallen in a special container, is automatically weighed, and its weight recorded in the memory. After this, the first exciter part is energized, which induces the motion of the larger granules remaining on the screens in the opposite direction, i.e., downward. During this process, after reaching the screen edges, the material falls into a container through the radial slots. After this, the whole sample is weighted, and the signal proportional to the ratio of the weight of the fine fraction to the total sample weight is displayed.

A photograph of a universal vibratory shaker table (55) with the control panel is shown in Fig. 29a. This table can be easily adjusted to perform vibrations of a varied character. This is very useful for studies of various vibratory processes, such as vibroconveyance, screening, vibroseparation, vibrocompaction, vibrocontainerization. Its principle of operation is illustrated by the sketch in part 15, Table 2. Let's assume that the rotational axes of the exciter rotors can be arbitrarily turned around on straight line *AB*, which is perpendicular to them, and then fixed. Then it is clear that, by an appropriate rotation of the axes, selection of appropriate rotation directions of the exciter rotors, and, perhaps, by some mass redistribution in the rigid body, it would be easy to transform the schematic of part 15, Table 2 into, for example, parts 2 and 12, Table 2. As a result, the following vibration patterns of the table could be achieved:

FIG. 29.

a) rectilinear translational vibrations at any angle to the horizon;

b) circular translational vibrations parallel to an arbitrarily oriented plane;

c) angular vibrations about an axis inclined at an arbitrary angle;

d) screwlike vibrations whose axis has an arbitrary direction in space, and whose pitch could be varied in a broad range;

e) complex vibrations at which trajectories are arbitrarily oriented ellipses.

The potential for generating the various vibratory fields for the table could be widened even more if rotation and fixation of the rotor axes could be performed not only about line *AB* but also about the axes perpendicular to both *AB* and the rotor axes. It can be noted that planetary exciters as well as various vibratory systems can be attached to the table, thus obtaining schematics similar to parts 9 and 21 of Table 2. For example, the photographs in Figs. 29b–d show installation on the table of mockups of vibroconveying devices as per the schematics of parts 12 and 15 of Table 2, and of the cone inertial crusher as per part 21, Table 2; the latter will be discussed below.

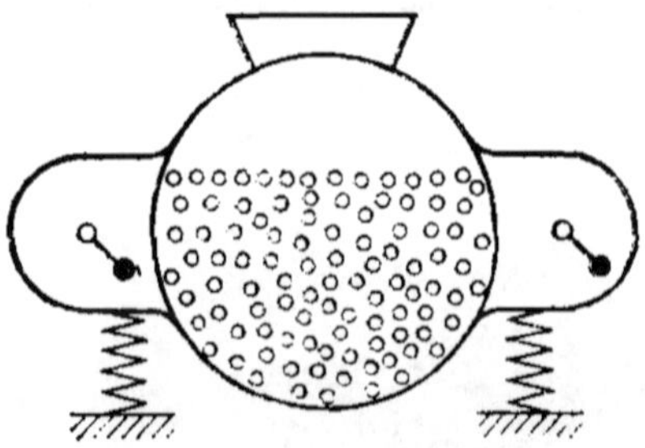

FIG. 30.

Variation of the frequency and amplitudes of the table vibrations can be achieved by the usual means.

The vibratory ball mill shown in Fig. 30 (37, 172) has a soft-suspended housing in which a cylindrical drum filled with crushing bodies is inserted. The housing is induced into circular translational vibrations by means of two identical debalanced exciters whose rotors rotate in the same direction (see part 2b, Table 2). These vibrations cause crushing of the material loaded in the drum with the balls. An advantage of this machine relative to mills driven from one centrally located exciter is in the fact that its chamber should not house this bulky exciter; in addition, loads on the exciter bearings can be reduced. For an analysis of the self-synchronization of vibro-exciters in the vibratory mill, usually neological characteristics of the local material being crushed should be considered (see part 6, Table 2).

The vibratory crushing-milling machines are a special group of devices in which an internal vibratory planetary vibro-exciter without a motor is used as a crushing body. Rotation of the roller of this exciter around its sleeve is achieved by the effect of vibratory maintenance of rotation (a special case of self-synchronization). One schematic of such a device is shown in part 21, Table 2: two identical debalanced vibro-exciters, whose rotors are rotating in the same direction, are symmetrically installed in the soft-suspended housing (machine frame); in the central part of the machine there is a sleeve into which a massive roller is inserted, the latter moving in a synchronous motion around the sleeve and crushing ore which is fed into the sleeve. The same schematic is used for several machines—mills, inertial crushers, etc. [see (57)].

The so-called inertial conical crusher belongs to the same group (139, 250–252). Its schematic cross section is shown (for one design) in Fig. 31a and the photograph in Fig. 32. Its simplified ("planar") schematic corresponds to one in part 20, Table 2. The crushing cone *1* of the machine (Fig. 31) includes a debalanced exciter *3* driven by motor *5* (attached to supporting frame *6*) through driveshaft *4* with cardan joints The housing *2* is attached to the frame via soft spring-cable suspensions *7*, *8*, resulting in the almost complete balancing of the machine. There are designs in which the same goal is achieved by installing a housing on the rubber-pneumatic isolators *6*, Fig. 31b.

Ore crushing is performed in an annular cavity between the external surface of the cone, and the internal surface of the housing.

In the regular steady regime of the crusher, the cone axis rotates about the housing axis with the angular velocity equal to the angular velocity of the exciter rotor.

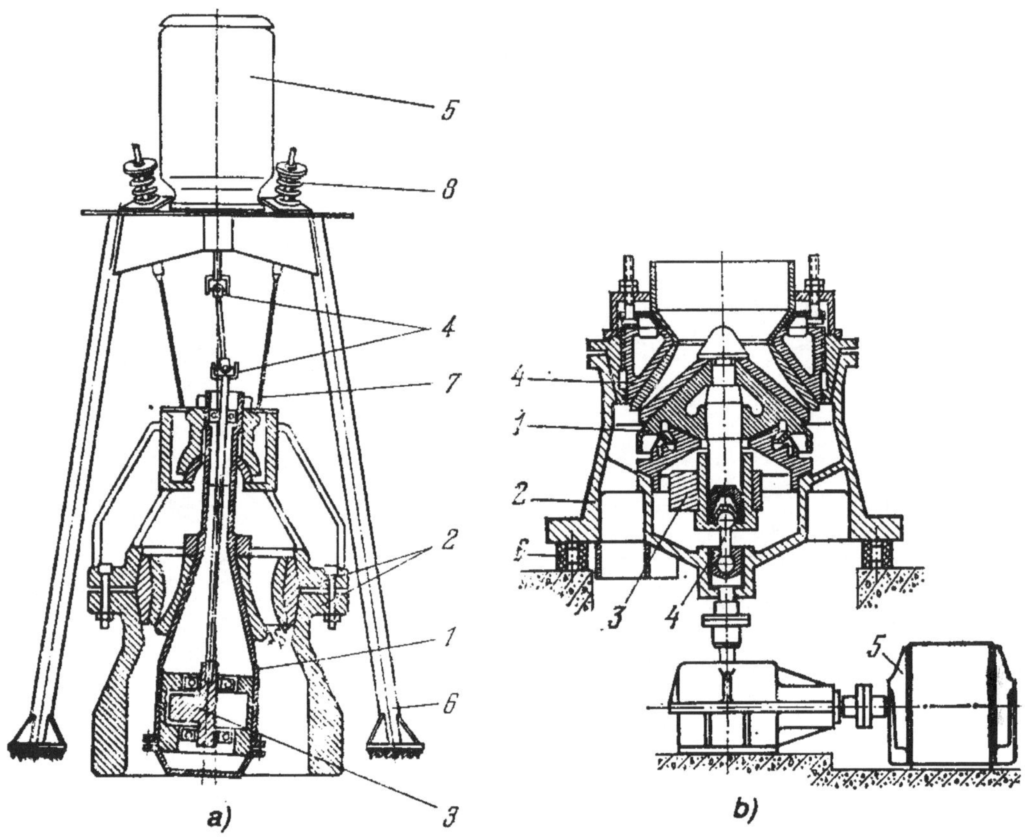

FIG. 31.

In this process, the cone comes in contact with the housing either directly or through a layer of ore in the working cavity. The direction of the inclination of the cone axis in the steady regime coincides (within accuracy up to a certain sharp angle) with the direction of the vector of eccentricity of debalance (unbalanced weight) of the exciter shaft. Thus, the total crushing force is not only due to the centrifugal force of the rotating debalance, but also to the centrifugal force generated by the precession motion of the cone.

The crusher operates at debalanced exciter rpms that are much higher than the frequency of the cone rocking in the traditional conical crushers, in which the cone is driven by a crank mechanism instead of the vibro-exciter. This, together with the design specifics of the machine, allows us to achieve much finer crushing than that feasible in traditional crushers.

A more detailed description of the design and processing characteristics of this crusher (developed in the "Mekhanobr" Institute) is given in (251, 252), and (57). One remark: the crushing cone can be considered a supporting rigid body, and the housing a ring of external planetary exciter without a driving motor. The fact that in this case the role of the exciter housing is played by an internal body (crushing cone) and the housing plays the role of the roller is not of critical importance in this case. The conditions of the existence and stability of a normal operating regime of the crusher are conditions of the existence and stability of the synchronous motion of both vibro-exciters of the crusher—debalanced and planetary ones or, since the latter does not have a motor—conditions of vibratory maintenance of the motion of the planetary exciter. Appropriate conditions derived for a "planar" crusher design are given in part 20, Table 2.

These examples do not exhaust the successful applications of the subject effects in vibratory technology. More detailed examples are given in (57), as well as in (42, 164, 166, 215).

FIG.32.

2. Application to group foundations for unbalanced machines. Recently, unified (group) foundations for several similar unbalanced machines which are rigidly attached to the foundation and driven by induction motors, gained popularity. Analysis of such foundations is based on the assumption that phases of unbalanced forces have a random character; thus unbalances of individual machines are mutually compensated to a certain degree (255, 290). However, there are known cases when synchronous and synphaseous operation of the machines developed, thus causing catastrophic consequences. From the aforesaid, such occurrences are understandable: under certain conditions, machines develop a tendency toward synchronous rotation with not a random, but a very specific phasing. In some cases, the phases could be close to each other or identical (this can explain the catastrophic occurrences mentioned above), while in other cases unbalanced forces would cancel each other.

The theory of synchronization as developed above allows us to decide when vibratory coupling between the machines is weak (thus self-synchronization is practically excluded and traditional methods of foundation design can be used), and when it is strong and the possibility of self-synchronization should be considered. In the

TABLE 3. General recommendations on use of the self-synchronization effect for the design of group foundations for unbalanced machines.

Correlation between frequencies p_{min}, p_{max}, and ω	Intervals of values $\varkappa_\omega$, and k_ω		
	$0 < \varkappa < 0.05$ $0 < k_\omega < 0.01$ (very weak vibratory coupling)	$0.05 < \varkappa_\omega < 1$ $0.01 < k_\omega < 0.2$ (relatively weak or relatively strong vibratory coupling)	$\varkappa_\omega > 1$ $k_\omega > 0.2$ (strong vibratory coupling)
$\omega > p_{max}$ (after resonance region)	—	−+0	+0
$p_{min} < \omega < p_{max}$ (between resonance region)	—	+0 or y	+0 or y
$\omega < p_{min}$ (before resonance region)	—	+y	+y

latter case, design of the foundation and the machine's layout should be done in such a way that cancellation, not summation, of unbalanced forces would occur.

Some guidance for such designs, based on subsection 4, § 9, and § 14, is given in Table 3, which is, in fact, an appropriately modified version of Table 1. The table is proposed in (64).

In Table 3, k_ω and χ_ω denote coefficients (introduced in subsection 2, § 9) characterizing the relative strength of vibratory coupling. More preferable is the use of coefficient χ_ω; only when necessary data is not available can one be limited with k_ω, which is very simple to calculate. As was noted in § 9, for a very weak vibratory coupling $\chi_\omega < 0.05$, $k_\omega < 0.01$, the probability of self-synchronization is less than 10%; in the case of a very strong coupling, $\chi_\omega > 1$, $k_\omega > 0.2$, this probability exceeds 90% (see Table 1). In Table 3, p_{min} and p_{max} are the lowest and highest natural frequencies of the foundation of the machines. The sign "—" indicates that computations are advisable without considering vibratory coupling, sign "+" means that the coupling must be considered, while "– +" shows that the calculations could be performed without considering vibratory coupling; however, at too high amplitudes, values might be obtained. The letters O and Y indicate cases when the occurrence of self-synchronization would attenuate or amplify the effects of individual unbalances. Attenuation or amplification is understood in terms of an average during a periodic value of the kinetic energy of the foundation vibrations. Due to the effect of mutual balancing of the exciters attached to a soft-suspended rigid body (subsection 6, § 14), attenuation would take place in the case of "soft" suspension of the foundation block on the supporting structure ($\omega > p_{max}$). In this case, if such a phasing is possible in a synchronous motion, when a complete mutual balancing takes place (so-called compensating phasing), the machinery would develop a tendency to such complete mutual balancing. In this case, the concept of "attenuation" in terms of the reduction of the average kinetic energy would approximately coincide with the traditional understanding. However, when compensating phasing is impossible (this case is characteristic for two machines which are analogous to debalanced exciters, see part 2a, Table 2), these concepts might not be equivalent.

One conclusion in designing group foundations for unbalanced machines is that it is important to achieve a situation when machines could mutually balance at a synchronous motion and make a soft connection between the foundation and the supporting structure.

The use of these recommendations on an example of a system whose model is given in part 14, Table 2 is illustrated in (57), where detailed computations and exemplary cases for machines driven by synchronous motors are given. The problem of self-synchronization of unbalanced machines developing polyharmonic forces and after being installed on a single DOF foundation, to which dynamic vibration absorbers are attached, is considered in (158).

Chapter 4

SYNCHRONIZATION OF OBJECTS SIMILAR TO MECHANICAL VIBRO-EXCITERS

§ 1) Introductory remarks.

This chapter considers the problems of dynamics of some engineering devices with rotating rotors, such as automatic balancers and flexible shafts with unbalanced discs. In all such devices, synchronization plays a very significant, if not the most important, role.

Despite substantial differences in appearance, such systems have many common features with systems containing mechanical vibro-exciters. As for the latter, they could be considered as orbital systems (see § 6, chap. 12) and as systems with quasi-uniform rotational motions (see § 4, chap. 12). Because of this, some basic results from chap. 3 are directly applicable here.

§ 2) Autobalancer.

One possible design schematic for an automatic balancer for rotating discs is shown in Fig. 33. Disc *2* is attached to flexible rotating shaft *1*, with the center of gravity C of the disc shifted from the shaft axis AO_1B. The disc has a cylindrical or toroidal cavity filled with oil. The axis of said cavity coincides with the line which is tangent to the shaft axis at the point O_1 of the disc attachment. There are two ball bearings *3* in the cavity which, under certain conditions, are set in the rotating disc in such a way that they compensate debalance of the disc and thus eliminate shaft vibrations and transmission of dynamic loads on its supports (bearings).

In other designs of the autobalancer, instead of a cavity and ball bearings there are either rings put on the shaft with a large clearance, or levers with one end attached to a sleeve that freely rotates on the shaft, while the other end carries an unbalanced weight.

Such autobalancers were invented by E. Searle (332, 333) and are used in machines such as centrifuges. They are especially effective if the rotor unbalance can vary somewhat during the machine operation.

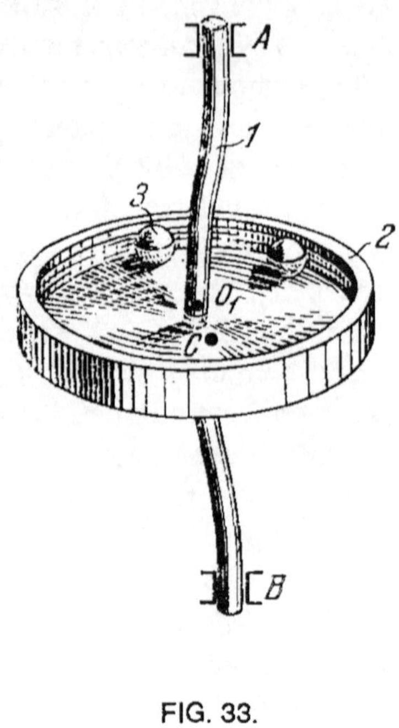

FIG. 33.

Elements of the theory of the autobalancer are given in (113), (119), (205), (225).

Equations of motion for the system with the ball-bearing autobalancer and one unbalanced disc at the center of a weightless shaft are as follows* (119)

---

* We use notation slightly different from (119); besides, the number k of ball bearings is arbitrary.

$$\ddot{\varphi}_s + \beta_0(\dot{\varphi}_s - \omega) = \frac{1}{R}(\ddot{x}\sin\varphi_s + \ddot{y}\cos\varphi_s)\ (s = 1, \ldots, k); \tag{2.1}$$

$$M\ddot{x} + \beta\dot{x} + cx = \mathcal{M}r\omega^2\cos\omega t + mR\sum_{s=1}^{k}(\ddot{\varphi}_s\sin\varphi_s + \dot{\varphi}_s^2\cos\varphi_s),$$
$$M\ddot{y} + \beta\dot{y} + cy = -\mathcal{M}r\omega^2\sin\omega t + mR\sum_{s=1}^{k}(\ddot{\varphi}_s\cos\varphi_s - \dot{\varphi}_s^2\sin\varphi_s). \tag{2.2}$$

In Fig. 34 x and y are coordinates of the disc center O_1 in a stationary frame xOy, whose origin is at the intersection point between the disc plane and the bearing axis; φ_s is the angle between the line connecting O_1 with the center of the s-th ball, and the direction of the axis Ox (counterclockwise); $\mathcal{M}$ = the disc mass; m = mass of a ball bearing; $M = \mathcal{M} + km$ is the mass of the whole system; r is the disc eccentricity; R is the distance between the centers of the ball bearings and the shaft axis; ω is the angular velocity of shaft rotation; β_0, β are the coefficients of the viscous resistance; c is the bending stiffness of the shaft relative to the force applied at O_1.

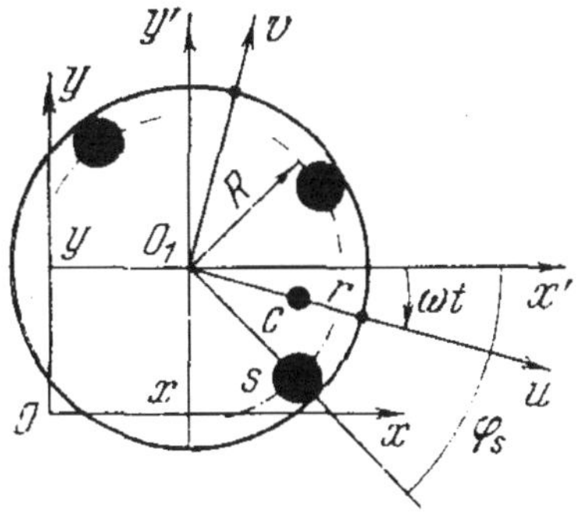

FIG. 34.

Especially interesting for the dynamics of autobalancers is the clarification of the conditions of existence and stability of such systems of differential equations (2.1), (2.2) in which $x(\omega t) \equiv y(\omega t) \equiv 0$ (or at least remain sufficiently small), i.e., solutions corresponding to the self-balancing of the system when shaft vibrations cease.

Such a formulated problem of the dynamics of autobalancers has been treated in (119) using methods different from the ones used here. There was also a much more complex case studied, in which the shaft had an unbalance distributed along its length and carried an arbitrary number n of discs, each having two ball bearings. This case is described by n groups of equations like (2.1), and equations (2.2) are replaced by partial differential equations.

Below, the solution of the problem is given as the problem of synchronization of the balls in the disc cavity (57). Equations (2.1) and (2.2) are presented as

$$\ddot{\varphi}_s + \beta_0(\dot{\varphi}_s - \omega) = \mu\omega^2\Phi(\varphi_s, \ddot{x}, \ddot{y}) \qquad (s = 1, \ldots, k), \tag{2.3}$$

$$M\ddot{x} + cx = mR\sum_{s=1}^{k}(\ddot{\varphi}_s\sin\varphi_s + \dot{\varphi}_s^2\cos\varphi_s) + \mathcal{M}r\omega^2\cos\omega t - \mu\beta'\dot{x},$$
$$M\ddot{y} + cy = mR\sum_{s=1}^{k}(\ddot{\varphi}_s\cos\varphi_s - \dot{\varphi}_s^2\sin\varphi_s) - \mathcal{M}r\omega^2\sin\omega t - \mu\beta'\dot{y}, \tag{2.4}$$

where denoted

$$\mu\Phi(\varphi_s, \ddot{x}, \ddot{y}) = \frac{1}{R\omega^2}(\ddot{x}\sin\varphi_s + \ddot{y}\cos\varphi_s), \quad \mu\beta' = \beta, \tag{2.5}$$

Parameter $\mu > 0$ will be considered as small with the same reasoning as in § 4, chap. 3 during analysis of the synchronization of vibro-exciters.

In accordance with the aforesaid, solutions of (2.3) and (2.4) will be sought having format as follows:

$$\varphi_s = \omega t + \alpha_s + \mu\psi_s^* (\omega t), \quad x = x(\omega t), \quad y = y(\omega t), \tag{2.6}$$

where ψ_s^*, x,y are periodic functions of time t with period $2\pi/\omega$, and α_s are constant. The most interesting solutions are those in which "initial phases" α_s are such that $x \approx 0$, $y \approx 0$; thus self-balancing occurs.

The problem is a special case of the problems of synchronization of objects with quasi-uniform rotational motions, considered in § 4, chap. 12, and § 8, chap. 3. Equations (2.3), (2.4) and solutions (2.6) are of the same type as equations (4.2), (4.3), and solutions (4.1) of chap. 12.

The fundamental system corresponding to equations (2.3) and (2.4) at $p^2 = c/M \neq \omega^2$ (the nonresonance case is considered) tolerates the following family of solutions in the format of (4.5), (4.6), chap. 12:

$$\begin{aligned} \varphi_s^0 &= \omega t + \alpha_s, \\ x^0 &= wR\left[\eta \cos \omega t + \sum_{j=1}^{k} \cos(\omega t + \alpha_j)\right], \\ y^0 &= -wR\left[\eta \sin \omega t + \sum_{j=1}^{k} \sin(\omega t + \alpha_j)\right]. \end{aligned} \tag{2.7}$$

It is a function of k arbitrary parameters $\alpha_1, \ldots, \alpha_k$.

It is denoted here:

$$\eta = \frac{\mathcal{M} r}{mR}, \quad w = \frac{m}{\mathcal{M}} \frac{\omega^2}{p^2 - \omega^2}, \quad p^2 = \frac{c}{M}. \tag{2.8}$$

Now the *basic equations* (4.6), chap. 12, will be written for determining the values of parameters $\alpha_1, \ldots, \alpha_k$, for which solutions of the format (2.6) may exist. In accordance with (2.5) and (2.7), these equations can be written as

$$\begin{aligned} P_s(\alpha_1, \ldots, \alpha_k, \omega) &\equiv \frac{\mu\omega^2}{\beta_0} \langle \Phi(\varphi_s^0, \ddot{x}^0, \ddot{y}^0) \rangle \equiv \\ &\equiv \frac{1}{\beta_0 R} \int_0^{2\pi/\omega} (\ddot{x}^0 \sin \varphi_s^0 + \ddot{y}^0 \cos \varphi_s^0)\, dt \equiv \\ &\equiv -\frac{1}{2} \frac{\omega^2}{\beta_0} w \left[\eta \sin \alpha_s + \sum_{j=1}^{k} \sin(\alpha_s - \alpha_j)\right] = 0 \\ &(s = 1, \ldots, k). \end{aligned} \tag{2.9}$$

These equations could undoubtedly tolerate solutions relative to $\alpha_1, \ldots, \alpha_k$, complying with conditions

$$\eta + \sum_{j=1}^{k} \cos \alpha_j = 0, \quad \sum_{j=1}^{k} \sin \alpha_j = 0, \tag{2.10}$$

When these conditions are satisfied, it follows from (2.7) that shaft vibrations in the first approximation do not exist,

$$x^0(\omega t) = y^0(\omega t) = 0. \tag{2.11}$$

Thus, the most important issues are, first, to formulate stability conditions corresponding to solutions of the (2.10) type, and, second, to find out if, besides these motions, some other solutions would be stable, described by solutions of (2.9) not complying with conditions (2.10). It is easy to see that the latter solutions always exist.

Solutions of equations (2.9) which comply with conditions (2.10), and the corresponding motions, would be called *principal*, and all other solutions (and motions), *secondary*.

Stability of the solutions of equations (2.9) depends on the signs of the real parts of the roots of the algebraic equation (2.12), chap. 12.

Firstly, the case of two ball bearings in the disc cavity will be considered. At $k = 2$, equations (2.9) become

$$\begin{aligned} P_1(\alpha_1, \alpha_2, \omega) &\equiv -\frac{1}{2}\frac{\omega^2}{\beta_0} w\, [\eta \sin \alpha_1 + \sin(\alpha_1 - \alpha_2)] = 0, \\ P_2(\alpha_1, \alpha_2, \omega) &\equiv -\frac{1}{2}\frac{\omega^2}{\beta_0} w\, [\eta \sin \alpha_2 - \sin(\alpha_1 - \alpha_2)] = 0. \end{aligned} \tag{2.12}$$

These equations allow, in general, for four essentially different solutions. Obviously, the solutions should not be distinguished, in which appropriate α_1 and α_2 differ by $2\pi n$, where n is a whole number and, since the ball bearings are considered identical, in which values of α_1 and α_2 are interchanged. The solutions are as follows:

$$\begin{aligned} &1) \quad \alpha_1^{(1)} = \pi - \gamma, \quad \alpha_2^{(1)} = -(\pi - \gamma); \\ &2) \quad \alpha_1^{(2)} = 0, \quad \alpha_2^{(2)} = 0; \\ &3) \quad \alpha_1^{(3)} = \pi, \quad \alpha_2^{(3)} = \pi; \\ &4) \quad \alpha_1^{(4)} = 0, \quad \alpha_1^{(4)} = \pi; \end{aligned} \tag{2.13}$$

where $\gamma = \arccos \frac{\eta}{2} = \arccos \frac{\mathcal{M} r}{2mR}$.

Only the first of these solutions satisfies the conditions of the absence of shaft vibrations (2.10). In other words, only the first solution is principal; the other ones are secondary. The condition for existence of the principal solution is

$$\frac{\eta}{2} = \frac{\mathcal{M} r}{2mR} < 1, \tag{2.14}$$

It expresses the obvious requirement that the maximal static moment of both ball bearings about the shaft axis should be larger than the static moment of the disc about the same axis.

The mutual positions of the ball bearing centers c_1, c_2, the shaft center in the undeformed condition O during motion O_1, and also the center of gravity C of the disc for each of the four solutions (2.13) are shown in Fig. 35a–d. The configurations are shown in accordance with formulae (2.13) and (2.7) for $t = 0$. In an arbitrary moment of time, the disc with the ball bearings and coordinate frame uOv would rotate by angle $\varphi_0 = \omega t$.

Let us analyze the stability of solutions (2.13). Equation (2.12), chap. 12, can be written in this case as

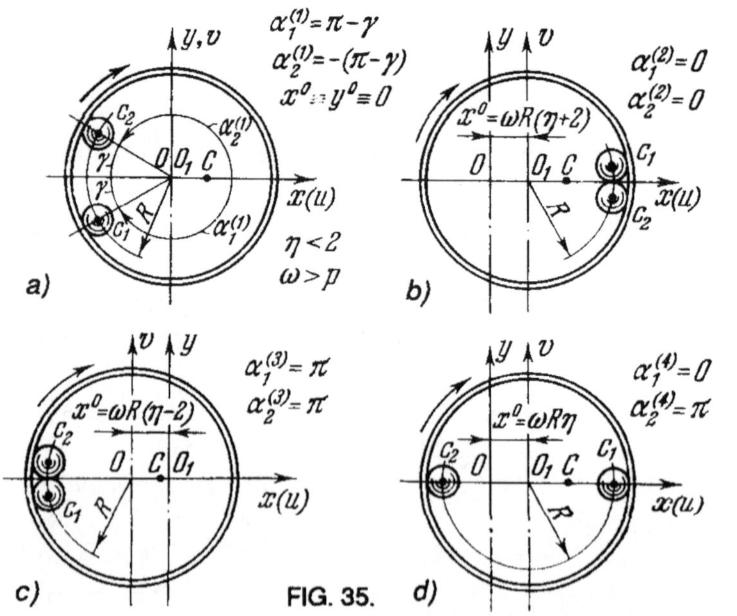

FIG. 35.

$$\varkappa^2 + A\varkappa + B = 0, \tag{2.15}$$

where, according to (2.12)

$$A = -\left(\frac{\partial P_1}{\partial\alpha_1} + \frac{\partial P_2}{\partial\alpha_2}\right) = \frac{\omega^2 w}{2\beta_0}\left[\eta(\cos\alpha_1 + \cos\alpha_2) + 2\cos(\alpha_1-\alpha_2)\right],$$

$$B = \left(\frac{\partial P_1}{\partial\alpha_1}\frac{\partial P_2}{\partial\alpha_2} - \frac{\partial P_1}{\partial\alpha_2}\frac{\partial P_2}{\partial\alpha_1}\right) = \tag{2.16}$$

$$= \frac{\omega^4 w^2\eta}{4\beta_0^2}\left[\eta\cos\alpha_1\cos\alpha_2 + (\cos\alpha_1 + \cos\alpha_2)\cos(\alpha_1-\alpha_2)\right]$$

Sufficient conditions for the existence and asymptotic stability of motion of the considered type, complying with solutions (2.13), are inequalities*

$$A > 0, \quad B > 0. \tag{2.17}$$

For the first, principal solution in (2.13), it can be found from (2.16) that:

$$A = A^{(1)} = \frac{\omega^2 w}{\beta_0}(-\eta\cos\gamma + \cos 2\gamma) = -\frac{\omega^2 w}{\beta_0},$$

$$B = B^{(1)} = \frac{\omega^4 w^2\eta}{4\beta_0^2}(\eta\cos^2\gamma - 2\cos\gamma\cos 2\gamma) = \left(1-\frac{\eta^2}{4}\right)\frac{\omega^4 w^2\eta^2}{4\beta_0^2}. \tag{2.18}$$

According to (2.14), parameter B is always positive inside the domain of existence of the considered solution. The condition for A to be positive leads to a requirement $w < 0$ or, according to (2.8), to inequality

* Strictly speaking, some additional stability conditions (see § 4, chap. 3) should be attached to inequalities (2.17), similar to the problem of synchronization of vibro-exciters. However, in this case these conditions are also not very important at a sufficient distance from the resonance.

$$\omega > p = \sqrt{c/M}. \tag{2.19}$$

Thus, the principal motion (one at which the shaft does not vibrate due to compensation of the disc unbalance due to ball-bearing effect) is stable in the postcritical range.

For two secondary motions (2.13), it follows from (2.16) that

$$\begin{aligned} A^{(2)} &= \frac{\omega^2 w}{\beta_0}(1+\eta), & B^{(2)} &= \frac{\omega^4 w^2 \eta}{4\beta_0^2}(2+\eta), \\ A^{(3)} &= \frac{\omega^2 w}{\beta_0}(1-\eta), & B^{(3)} &= \frac{\omega^4 w^2 \eta}{4\beta^2}(\eta-2), \\ A^{(4)} &= -\frac{\omega^2 w}{\beta_0}, & B^{(4)} &= \frac{\omega^4 w^2 \eta^2}{4\beta_0^2}. \end{aligned} \tag{2.20}$$

It follows that the second solution is stable only in the precritical range ($\omega < p$); the third solution is stable only in the postcritical range ($\omega > p$), and only if $\eta > 2$; the fourth solution is altogether unstable. Thus, if (2.14) is satisfied, which guarantees existence of the principal solution, this solution in the postcritical zone is the only stable solution of the considered type. However, at precritical speeds this solution is unstable. This defines the domain of the effective use of the studied autobalancers.

Thus, the problem is in effect covered by the well-known solution of the classical problem of self-balancing an unbalanced disc without ball bearings [e.g., (30)]. This was previously called the "Laval Principle" (see subsection 6, § 14, chap. 3).

The obtained results correlate with (and, to a certain degree, complement) the results obtained by different methods in (119). Since the autobalancer already performs with two ball bearings, there is no need to use a larger number of ball bearings. On the other hand, at $k > 2$ the study becomes more complex, since then solutions of (2.10), and thus of basic equations (2.9), cease to be isolated. It reflects the facts that with $k > 2$ and with satisfaction of a condition equivalent to inequality (2.14), balancing of the disc is possible at any fixed configuration of $k-2$ ball bearings by the selection of the positions of two remaining ball bearings.

Finally, in the considered problems it is easy to derive a simple integral stability criterion (extremal property) of synchronous motions from the general correlations of § 4, chap. 12 and also § 8, chap. 3. The role of potential function is played by the Lagrange function corresponding to the vibrations relative to the shaft of the fully balanced disc, to whose axis the ball-bearing masses are attached. In this case, the Lagrange function is calculated in a fundamental approximation:

$$\begin{aligned} D = \Lambda^{(I)} &= \langle (T^{(I)} - \Pi^{(I)}) \rangle = \\ &= \left\langle \frac{1}{2} M [(\dot{x}^0)^2 + (\dot{y}^0)^2] - \frac{1}{2} c [(x^0)^2 + (y^0)^2] \right\rangle. \end{aligned} \tag{2.21}$$

Averaging, considering (2.7) and expressions (4.17) of chap. 3, we arrive at

$$D = \Lambda^{(I)} = -M \frac{m\omega^2 R^2}{\mathcal{M}} w \left[\eta \sum_{s=1}^{k} \cos \alpha_s + \frac{1}{2} \sum_{s=1}^{k} \sum_{j=1}^{k} \cos(\alpha_s - \alpha_j) \right]. \tag{2.22}$$

Comparing this result with the one obtained for the problem of vibro-exciters' synchronization [see (5.23), chap. 3], it should be understood that in this case there are $k+1$ equivalent vibro-exciters—one unbalanced disc, and k ball bearings.

According to the integral stability criterion, stable motions correspond to points of rough minima of the function D. Both basic equations (2.9) and the respective stability conditions are easily derived from this concept.

It is easy to see that a conclusion about the self-balancing of the system far in the postcritical range, where potential energy $\Pi^{(I)}$ in (2.21) can be neglected compared with kinetic energy $T^{(I)}$, follows directly from the formulated integral stability criterion. This conclusion correlates with the conclusion from the integral stability criterion in the problem of synchronization of vibro-exciters described in § 5 and in subsection 6, § 14, chap. 3, or with the concept previously called the *generalized Laval Principle.*

§ 3) Rotor with pendulums.

For the theory of some devices, the problem about steady motion of a disc on a vertical shaft with two physical pendulums having pivoting axes perpendicular to the disc plane (Fig. 36a) attached to the disc is important. Especially important is finding the conditions of the existence and stability for such motion of the system's disc-pendulums, at which the pendulums are mutually balancing each other. As a result, the disc center remains stationary in space and no unbalanced forces are transmitted to the shaft bearings and to the stationary support structure.

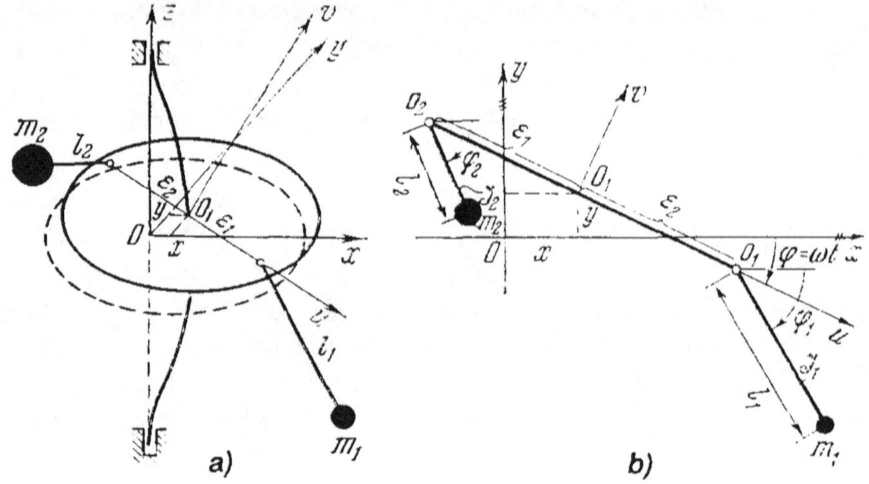

FIG. 36.

Such a problem was treated in (190) by reducing it to the problem of synchronization of objects with almost uniform rotational motions. The results of this paper, with minor changes and corrections, are given below.

The dynamic model of the system is given in Fig. 36b. It is assumed that the disc itself is statically and dynamically balanced and attached halfway between the bearings. Thus, center O_1 of the disc can vibrate only in the disc plane. Then, assuming uniform rotation of the disc, the system has four DOF, and its motion can be described in the stationary coordinate frame xOy by two equations of pendulum oscillations considering the mobility of their pivots, and two equations of small oscillations of the disc considering the reactions due to the pendulums:

$$I_1\ddot{\varphi}_1 + k_1\dot{\varphi}_1 - m_1 l_1 (\ddot{x}\sin\varphi_1 + \ddot{y}\cos\varphi_1) + \\ + m_1\varepsilon_1 l_1\omega^2 \sin(\varphi_1 - \omega t) = 0, \qquad (3.1)$$

$$I_2\ddot{\varphi}_2 + k_2\dot{\varphi}_2 - m_2 l_2 (\ddot{x}\sin\varphi_2 + \ddot{y}\cos\varphi_2) - \\ - m_2\varepsilon_2 l_2\omega^2 \sin(\varphi_2 - \omega t) = 0,$$

$$M\ddot{x} + cx - s\omega^2\cos\omega t - m_1 l_1 (\ddot{\varphi}_1 \sin\varphi_1 + \dot{\varphi}_1^2\cos\varphi_1) - \\ - m_2 l_2 (\ddot{\varphi}_2\sin\varphi_2 + \dot{\varphi}_2^2\cos\varphi_2) = 0. \qquad (3.2)$$

$$M\ddot{y} + cy + s\omega^2\sin\omega t - m_1 l_1 (\ddot{\varphi}_1\cos\varphi_1 - \dot{\varphi}_1^2\sin\varphi_1) - \\ - m_2 l_2 (\ddot{\varphi}_2\cos\varphi_2 - \dot{\varphi}_2^2\sin\varphi_2) = 0.$$

Here x,y are the coordinates of the center of gravity of the disc; φ_1, φ_2 are the swinging angles of the pendulums, counterclockwise from the positive direction of the axis Ox; m, m_1, m_2 are the masses of the disc and the pendulums, respectively; $\mathcal{J}_1, \mathcal{J}_2$ are the moments of inertia of the pendulums about their axes; l_1, l_2 are the distances from the pendulum pivots to their centers of gravity; ε_1, ε_2 are the distances from the center of the disc to the axes of the pendulums (eccentricities of the pendulum suspensions); k_1, k_2 are the coefficients of the viscous resistance for the pendulum rotations; c is the bending stiffness of the shaft; ω is the angular velocity of the disc; $M = m + m_1 + m_2$is the overall mass of the system.

$$I_1 = \mathcal{J}_1 + m_1\varepsilon_1^2, \quad I_2 = \mathcal{J}_2 + m_2\varepsilon_2^2, \qquad s = m_1\varepsilon_1 - m_2\varepsilon_2. \qquad (3.3)$$

Introducing expressions

$$\mu\omega^2\Phi_1(\varphi_1, \ddot{x}, \ddot{y}, \omega t) = \\ = m_1 l_1(\ddot{x}\sin\varphi_1 + \ddot{y}\cos\varphi_1) - m_1\varepsilon_1 l_1\omega^2\sin(\varphi_1 - \omega t) - k_1\omega. \\ \mu\omega^2\Phi_2(\varphi_2, \ddot{x}, \ddot{y}, \omega t) = \\ = m_2 l_2(\ddot{x}\sin\varphi_2 + \ddot{y}\cos\varphi_2) - m_2\varepsilon_2 l_2\omega^2\sin(\varphi_2 - \omega t) - k_2\omega. \qquad (3.4)$$

where μ is a positive small parameter (introduced in the same manner as many times before), equations of motion (3.1) for the pendulums can be written as

$$I_s\ddot{\varphi}_s + k_s(\dot{\varphi}_s - \omega) = \mu\omega^2\Phi_s(\varphi_s, \ddot{x}, \ddot{y}, \omega t) \qquad (s = 1, 2). \qquad (3.5)$$

Especially interesting are solutions like

$$\varphi_s = \omega t + \alpha_s + \mu\psi_s^*(\omega t) \qquad (s = 1, 2). \\ x = x(\omega t), \quad y = y(\omega t), \qquad (3.6)$$

where $\psi_s^*(\omega t)$, $x(\omega t)$, $y(\omega t)$ are periodic time functions with period $2\pi/\omega$. Since equations (3.2), (3.5) and solutions (3.6) correspond by their format to equations (4.2), (4.3) and solutions (4.1) of chap. 12, then, as in the preceding paragraph, the results for the synchronization in the systems with almost uniform rotational motions can be used.

The fundamental system for equations (3.2) and (3.5) allows a family of synchronous solutions

$$\varphi_1^0 = \omega t + \alpha_1, \quad \varphi_2^0 = \omega t + \alpha_2;$$
$$x^0 = w[a\cos\omega t + b_1\cos(\omega t + \alpha_1) + b_2\cos(\omega t + \alpha_2)],$$
$$y^0 = \dot{x}^0/\omega = -w[a\sin\omega t + b_1\sin(\omega t + \alpha_1) + b_2\sin(\omega t + \alpha_2)], \tag{3.7}$$

which is a function of two arbitrary constants α_1 and α_2. These constants are obviously angles of the deviation of the pendulums from the line o_1o_2, which is uniformly rotating with angular velocity ω.

In expressions (3.7)

$$w = \frac{\omega^2}{p^2 - \omega^2}, \quad p^2 = \frac{c}{M}, \quad a = \frac{s}{M},$$
$$b_1 = \frac{m_1 b_1}{M}, \quad b_2 = \frac{m_2 l_2}{M}. \tag{3.8}$$

Let us now construct *basic equations* (4.6) of chap. 12 to determine the values of α_1 and α_2 which would comply with the solutions of equations (3.5) and (3.2) having the format of (3.6). According to expressions (3.4) and (3.7), these equations can be written as

$$P_1(\alpha_1, \alpha_2) \equiv \frac{\mu\omega^2}{k_1}\langle \Phi_1(\varphi_1^0, \ddot{x}^0, \ddot{y}^0, \omega t)\rangle \equiv$$
$$\equiv \frac{1}{k_1}\langle m_1 l_1[\ddot{x}_1^0 \sin(\omega t + \alpha_1) + \ddot{y}_1^0\cos(\omega t + \alpha_1)] -$$
$$- m_1 l_1 \varepsilon_1 \omega^2 \sin\alpha_1 - k_1\omega\rangle \equiv -\frac{1}{k_1}\{m_1 l_1\omega^2[(wa + \varepsilon_1)\sin\alpha_1 +$$
$$+ b_2 w\sin(\alpha_1 - \alpha_2)] - k_1\omega\} = 0,$$
$$P_2(\alpha_1, \alpha_2) \equiv \frac{\mu\omega^2}{k_2}\langle \Phi_2(\varphi_1^0, \ddot{x}^0, \ddot{y}^0, \omega t)\rangle \equiv \tag{3.9}$$
$$\equiv \frac{1}{k_2}\langle m_2 l_2[\ddot{x}^0 \sin(\omega t + \alpha_2) + \ddot{y}^0\cos(\omega t + \alpha_2)] +$$
$$+ m_2 l_2 \varepsilon_2 \omega^2 \sin\alpha_2 - k_2\omega\rangle \equiv -\frac{1}{k_2}\{m_2 l_2\omega^2[(wa - \varepsilon_2)\sin\alpha_2 -$$
$$- b_1 w\sin(\alpha_1 - \alpha_2)] - k_2\omega\} = 0.$$

A study of the stability of the motion corresponding to one of the solutions $\alpha_1 = \alpha_1^*, \alpha_2 = \alpha_2^*$ of the latter equations becomes a study of the signs of the real parts of the roots of the quadratic equation [see (2.12) in chap. 12]

$$\begin{vmatrix} \dfrac{\partial P_1}{\partial \alpha_1} - \varkappa & \dfrac{\partial P_1}{\partial \alpha_2} \\ \dfrac{\partial P_2}{\partial \alpha_1} & \dfrac{\partial P_2}{\partial \alpha_2} - \varkappa \end{vmatrix}_{\alpha_j = \alpha_j^*} = 0. \tag{3.10}$$

In actual systems, resistance moments $k_1\omega$ and $k_2\omega$ are small as compared with other averaged moments in equations (3.9). Thus, it is sufficient to solve these equations with accuracy up to the terms containing $k_1\omega$ and $k_2\omega$ in powers not higher than one.

It is easy to establish that there exist (with accuracy up to the nonsignificant constant $2\pi n$, where n is an arbitrary whole number) six solutions of system (3.9):

$$
\begin{aligned}
&1)\ \alpha_1^{(1)} \approx 0 + k_1\omega A_1^{(1)} + k_2\omega B_1^{(1)}, && \alpha_2^{(1)} \approx 0 + k_1\omega A_2^{(1)} + k_2\omega B_2^{(1)};\\
&2)\ \alpha_1^{(2)} \approx 0 + k_1\omega A_1^{(2)} + k_2\omega B_1^{(2)}, && \alpha_2^{(2)} \approx \pi + k_1\omega A_2^{(2)} + k_2\omega B_2^{(2)};\\
&3)\ \alpha_1^{(3)} \approx \pi + k_1\omega A_1^{(3)} + k_2\omega B_1^{(3)}, && \alpha_2^{(3)} \approx \pi + k_1\omega A_2^{(3)} + k_2\omega B_2^{(3)};\\
&4)\ \alpha_1^{(4)} \approx \pi + k_1\omega A_1^{(4)} + k_2\omega B_1^{(4)}, && \alpha_2^{(4)} \approx 0 + k_1\omega A_2^{(4)} + k_2\omega B_2^{(4)};\\
&5)\ \alpha_1^{(5)} \approx \beta + k_1\omega A_1^{(5)} + k_2\omega B_1^{(5)}, && \alpha_2^{(5)} \approx \gamma + k_1\omega A_2^{(5)} + k_2\omega B_2^{(5)};\\
&6)\ \alpha_1^{(6)} \approx -\beta + k_1\omega A_1^{(6)} + k_2\omega B_1^{(6)}, && \alpha_2^{(6)} \approx -\gamma + k_1\omega A_2^{(6)} + k_2\omega B_2^{(6)}.
\end{aligned}
\tag{3.11}
$$

Here

$$
\gamma = \arccos \frac{1}{2\eta_1}\left(\frac{\eta_1^2}{\eta_2^2} - 1 - \eta_1^2\right), \quad \eta_1 = \frac{wa - \varepsilon_1}{b_2 w},
$$

$$
\eta_2 = \frac{wa - \varepsilon_2}{b_1 w}, \quad \beta = \arccos\left\{-\frac{\eta_2}{\eta_1}\left[\eta_1 + \frac{1}{2\eta_1}\left(\frac{\eta_1^2}{\eta_2^2} - 1 - \eta_1^2\right)\right]\right\}. \tag{3.12}
$$

Constants $A_j^{(k)}$ and $B_j^{(k)}$ ($j = 1, 2$; $k = 1, \ldots, 6$) are easy to find from equations(3.9) by substituting into them the appropriate expressions (3.11) and neglecting terms $k_j\omega$ in powers higher than one.

From all the solutions (3.11), the most important for the considered problem is the second solution. If the system parameters are chosen to satisfy correlation

$$
m_1(\varepsilon_1 + l_1) = m_2(\varepsilon_2 + l_2), \tag{3.13}
$$

or, in notation of (3.8),

$$
a + b_1 - b_2 = 0, \tag{3.14}
$$

then for this solution deviations x, y of the coordinates of the center of the disc are zeros, with an accuracy up to the small terms $k_j\omega$. This means that a practical self-balancing of the system takes place.

To prove it, it follows from (3.7), (3.11), and (3.14) at $\alpha_1 = \alpha_1^{(2)}$ and $\alpha_2 = \alpha_2^{(2)}$

$$
\begin{aligned}
x^0/w &\approx a\cos\omega t + b_1\cos\left(\omega t + k_1\omega A_1^{(2)} + k_2\omega B_1^{(2)}\right) + \\
&\quad + b_2\cos\left(\omega t + \pi + k_1\omega A_2^{(2)} + k_2\omega B_2^{(2)}\right) \approx (a + b_1 - b_2)\cos\omega t - \\
&\quad - b_1\left(k_1\omega A_1^{(2)} + k_2\omega B_1^{(2)}\right)\sin\omega t + b_2\left(k_1\omega A_2^{(2)} + k_2\omega B_2^{(2)}\right)\sin\omega t = \\
&\quad = \left[b_2\left(k_1\omega A_2^{(2)} + k_2\omega B_2^{(2)}\right) - b_1\left(k_1\omega A_1^{(2)} + k_2\omega B_1^{(2)}\right)\right]\sin\omega t \approx 0,\\
y^0/w &\approx -\left[b_2\left(k_1\omega A_2^{(2)} + k_2\omega B_2^{(2)}\right) - \right.\\
&\quad \left. - b_1\left(k_1\omega A_1^{(2)} + k_2\omega B_1^{(2)}\right)\right]\cos\omega t \approx 0.
\end{aligned}
\tag{3.15}
$$

For all other solutions (3.11), parameters x and y would contain free terms in their expansions by $k_1\omega$ and $k_2\omega$; thus self-balancing (in the considered approximation) would not take place. Thus, stability of the specific solution corresponding to $\alpha_1 = \alpha_1^{(2)}$ and $\alpha_2 = \alpha_2^{(2)}$ would be analyzed.

Up to the terms containing nonnegative powers of $k_1\omega$ and $k_2\omega$, for this solution, considering (3.9) and (3.14), it can be found that

$$\frac{\partial P_1}{\partial \alpha_1} = \frac{1}{k_1} m_1 l_1 \omega^2 (wb_1 - \varepsilon_1), \quad \frac{\partial P_2}{\partial \alpha_1} = -\frac{1}{k_2} m_2 l_2 \omega^2 b_1 w,$$
$$\frac{\partial P_1}{\partial \alpha_2} = -\frac{1}{k_1} m_1 l_1 \omega^2 b_2 w, \quad \frac{\partial P_2}{\partial \alpha_2} = \frac{1}{k_2} m_2 l_2 \omega^2 (wb_2 - \varepsilon_2). \tag{3.16}$$

Thus, the conditions for the nonnegative sign of the real parts of the roots of the quadratic equation (3.10), coinciding with the condition for its coefficients being all positive, can be presented as

$$-\left(\frac{\partial P_1}{\partial \alpha_1} + \frac{\partial P_2}{\partial \alpha_2}\right) = -\frac{m_1 l_1 \omega^2}{k_1}(wb_1 - \varepsilon_1) - \frac{m_2 l_2 \omega^2}{k_2}(wb_2 - \varepsilon_2) > 0,$$
$$\left(\frac{\partial P_1}{\partial \alpha_1}\frac{\partial P_2}{\partial \alpha_2} - \frac{\partial P_1}{\partial \alpha_2}\frac{\partial P_2}{\partial \alpha_1}\right) = \frac{m_1 l_1 m_2 l_2 \omega^4}{k_1 k_2}[(wb_1 - \varepsilon_1)(wb_2 - \varepsilon_2) - b_1 b_2 w^2] =$$
$$= \frac{m_1 l_1 m_2 l_2 \omega^4}{k_1 k_2}[\varepsilon_1 \varepsilon_2 - w(b_1 \varepsilon_2 + b_2 \varepsilon_1)] > 0. \tag{3.17}$$

After transformations considering (3.8), the latter inequalities become

$$\frac{\omega^2}{p^2 - \omega^2} < \frac{(m_1 l_1 \varepsilon_1 k_2 + m_2 l_2 \varepsilon_2 k_1) M}{m_1^2 l_1^2 k_2 + m_2^2 l_2^2 k_1} = R_1,$$
$$\frac{\omega^2}{p^2 - \omega^2} < \frac{M \varepsilon_1 \varepsilon_2}{m_1 l_1 \varepsilon_2 + m_2 l_2 \varepsilon_1} = R_2. \tag{3.18}$$

Conditions (3.18) are satisfied at

$$\omega < p\sqrt{R/(R+1)} < p,$$

where $R = \text{Inf}(R_1, R_2)$ is the smallest of the two positive numbers R_1, R_2, which are determined from (3.18). Besides, regardless of the values of the other parameters of the system, stability conditions (3.18) are satisfied if

$$\omega > p = \sqrt{c/M}, \tag{3.19}$$

This means that the angular velocity ω of the shaft exceeds the natural frequency of the disc with the pendulums attached to it, or the so-called first critical speed. Thus it can be admitted that the principal condition for self-balancing of the system is the same as for the system in § 2, and it coincides with the respective conditions for the classical case—unbalanced disc without pendulums. In other words, it is also the case of the *generalized Laval Principle*.

§ 4) Rotating flexible shaft with unbalanced discs.

Traditionally, the problem of bending/torsional vibrations of a rotating elastic shaft with several unbalanced discs is considered without referring to synchronization phenomena [see, e.g., (30, 120)]. However, the vibratory moments are developing in the system, which are characteristic for systems with self-synchronizing vibro-exciters (see §§ 5 and 8 of chap. 3). Under certain conditions, these moments could substantially influence the motion of the shaft with discs attached (57).

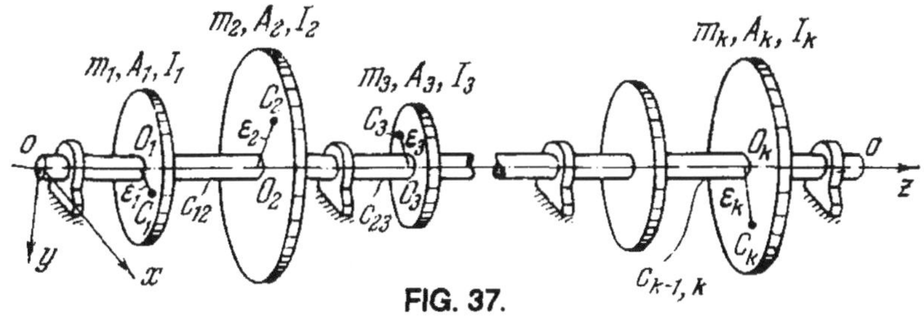

FIG. 37.

In the most general case, the system in Fig. 37 is a multisupported shaft with a certain number k of statically and dynamically unbalanced discs driven by some motors. It is assumed that during the motion the shaft can participate not only in bending but also in torsional vibrations; thus torsional stiffnesses of the shaft segment are not infinite. Shaft supports (bearings) can be rigid or elastic, with different stiffnesses in different directions.

It will be assumed that some of the discs are rotors of the driving motors.

Let $Oxyz$ be a stationary cartesian frame with Oz being directed along the axes of the shaft bearings, while $O_s'x_s'y_s'z_s'$ are coordinate frames with axes parallel to the respective axes $Oxyz$ and having origins in the intersection points O_s' of the disc planes with the deflected shaft axis oo (Fig. 38). Further, $O_s'u_sv_sw_s$ are coordinate frames rigidly attached to the discs with their axes coinciding with their principal inertia axes.

The position of each disc is determined by five generalized coordinates—the cartesian coordinates x_s and y_s of the point O_s' in the stationary frame $Oxyz$ and three Euler angles α_s, β_s, and φ_s, chosen in accordance with Fig. 38: α_s is the angle between axis $O_s'w_s$ and plane $x_s'O_s'z_s'$; β_s is the angle between axis $O_s'z_s'$ and the projection of axis $O_s'w_s$ on plane $x_s'O_s'z'$; and φ_s is the angle between the line n of the intersection of the planes $x_s'O_s'z_s'$ and $u_sO_s'v_s$, and axis $O_s'u_s$ (this line is usually called the modal line, and angle φ_s is the angle of proper rotation).

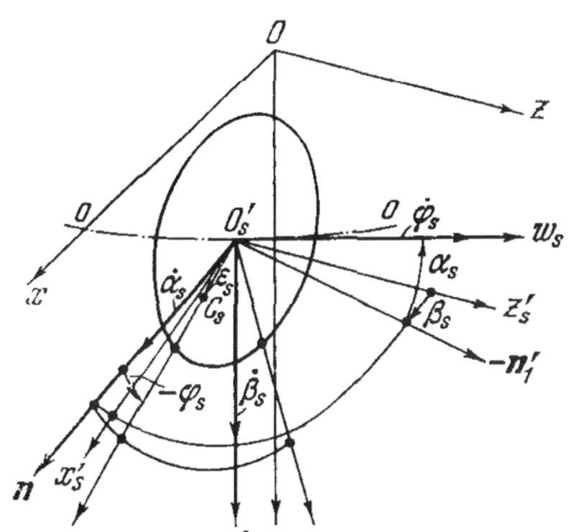

FIG. 38.

Sticking to the case of static unbalance of the discs, let's assume their centers of gravity C_s are located on axes Ou_s.

If giroscopic terms are neglected, as well as the terms having higher than second order relative to coordinates x_s, y_s, α_s, β_s and corresponding velocities, and if gravity forces are not considered, then the expressions for the potential and kinetic energies become

$$\Pi = \Pi^{(I)} + \Pi^{(II)},$$

$$\Pi^{(I)} = \frac{1}{2}\sum_{s=1}^{k}\sum_{j=1}^{k}\left(c_{sj}^{(x)}x_s x_j + c_{sj}^{(\alpha)}\alpha_s\alpha_j + 2c_{sj}^{(x\alpha)}x_s\alpha_j + \right.$$
$$\left. + c_{sj}^{(y)}y_s y_j + c_{sj}^{(\beta)}\beta_s\beta_j + 2c_{sj}^{(y\beta)}y_s\beta_j\right),$$

$$\Pi^{(II)} = \frac{1}{2}\sum_{s=1}^{k-1} c_{s,s+1}(\varphi_{s+1} - \varphi_s - \varkappa_{s+1} + \varkappa_s)^2, \qquad (4.1)$$

$$T = T^{(I)} + \Delta T^* + \sum_{s=1}^{k} T_s,$$

$$T^{(I)} = \frac{1}{2}\sum_{s=1}^{k}\left[m_s(\dot{x}_s^2 + \dot{y}_s^2) + A_s(\dot{\alpha}_s^2 + \dot{\beta}_s^2)\right], \qquad (4.2)$$

$$\Delta T^* = -\sum_{s=1}^{k} m_s\varepsilon_s(\dot{x}_s\cos\varphi_s + \dot{y}_s\sin\varphi_s)\dot{\varphi}_s, \quad T_s = \frac{1}{2}I_s\dot{\varphi}_s^2.$$

Expression (4.2) can be obtained in the most simple manner from the formula for the kinetic energy of a rigid body in a general case when an arbitrary point in this body, but not necessarily its center of gravity, is taken as a pole (174).

In (4.1) and (4.2) it is denoted as follows: m_s, ε_s, A_s, I_s are, respectively, mass, eccentricity, equatorial and polar moments of inertia of the s-th disc; $c_{s,\,s+1}$ is torsional stiffness of the shaft segment between s-th and $(s+1)$-th discs, with $c_{01} = c_{k,\,k+1} = 0$; $c_{sj}^{(x)}, c_{sj}^{(y)}, c_{sj}^{(\alpha)}, c_{sj}^{(\beta)}, c_{sj}^{(x\alpha)}, c_{sj}^{(y\beta)}$ are the respective bending stiffnesses of the shaft calculated with consideration of bearing compliances; χ_s are the values of the angles φ_s at which elastic torques in the shaft spans are zeros; these angles, determined up to a constant rotation χ_0, characterize directions of eccentricity vectors ε_s of discs with nontwisted shaft.

The equations of motion of the system, derived in accordance with (4.1) and (4.2) as Lagrange equations of the second type, are as follows:

$$I_s\ddot{\varphi}_s = m_s\varepsilon_s(\ddot{x}_s\cos\varphi_s + \ddot{y}_s\sin\varphi_s) - c_{s-1,s}(\varphi_s - \varphi_{s-1} - \varkappa_s + \varkappa_{s-1}) +$$
$$+ c_{s,s+1}(\varphi_{s+1} - \varphi_s - \varkappa_{s+1} + \varkappa_s) + L_s - R_s;$$

$$m_s\ddot{x}_s + \sum_{j=1}^{k}\left(c_{sj}^{(x)}x_j + c_{sj}^{(x\alpha)}\alpha_j\right) = m_s\varepsilon_s\left(\ddot{\varphi}_s\cos\varphi_s - \dot{\varphi}_s^2\sin\varphi_s\right) + Q_s^{(x)},$$

$$(4.3)$$

$$m_s\ddot{y}_s + \sum_{j=1}^{k}\left(c_{sj}^{(y)}y_j + c_{sj}^{(y\beta)}\beta_j\right) = m_s\varepsilon_s(\ddot{\varphi}_s\sin\varphi_s + \dot{\varphi}_s^2\cos\varphi_s) + Q_s^{(y)};$$

$$A_s\ddot{\alpha}_s + \sum_{j=1}^{k}\left(c_{sj}^{(\alpha)}\alpha_j + c_{sj}^{(x\alpha)}x_j\right) = Q_s^{(\alpha)},$$

$$A_s\ddot{\beta}_s + \sum_{j=1}^{k}\left(c_{sj}^{(\beta)}\beta_j + c_{sj}^{(y\beta)}y_j\right) = Q_s^{(\beta)} \qquad (s = 1, \ldots, k).$$

Here $Q_s^{(x)}$, $Q_s^{(y)}$, $Q_s^{(\alpha)}$, $Q_s^{(\beta)}$ are forces and moments of internal and external resistance to the shaft vibrations which can depend on all generalized coordinates and velocities of the system, with coordinates φ_s and velocities $\dot{\varphi}_s$ entering the expressions for Q only as differences $\varphi_s - \varphi_j$ *and* $\dot{\varphi}_s - \dot{\varphi}_j$. L_s and R_s are, respectively, driving torques of motors and resistance torques. Further, it is assumed that the shaft is driven by inductionlike motors and that torques L_s and R_s are determined by expressions (4.10) and (4.11) of chap. 3. Also it is assumed that all natural frequencies of the torsional vibrations of the shaft with the fully balanced discs are substantially lower than the angular velocity ω of the steady rotation, and the natural bending frequencies are sufficiently different from $n\omega$, where n is the positive integer.* Then, denoting

$$\mu\Phi_s = \\ = m_s\varepsilon_s(\ddot{x}_s\cos\varphi_s + \ddot{y}_s\sin\varphi_s) - c_{s-1,s}(\varphi_s - \varphi_{s-1} - \varkappa_s + \varkappa_{s-1}) + \\ + c_{s+1,s}(\varphi_{s+1} - \varphi_s - \varkappa_{s+1} + \varkappa_s) + L_s(\omega) - R_s(\omega), \quad (4.4)$$

$$Q_s^{(x)} = \mu\widetilde{Q}_s^{(x)}, \quad Q_s^{(y)} = \mu\widetilde{Q}_s^{(y)}, \quad Q_s^{(\alpha)} = \mu\widetilde{Q}_s^{(\alpha)}, \quad Q_s^{(\beta)} = \mu\widetilde{Q}_s^{(\beta)},$$

equations (4.3) can be rewritten as

$$\begin{gathered} I_s\ddot{\varphi}_s + k_3(\dot{\varphi}_s - \omega) = \mu\Phi_s(\varphi_{s-1}, \varphi_s, \dot{\varphi}_s, \ddot{x}_s, \ddot{y}_s); \\ m_s\ddot{x}_s + \sum_{j=1}^{k}\left(c_{sj}^{(x)}x_j + c_{js}^{(x\alpha)}\alpha_j\right) = m_s\varepsilon_s\left(\ddot{\varphi}_s\cos\varphi_s - \dot{\varphi}_s^2\sin\varphi_s\right) + \mu\widetilde{Q}_s^{(x)}, \\ m_s\ddot{y}_s + \sum_{j=1}^{k}\left(c_{sj}^{(y)}y_j + c_{sj}^{(y\beta)}\beta_j\right) = m_s\varepsilon_s\left(\ddot{\varphi}_s\sin\varphi_s + \dot{\varphi}_s^2\cos\varphi_s\right) + \mu\widetilde{Q}_s^{(y)}; \\ A_s\ddot{\alpha}_s + \sum_{j=1}^{k}\left(c_{sj}^{(\alpha)}\alpha_j + c_{sj}^{(x\alpha)}x_j\right) = \mu\widetilde{Q}_s^{(\alpha)}, \\ A_s\ddot{\beta}_s + \sum_{j=1}^{k}\left(c_{sj}^{(\beta)}\beta_j + c_{sj}^{(y\beta)}y_j\right) = \mu\widetilde{Q}_s^{(\beta)} \\ (s = 1, \ldots, k). \end{gathered} \quad (4.5)$$

Parameter $\mu > 0$ is still considered as small, and can be argumented by the adopted assumptions as in § 4, chap. 3.

Equations (4.5) look like equations (4.2), (4.3) from chap. 12 which describe synchronization in the system with almost uniform rotations. They also correspond to similar equations of the generalized problem of synchronization of vibro-exciters (see § 8, chap. 3). Moreover, it can be said that the problem considered here is a special case of the generalized synchronization problem. As in the latter, the main goal of the analysis is to establish conditions of existence and stability, as well as the computation, with some degree of accuracy, of the synchronous motions of the system, i.e., motions like

$$\begin{gathered} \varphi_s = \omega t + \overset{*}{\alpha}_s + \mu\psi_s(\omega t); \\ x_s = x_s(\omega t), \quad y_s = y_s(\omega t); \\ \alpha_s = \alpha_s(\omega t), \quad \beta_s = \beta_s(\omega t), \end{gathered} \quad (4.6)$$

---

* Some of the assumptions are not necessary for correctness of the principal conclusions and are adopted only for simplification of the derivations.

Here φ_s, x_s, y_s, α_s, and β_s are periodic time functions with period $2\pi/\omega$, and α_s^* are constants. The value of the synchronous frequency ω is, generally speaking, not known in advance and should be determined during the solution process (the case of internal synchronization).

The problem will be considered in the spirit of the analysis in § 8, chap. 3. The fundamental system complying with equations (4.5) would allow for the following family of solutions at the conditions formulated above, which depends on k arbitrary constants α_s^*

$$\varphi_s^0 = \omega t + \alpha_s^*; \quad x_s^0 = -\sum_{j=1}^{k} \mathscr{M}_{sj} \sin(\omega t + \alpha_j^*),$$

$$\alpha_s^0 = -\sum_{j=1}^{k} U_{sj} \sin(\omega t + \alpha_j^*); \quad y_s^0 = \sum_{j=1}^{k} \mathscr{N}_{sj} \cos(\omega t + \alpha_j^*), \tag{4.7}$$

$$\beta_s^0 = -\sum_{j=1}^{k} V_{sj} \cos(\omega t + \alpha_j^*) \qquad (s = 1, \ldots, k),$$

In (4.7), $\mathscr{M}_{sj}$, U_{sj}, $\mathscr{N}_{sj}$, and V_{sj} designate constants determined from the solution of the linear algebraic equations which would be obtained at $\mu = 0$ from the last $4k$ equations (4.5) after substitution of expressions (4.6), and equating coefficients at $\sin(\omega t + \alpha_j^*)$ and $\cos(\omega t + \alpha_j^*)$ in both parts of the resulting expressions.

Let's derive *basic equations* (4.6) of chap. 12 or, which is the same thing, equations (8.4) of chap. 3 for determining the values of $\alpha_1^*, \ldots, \alpha_k^*$, which correspond to the solutions of equations (4.5) in the format of (4.6). According to (4.4) and (4.7), these equations are as follows (superscript "*" at α_s is further omitted):

$$P_s(\alpha_1, \ldots, \alpha_k) \equiv \frac{\mu}{k_s} \langle \Phi_s(\varphi_{s-1}^0, \varphi_s^0, \varphi_{s+1}^0, \dot{\varphi}_s^0, \ddot{x}_s^0, \ddot{y}_s^0) \rangle \equiv$$

$$\equiv \frac{1}{k_s}\Big[-\frac{m_s \varepsilon_s \omega^2}{2} \sum_{j=1}^{k} (\mathscr{M}_{sj} + \mathscr{N}_{sj}) \sin(\alpha_s - \alpha_j) - \tag{4.8}$$

$$-c_{s-1,s}(\alpha_s - \alpha_{s-1} - \varkappa_s + \varkappa_{s+1}) + c_{s,s+1}(\alpha_{s+1} - \alpha_s - \varkappa_{s+1} + \varkappa_s) +$$

$$+ L_s(\omega) - R_s(\omega)\Big] = 0$$

$$(s = 1, \ \ldots, \ k).$$

To interpret the latter equations, the arguments in subsection 2, § 8, chap. 3 can be repeated almost verbatim. Thus, parameters

$$W_s^{(K)}(\alpha_1, \ldots, \alpha_k) = \frac{m_s \varepsilon_s \omega^2}{2} \sum_{j=1}^{k} (\mathscr{M}_{sj} + \mathscr{N}_{sj}) \sin(\alpha_s - \alpha_j) \tag{4.9}$$

can be called vibratory torques, analogously to the problem of synchronization of vibro-exciters. Parameters

$$S_s^{(K)}(\alpha_{s-1}, \alpha_s, \alpha_{s+1}) = c_{s-1,s}(\alpha_s - \alpha_{s-1} - \varkappa_s + \varkappa_{s-1}) -$$
$$- c_{s,s+1}(\alpha_{s+1} - \alpha_s - \varkappa_{s+1} + \varkappa_s) \tag{4.10}$$

averaged during the period $2\pi/\omega$, are the values of the elastic torques, and $L_s(\omega)$ and $R_s(\omega)$ are the average values of the motor torque and resistance torque acting on the s-th disc. Frequently, the torque is applied only to the end disc, then

$$L_1(\omega) \neq 0, \quad L_2(\omega) = \ldots = L_k(\omega) = 0.$$

In accordance with equalities (8.34) and (8.35) of chap. 3, the following correlations are correct:

$$W_s^{(K)} = \frac{\partial \Lambda^{(I)}}{\partial \alpha_s}, \qquad S^{(K)} = -\frac{\partial \Lambda^{(II)}}{\partial \alpha_s},$$
$$\Lambda^{(I)} = \langle (T^{(I)} - \Pi^{(I)}) \rangle, \quad \Lambda^{(II)} = -\langle (\Pi^{(II)}) \rangle = -(\Pi^{(II)}). \tag{4.11}$$

Considering expressions (4.9) to (4.11), equations (4.8) can be presented as

$$P_s(\alpha_1, \ldots, \alpha_k) \equiv$$
$$\equiv \frac{1}{k_s}[L_s(\omega) - R_s(\omega) - S_s^{(K)}(\alpha_{s-1}, \alpha_s, \alpha_{s+1}) - \tag{4.12}$$
$$- W_s^{(K)}(\alpha_1, \ldots, \alpha_k)] = 0$$
$$(s = 1, \ldots, k)$$

or as

$$P_s(\alpha_1, \ldots, \alpha_k) \equiv \frac{1}{k_s}\left[-\frac{\partial(\Lambda^{(I)} - \Lambda^{(II)})}{\partial \alpha_s} + L_s(\omega) - R_s(\omega)\right] \equiv$$
$$\equiv \frac{1}{k_s}\left[-\frac{\partial \langle (T^{(I)} - \Pi^{(I)} + \Pi^{(II)}) \rangle}{\partial \alpha_s} + L_s(\omega) - R_s(\omega)\right] = 0 \tag{4.13}$$
$$(s = 1, \ldots, k).$$

The latter equations are quite analogous to the corresponding equations of § 8, chap. 3, related to the problem of synchronization of the vibro-exciters.

Summing up all equations (4.8), premultiplied by ωk_s, the expression would be the same as in § 4, chap. 3

$$\omega \sum_{s=1}^{k} L_s(\omega) = \sum_{s=1}^{k} R_s(\omega)\,\omega, \tag{4.14}$$

which is an equation of energy balance in the system (in the initial approximation) and serves to determine ω. Fundamental phases α_s are determined from (4.8) or, which is the same thing, from equations (4.12) or (4.13) only up to the additive arbitrary constant. All this, as has been repeatedly noted, is a consequence of the autonomous character of the original system of equations.

The conclusion about the stability of a motion corresponding to a certain solution $\alpha_1 = \alpha_1^*, \ldots, \alpha_k = \alpha_k^*$ of transcendant equations (4.8) depends on the signs of the roots of the algebraic equation (8.31) of chap. 3 having power $k - 1$.

For the considered system, the integral stability criterion stands, with the potential function expressed as

$$D = \Lambda^{(I)} - \Lambda^{(II)} - B = \langle (T^{(I)} - \Pi^{(I)} + \Pi^{(II)}) \rangle - B, \tag{4.15}$$

where

$$B = B(\alpha_1, \ldots, \alpha_k, \omega) = \sum_{j=1}^{k} [L_j(\omega) - R_j(\omega)] \alpha_j. \tag{4.16}$$

Our goal is not a detailed study of the bending/torsional vibrations of a rotating shaft with unbalanced discs. Accordingly, after the basic correlations derived above, the main qualitative conclusions will be briefly addressed.

One conclusion is that due to the unbalanced condition of the discs, bending and torsional vibrations are coupled. In some cases this coupling cannot be ignored, and the problem should be considered in the presented nonlinear formulation. A specific indication of the coupling is the appearance of vibratory moments $W_s^{(K)}$ in equations (4.12). As a result, at certain conditions during motion, shaft segments might get twisted for significant additional angles, and torsional vibrations would develop around these angles as mean positions. In such cases, there will be additional stresses in the shaft cross section, and the character of the interaction between the shaft and its bearings might change significantly.

Such effects are more pronounced with increasing of maximal values of vibratory moments $W_s^{(K)}$ in relation to the maximum elastic torques $S_s^{(K)}$, or in other words, with diminishing torsional stiffnesses $c_{s,\, s+1}$ of the shaft segments, increasing unbalances $m_s\varepsilon_s$, and increasing angular velocity ω. These effects can be, of course, alleviated as desired by assigning sufficiently large stiffnesses $c_{s,\, s+1}$.

Again, at $L_s(\omega) \approx R_s(\omega)$ this system can be described as a confirmation of the *generalized Laval Principle* (see subsection 6, § 14, chap. 3): if angular velocity ω of a shaft is much higher than the natural frequency of the system, then the system tends to self-balance.

§ 5) Cages of antifriction bearings.

An assumption about the possibility of self-synchronization in rolling friction bearings was formulated in (1). Later in (127) synchronization of the cages of the two bearings supporting a rotating rigid shaft (Fig. 39) was studied. The study was performed using an asymptotic method not related to the problems of the synchronization of exciters and other objects with almost uniform rotations. A case of near-resonance motion when synchronous angular velocity ω of the cages is near the natural frequency p of the shaft in its supports was considered. It will be shown that the conditions of existence and stability of synchronous motions of cages both near to and far from resonance are easy to derive from general expressions of the synchronization theory, and that at $\omega \approx p$ they coincide with the results of (127). The conditions will be analyzed.

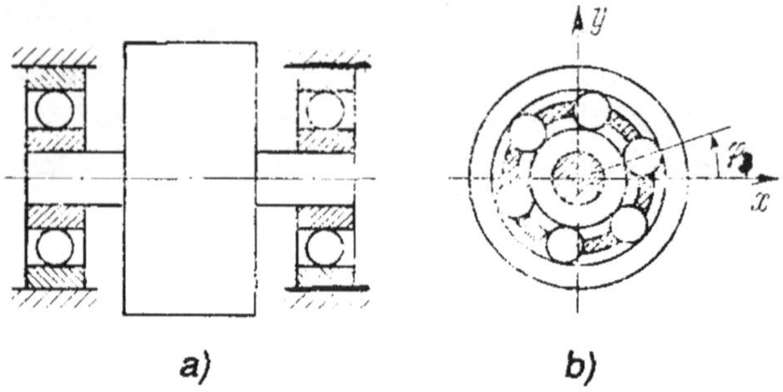

FIG. 39.

The problem is considered in the same formulation as in (127). The main assumption is that of the possibility of the slippage of the ball bearings in a hydrodynamic contact regime, and as a consequence the case having the "proper" degree of freedom (DOF). The angular velocity $\dot{\varphi}_s$ $(s = 1, 2)$ of the cages can differ from the "kinematic" values Ω_s corresponding to the rolling of the ball bearings without slippage. Equations of translational vibrations of the rotor and equations of the motion of the cages are written as

$$M\ddot{x} + \beta\dot{x} = -\frac{\partial \Pi}{\partial x}, \quad M\ddot{y} + \beta\dot{y} = -\frac{\partial \Pi}{\partial y};$$
$$I_s\ddot{\varphi}_s = k_s(\Omega_s - \dot{\varphi}_s) + B_s(\Omega_s - \dot{\varphi}_s) - \frac{\partial \Pi}{\partial \varphi_s} \qquad (s = 1, 2). \tag{5.1}$$

Here Π is the potential energy of the elastic forces acting on the bearings, and is approximately expressed as

$$\Pi = \frac{1}{2} c(x^2 + y^2) - y(q_1 \cos\varphi_1 + q_2 \cos\varphi_2) -$$
$$- x(q_1 \sin\varphi_1 + q_2 \sin\varphi_2). \tag{5.2}$$

In (5.1) and (5.2) x, y are rotor displacements; $\dot{\varphi}_s$ are the angular displacements of the cages; M is the mass of the rotor; β is the coefficient of viscous resistance to the rotor vibrations; I_s are the moments of inertia of the cages with the ball bearings relative to the rotor axis; $k_s(\Omega_s - \dot{\varphi}_s)$ and $B_s(\Omega_s - \dot{\varphi}_s)$ respectively, moments of viscous and dry friction, with

$$B_s(\Omega_s - \dot{\varphi}_s) = \begin{cases} b_s \operatorname{sgn}(\Omega_s - \dot{\varphi}_s) & \text{for} \quad \Omega_s \neq \dot{\varphi}_s, \\ \eta b_s & \text{for} \quad \Omega_s = \dot{\varphi}_s, \end{cases} \tag{5.3}$$

where k_s and b_s are positive constants and $|\eta| < 1$. The maximum moment of the dry friction forces without slippage of the ball bearings is assumed to be equal to the friction moment at slippage. The radial stiffness of the bearings is denoted by c, while q_s are the coefficients characterizing bearing defects (manufacturing errors). Without a loss of generality, it can be assumed that $q_s \geq 0$, with $q_s = 0$ with no defects. It is adopted in (5.2) that angles φ_s are counted from the Ox axis to such directions on the cages, in relation to which these defects are "identically situated" on both bearings.

If there is no slippage of the ball bearings, then the cages are rotating with kinematic values of angular velocities Ω_1, Ω_2. It is assumed that $\Omega_1 \neq \Omega_2$, e.g., due to manufacturing errors; if $\Omega_1 = \Omega_2$, then there is no synchronization problem.

It will be clear later that velocities Ω_1 and Ω_2 are largely analogous to the partial angular velocities used for the analysis of self-synchronization of vibro-exciters (see § 5, chap. 3).

With the slippage of the ball bearings and satisfying certain conditions, at least two regimes of synchronous rotation of the cages are possible: a regime with slippage of the balls in both bearings, when the synchronous angular velocity ω of the cages is different from both Ω_1 and Ω_2, and a regime with slippage only in one bearing. In the latter case, e.g., $\omega = \Omega_1$ and $\Omega_2 \neq \omega$.

The first regime corresponds to *the problem of internal (mutual) synchronization of cages*. In this case, considering (5.2), (5.3) and denoting

$$z = y + ix,$$
$$\mu\Phi_s = k_s(\Omega_s - \omega) + b_s \operatorname{sgn}(\Omega_s - \dot{\varphi}_s) + q_s \operatorname{Im}\left(ze^{-i\varphi_s}\right) \qquad (5.4)$$

equations (5.1) can be written as

$$I_s\ddot{\varphi}_s + k_s(\dot{\varphi}_s - \omega) = \mu\Phi_s(\varphi_1, \varphi_2, z) \qquad (s = 1, 2), \qquad (5.5)$$

$$M\ddot{z} + \beta\dot{z} + cz = \sum_{s=1}^{2} q_s e^{i\varphi_s}, \qquad (5.6)$$

This corresponds to the format of the equations for the simple (nonmultiple) problem of objects with almost uniform rotations [see equations (4.2), (4.3) of chap. 12] and, specifically, to the equations of the problem of self-synchronization of vibro-exciters [see equations (4.14) of chap. 3]. To obtain a result applicable both in resonance and nonresonance regimes, viscous friction forces in (5.6) are not considered proportional to the small parameter μ, as was done in § 4, chap. 3 for some simplification of derivation.

The fundamental system corresponding to equations (5.5), (5.6) allows a family of synchronous solutions

$$\varphi_s^0 = \omega t + \alpha_s \qquad (s = 1, 2),$$
$$z^0 = \frac{1}{M\Delta}\sum_{j=1}^{2} q_j e^{i(\omega t + \alpha_j - \varepsilon)} \qquad (5.7)$$

where

($\Delta = \sqrt{(p^2 - \omega^2)^2 + 4n^2\omega^2}$, $p^2 = c/m$, $2n = \beta/m$, $\cos\varepsilon = (p^2 - \omega^2)/\Delta$, $\sin\varepsilon = 2n\omega/\Delta$). The family depends on two arbitrary parameters α_1 and α_2, which play the role of initial phases of the cages in rotation.

The basic equations of the problem [see equations (4.6) of chap. 12] are

$$P_s \equiv \frac{\mu}{k_s}\langle\Phi_s(\varphi_1^0, \varphi_2^0, z^0)\rangle = 0 \qquad (s = 1, 2)$$

and after averaging, (5.4) and (5.7) are written as

$$P_1(\alpha, \omega) \equiv \frac{1}{k_1}[k_1(\Omega_1 - \omega) + b_1 \operatorname{sgn}(\Omega_1 - \omega) - W_1(\alpha, \omega)] = 0,$$
$$P_2(\alpha, \omega) \equiv \frac{1}{k_2}[k_2(\Omega_2 - \omega) + b_2 \operatorname{sgn}(\Omega_2 - \omega) - W_2(\alpha, \omega)] = 0$$
$$(\alpha = \alpha_2 - \alpha_1). \qquad (5.8)$$

Here expressions

$$W_1(\alpha, \omega) = -q_1 \operatorname{Im}\left\langle z^0 e^{-i\varphi_1^0}\right\rangle = -\frac{q_1}{M\Delta}\operatorname{Im}\sum_{j=1}^{2} q_j e^{i(\alpha_j-\alpha_1-\varepsilon)} =$$

$$= \frac{q_1^2}{M\Delta}\sin\varepsilon - \frac{q_1 q_2}{M\Delta}\sin(\alpha-\varepsilon) = A_1[q_1\sin\varepsilon - q_2\sin(\alpha-\varepsilon)], \tag{5.9}$$

$$W_2(\alpha, \omega) = A_2[q_2\sin\varepsilon + q_1\sin(\alpha+\varepsilon)]$$

$$(A_s = q_s/M\Delta)$$

correspond to the expressions for the vibratory moments in the appropriate problems of self-synchronization of the vibro-exciters. In this case self-synchronization can occur specifically due to these moments characterizing vibratory coupling between the cages.

In (5.9), according to (5.7), A_s denote approximate amplitudes of steady forced vibrations of the shaft due to the defects in the s-th bearing.

Equation (5.8), like the vibro-exciters' problem, can be described as equilibrium equations for the average moments acting on the cages. From these equations, phase difference $\alpha = \alpha_2 - \alpha_1$ should be determined, as well as the approximate value of the synchronous angular velocity of the cages, which in this case is not known beforehand.

The conditions at which (5.8) have real solutions relative to α and ω would be conditions of the possibility of self-synchronization of the cages.

The necessary and sufficient condition of the asymptotic stability of the synchronous motion, corresponding to certain values of α and ω determined from equations (5.8), is inequality

$$\varkappa = \frac{\partial(P_1 - P_2)}{\partial\alpha_1} = -\frac{d(P_1-P_2)}{d\alpha} = -\frac{q_1 q_2}{M\Delta}\left[\frac{\cos(\alpha-\varepsilon)}{k_1} + \frac{\cos(\alpha+\varepsilon)}{k_2}\right] < 0, \tag{5.10}$$

It follows from equations of the type of (4.24) of chap. 3 [sufficiency of this condition in this case is due to the consideration of the dissipative forces along the vibratory coordinates in the fundamental approximation; see equation (5.6)].

In the case of the second regime, when the balls in one of the bearings (let it be the first one) do not slip $\omega = \Omega_1$, the synchronous velocity ω is known from the beginning and is equal to the kinematic velocity of the first cage. Here it is *the problem of external synchronization (entrainment)* of the second cage by the rotation of the first cage. A special analysis that is easily performed using the results of subsection 2, § 8, chap. 3 or §§ 2 and 7, chap. 10 is needed here. The phase difference of the cage rotation $\alpha = \alpha_2 - \alpha_1$ is determined in this case from equation

$$P_2(\alpha, \Omega_1) \equiv \frac{1}{k_2}[k_2(\Omega_2 - \Omega_1) + b_2\operatorname{sgn}(\Omega_2 - \Omega_1) - W_2(\alpha, \Omega_1)] = 0, \tag{5.11}$$

which coincides with the second equation (5.8) at $\omega = \Omega_1$.

A condition at which a real value of α is determined from (5.11) would be the condition of the possibility of synchronization. An asymptotically stable synchronous motion would actually correspond to this value if

$$\varkappa = \frac{dP_2(\alpha, \Omega_1)}{d\alpha_2} = \frac{dP_2(\alpha, \Omega_1)}{d\alpha} = -\frac{q_1 q_2}{k_2 M\Delta}\cos(\alpha+\varepsilon) < 0. \quad (5.12)$$

As could be expected, expressions (5.11) and (5.12) are formally derived from (5.8) and (5.10) at $k_1 \longrightarrow \infty$.

For the actual development of this regime it is necessary that, for values α satisfying (5.11) and (5.12), a condition for the non-slippage of the balls of the second bearing should be satisfied.

$$|W_1(\alpha, \Omega_1)| < b_1. \quad (5.13)$$

Firstly, the results related to the simple second regime will be analyzed, when the balls of the first bearing do not slip.

Considering (5.9), equation (5.12) becomes

$$\sin(\alpha+\varepsilon) = Q_2$$
$$\left(Q_2 = \frac{k_2(\Omega_2-\Omega_1) + b_2 \operatorname{sgn}(\Omega_2-\Omega_1) - q_1 A_1 \sin\varepsilon}{q_2 A_1}\right). \quad (5.14)$$

The condition for the possibility of synchronization is inequality

$$|Q_2| < 1, \quad (5.15)$$

If it is satisfied, equation (5.14) has two essentially different real solutions

$$\alpha^{(1)} = -\varepsilon + \arcsin Q_2, \qquad \alpha^{(2)} = -\varepsilon + \pi - \arcsin Q_2,$$

According to (5.12), the first corresponds to a stable, and the second to an unstable motion.

It is easy to prove that at $\Omega_2 > \Omega_1$, and near resonance, when detuning $\delta = \omega - p = \Omega_1 - p$ is small, it can be assumed that

$$\Delta = \sqrt{(p^2-\omega^2)^2 + 4n^2\omega^2} \approx 2p\sqrt{\delta^2+n^2}$$

condition (5.15), up to rotation differences, coincides with condition (2.2) from (127), where it was derived by another technique and analyzed in defects.

However, condition (5.15) is correct also at any ω/p. If ω/p is sufficiently different from one (motion in the area far from resonance), then friction in the vibratory part of the system can be neglected, $\beta \approx 0$, $n \approx 0$, $\sin\varepsilon \approx 0$. If, in addition, the moment of dry friction forces b_2 is so small that it can be neglected, then (5.15) simplifies as

$$k_2|\Omega_2 - \Omega_1| < q_2 A_1 = q_1 A_2 = q_1 q_2 / M\Delta. \quad (5.16)$$

For the regime with slippage of the balls of both bearings, only the motion far from the resonance will be considered, where it can be assumed that $\beta \approx 0$, $n \approx 0$. Under these conditions, expressions (5.9) for the vibratory moments become

$$W_1 = -W_2 = -\frac{q_1 q_2}{M(p^2-\omega^2)}\sin\alpha. \quad (5.17)$$

Assume that

$$\Omega_1 < \omega < \Omega_2. \tag{5.18}$$

Then, considering (5.17), from equations (5.8) it can be found

$$\omega = (k_1\Omega_1 + k_2\Omega_2 + b_2 - b_1)/(k_1 + k_2), \quad \sin\alpha = Q$$
$$\left(Q = -\frac{M(p^2 - \omega^2)\,[k_1k_2(\Omega_2 - \Omega_1) + k_1b_2 + k_2b_1]}{(k_1 + k_2)\,q_1q_2}\right) \tag{5.19}$$

Thus, correlation (5.18) is actually satisfied if

$$-k_2(\Omega_2 - \Omega_1) < b_2 - b_1 < k_1(\Omega_2 - \Omega_1). \tag{5.20}$$

It follows from (5.19) that the following inequality is a condition for the possibility of the considered synchronous regime of the cage rotations

$$\left|\frac{k_1k_2(\Omega_2 - \Omega_1) + k_1b_2 + k_2b_1}{k_1 + k_2}\right| < q_1A_2 = q_2A_1 = \frac{q_1q_2}{M\,|p^2 - \omega^2|}, \tag{5.21}$$

If it is satisfied, then equation (5.19) allows for two essentially different solutions:

$$\alpha^{(1)} = \arcsin Q, \qquad \alpha^{(2)} = \pi - \arcsin Q. \tag{5.22}$$

The first solution is conditionally named synphaseous, and the second one antiphaseous, similar to the vibro-exciter problem.

Stability condition (5.10) in this case becomes

$$(\omega^2 - p^2)\cos\alpha < 0. \tag{5.23}$$

It follows from this condition that in the preresonance (precritical) range $\omega < p$, the synphaseous motion would be stable, while in the postresonance (postcritical) range $\omega > p$ antiphaseous synchronous motion would be stable. According to (5.7), the antiphaseous motion corresponds to a mutual compensation of rotor vibrations due to bearing defects. Specifically, for $q_1 = q_2$ and $Q \approx 0$, in such a case it will be $z^0 \approx 0$, or also $x = 0$ and $y = 0$. On the contrary, synphaseous motion corresponds to a mutual amplification of vibrations. It is easy to note that the same features are also characteristic for the considered regime with slippage of the balls of only one bearing.

Thus, for this system with the previously formulated assumptions, the *generalized Laval Principle* also holds: in the postcritical range there occurs some semblance of automatic balancing of the rotor.

In the case when dry friction can be neglected ($b_1 \approx 0$, $b_2 \approx 0$), the condition of the possibility for the synchronization of the cages (5.21) becomes

$$\frac{k_1k_2\,|\Omega_2 - \Omega_1|}{k_1 + k_2} < q_1A_2 = q_2A_1 = \frac{q_1q_2}{M\,|p^2 - \omega^2|}, \tag{5.24}$$

This differs from (5.16) only by replacing viscous friction coefficent k_2 with an equivalent friction coefficient $k_1k_2/(k_1 + k_2)$.

From inequalities (5.15), (5.16), (5.21), and (5.24) the following practical conclusions can be made:

1. The possibility of the self-synchronization of the cages of nominally identical bearings without considering the dry friction forces is determined by a correlation of the parameters characterizing the defects (manufacturing errors) of the bearings. The synchronization-inducing factors are defects characterized by parameters q_1, q_2 while synchronization-preventing factors are defects causing scatter of the kinematic values of the velocities $\Omega_2 - \Omega_1$.

Thus, self-synchronization of the cages is substantially different from that of the vibro-exciters. For the latter, only the desynchronizing factors depend on manufacturing errors, but the possibility of synchronization depends on a systemic factor-presence of rotor unbalances.

2. Self-synchronization is not possible if moments of dry friction forces are sufficiently large, thus preventing slippage of the cages.

3. An important synchronizing factor is rotor vibrations caused by bearing defects, whose amplitudes A_s are proportional to values q_s. Thus, approaching the resonance is beneficial for the development of the self-synchronization of the cages, while enhancement of viscous dissipative forces is not.

What is described above shows that self-synchronization of the cages of nominally identical bearings is possible only at some combinations of parameters q_1, q_2, $\Omega_2 - \Omega_1$, b_1, b_2, k_1 and k_2, and also p and ω. To decide if self-synchronization is actually possible, a thorough analysis of these parameters together with special observations and experiments are necessary.

It is important to remember that self-synchronization of the cages does not always cause negative consequences, such as a high wear rate of the rolling bodies and races or the development of some types of vibrations. Under certain conditions, a synchronous regime is characterized by a reduction in the vibration level. Then, if it is technically useful, the trend toward synchronization can be enhanced by, for example, an artificial formatting of defects characterized by parameters q_1 and q_2.

The considered mathematical model is quite crude since it does not consider some factors which can actually significantly influence self-synchronization of the cages. These factors include the nonlinearity of the elastic characteristics of the bearings; the angular vibrations of rotor axes; the giroscopic effect; "higher order" defects not considered in the expression (5.2) for potential energy.

Chapter 5

SYNCHRONIZATION OF PENDULUM CLOCKLIKE OBJECTS

§ 1) Mutual synchronization of pendulum clocks on an elastically supported platform (Huygens' problem).

As was noted in the introduction, synchronization of technical objects was discovered by Huygens in the example of self-synchronization and phase alignment of two pendulum clocks hanging from the same light beam.

Analysis of this problem is interesting both as an attempt to theoretically explain this *Huygens' problem,* and also for comparison with the laws of synchronization of other objects, such as mechanical vibro-exciters. It will be shown that there are both common features and substantial differences.

This problem is considered for one of the simplest but frequently adequate models of single-degree-of-freedom (DOF) clock mechanisms.* It is assumed that the clocks are hanging on an elastically supported rigid platform, also having a single DOF (Fig. 40). Analysis of this problem was given (42, 57), and is described below. A different solution (using a variational correlation, see § 6, chap. 11) is given in (175). A more simplistic formulation is analyzed in (194). Synchronization of clocks on a three DOF platform is analyzed in (60).

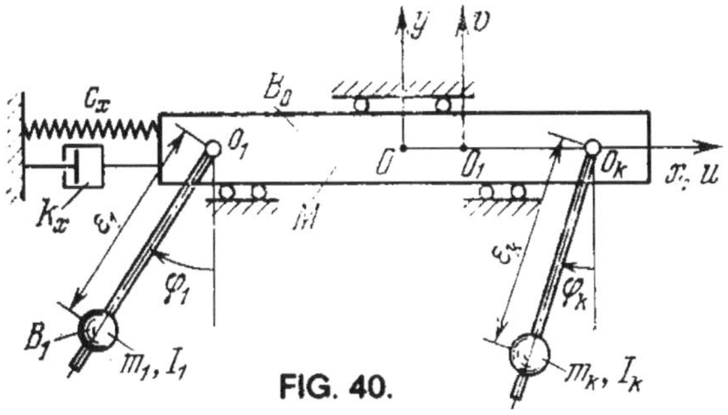

FIG. 40.

Equations of motion for this system can be derived from equations (4.6) and (4.7), chap. 3, describing the motion of an analogous system with vibro-exciters (compare Figs. 6 and 40). In the clock problem, only the meaning and expressions for the driving moments L_s are changing, which, under the assumptions adopted above, should be considered as depending on angles φ_s and the angular velocities $\dot{\varphi}_s$ of the pendulums. The action of moment L_s ($\dot{\varphi}_s$, φ_s) on a pendulum to simplify analysis, is sometimes considered an instantaneous pulse. In such a case, L_s ($\dot{\varphi}_s$, φ_s) is a nonanalytical function of its arguments. In the considered problem, a somewhat different character might also have expressions for moments of the resistance forces R_s. They are assumed, like L_s , to be dependent not only on angular velocity φ_s but also on angle $\dot{\varphi}_s$, i.e., $R_s = R_s$ ($\dot{\varphi}_s$, φ_s). Then the equations of motion of the system clocks platform will be written as

$$
\begin{gathered}
I_s\ddot{\varphi}_s + m_s g \varepsilon_s \varphi_s = m_s \varepsilon_s \ddot{x} + L_s(\dot{\varphi}_s, \varphi_s) - R_s(\dot{\varphi}_s, \varphi_s) \\
(s = 1, \ldots, k); \\
M\ddot{x} + k_x\dot{x} + c_x x = = \sum_{s=1}^{k} m_s \varepsilon_s \ddot{\varphi}_s .
\end{gathered}
\tag{1.1}
$$

Here, the same nomenclature is used as in § 4, chap. 3. Contrary to the vibro-

*Various dynamic models of clock mechanisms are considered in (3.4) and (20). Impact and nonimpact single DOF models also considered in (10).

exciters problem, angular deflections φ_s of the pendulums are considered to be small and are measured not from the horizontal direction of the platform vibrations, but from the vertical direction.

From observations by Huygens (see quotation in § 1 of the introduction), it can be expected that synchronization of the clocks is possible, for example, if their *"partial frequencies"*

$$\tilde{\omega}_s = \sqrt{m_s g \varepsilon_s / I_s}, \tag{1.2}$$

or natural frequencies of small oscillations of the pendulums with stationary pivoting points,* are close to each other. Thus, let's assume

$$\tilde{\omega}_s^2 = \tilde{\omega}^2 (1 - \chi_s), \tag{1.3}$$

Here χ_s are some small parameters, and $\tilde{\omega}$ can have a meaning of one of $\tilde{\omega}_s$, or some average value of $\tilde{\omega}_s$. In the process of solution, $\tilde{\omega}$ can also be defined from a condition of closeness between the first and the fundamental approximation as described in comment 5) of § 5, chap. 10. Then, transferring into dimensionless variables and introducing notations as follows:

$$\tau = \tilde{\omega} t, \quad x_0 = x/A, \quad 2n_x = k_x / M\tilde{\omega}, \quad n_x = \mu n_x',$$

$$\lambda_x^2 = p_x^2/\tilde{\omega}^2, \quad p_s = \frac{m_s \varepsilon_s A}{I_s}, \quad q_s = \frac{m_s \varepsilon_s}{MA}, \quad p_x^2 = c_x/M,$$

$$\mu\Phi_s\left(\frac{d\varphi_s}{d\tau}, \varphi_s, \frac{d^2 x_0}{d\tau^2}\right) = \tag{1.4}$$

$$= p_s \frac{d^2 x_0}{d\tau^2} + \chi_s \varphi_s + \frac{1}{I_s \tilde{\omega}^2}\left[L_s\left(\tilde{\omega}\frac{d\varphi_s}{d\tau}, \varphi_s\right) - R_s\left(\tilde{\omega}\frac{d\varphi_s}{d\tau}, \varphi_s\right)\right],$$

equations (1.1) can be presented as

$$\frac{d^2\varphi_s}{d\tau^2} + \varphi_s = \mu\Phi_s\left(\frac{d\varphi_s}{d\tau}, \varphi_s, \frac{d^2 x_0}{d\tau^2}\right) \quad (s = 1, \ldots, k),$$

$$\frac{d^2 x_0}{d\tau^2} + \lambda_x^2 x_0 = \sum_{j=1}^{k} q_j \frac{d^2\varphi_j}{d\tau^2} - 2\mu n_x' \frac{dx_0}{d\tau}. \tag{1.5}$$

Here A denotes an arbitrary parameter having dimension of length (it is convenient to assume that A is of the order of x_{max}), and $\mu \geq 0$ is a small parameter, as before. The algorithm of introduction of this small parameter corresponds to the assumptions that the clocks are quasi-identical, weakly coupled, quasi-conservative and quasi-linear autogenerators, and the viscous resistance force to oscillations of the platform is small.

* It should be noted that together with the similarity, there are some substantial differences between the introduced partial frequencies of pendulums (and of objects like quasi-harmonic autogenerators in general) and partial angular velocities ω_s of vibro-exciters (and objects with rotors in general) as introduced in § 5, chap. 3. For example, partial frequencies $\tilde{\omega}_s$ are nonnegative parameters, while ω_s can be negative.

The problem consists in studying conditions of the existence and stability of the periodic (synchronous) solutions of equations (1.5) having period 2π at $\mu = 0$.

A nonresonant case is considered with an assumption that the natural frequencies of pendulums on a stationary foundation $\tilde{\omega}_s \approx \tilde{\omega}$ are close to p_x/n, where p_x is the natural frequency of the platform and n is an integer. With such an assumption $\lambda_x = p_x/\tilde{\omega}$ is not an integer.

Another assumption is about the nondimensionless moment $(L_s - R_s)/I_s\,\tilde{\omega}^2$ being expressed as

$$\frac{1}{I_s\tilde{\omega}^2}(L_s - R_s) = a_s\left(1 - \nu_s\varphi_s^2\right)\frac{d\varphi_s}{d\tau}, \tag{1.6}$$

where a_s and ν_s are positive constants.

In other words, instead of a model with an impulse supply of energy to the clock pendulum, a van-der-Pol oscillatorlike model with a continuous supply of energy is considered (see also subsection 2, § 2, chap. 6). Such an idealization was supported in (194). The latter publication did not consider the clock problem, but was limited to the analysis of two inductance-coupled electronic generators (see § 2, chap. 6). Notwithstanding substantial differences of the initial equation systems, these problems are analogous, although only when there is a certain correlation between the parameters of the more complex system, representing the clock problem (shown below).

It is easy to find out, after a nonsingular linear transformation of variables

$$\begin{gathered}\varphi_s = y_s + y_{k+s}, \quad \dot{\varphi}_s = i\,(y_s - y_{k+s}) \qquad (s = 1, \ldots, k);\\ x_0 = y_{2k+1} + y_{2k+2} - \frac{1}{\lambda_x^2 - 1}\sum_{s=1}^{k} q_s(y_s + y_{k+s}),\\ \frac{dx_0}{d\tau} = i\lambda_x\,(y_{2k+1} - y_{2k+2}) - \frac{i}{\lambda_x^2 - 1}\sum_{s=1}^{k} q_s(y_s - y_{k+s})\end{gathered} \tag{1.7}$$

system (1.5) is reduced to a canonical format

$$\begin{aligned}\frac{dy_s}{d\tau} &= \lambda_s y_s + \mu F_s \qquad (s = 1, \ldots, 2k),\\ \frac{dy_{2k+1}}{d\tau} &= i\lambda_x y_{2k+1} + \mu F_{2k+1},\\ \frac{dy_{2k+2}}{d\tau} &= -i\lambda_x y_{2k+2} + \mu F_{2k+2}.\end{aligned} \tag{1.8}$$

This corresponds to the format of (3.1), chap. 10. The following notations are made in equations (1.8)

$$\begin{gathered}\lambda_s = i, \quad \lambda_{k+s} = -i; \quad F_s = -\frac{i}{2}\Phi_s, \quad F_{k+s} = \frac{i}{2}\Phi_s\\ (s = 1, \ldots, k),\\ F_{2k+1} = -F_{2k+2} = i\left[\frac{n'_x}{\lambda_x}\frac{dx_0}{d\tau} - \frac{1}{2}\frac{\lambda_x}{\lambda_x^2 - 1}\sum_{s=1}^{k} q_s\Phi_s\right].\end{gathered} \tag{1.9}$$

The theorem from § 3, chap. 10 on the existence and stability of the periodic solutions of the quasi-linear autonomous systems can be applied to system (1.8).

A fundamental system corresponding to equations (1.8) is as follows

$$\frac{dy_s^0}{d\tau} = \lambda_s y_s^0 \quad (s = 1, \ldots, 2k),$$
$$\frac{dy_{2k+1}^0}{d\tau} = -i\lambda_x y_{2k+1}^0, \quad \frac{dy_{2k+2}^0}{d\tau} = i\lambda_x y_{2k+2}^0 \tag{1.10}$$

It allows for a family of periodic (period 2π) solutions that depends on $2k$ arbitrary constants α_s,

$$y_s^0 = \alpha_s e^{\lambda_s \tau} = \begin{cases} \alpha_s e^{i\tau} & (s = 1, \ldots, k), \\ \alpha_s e^{-i\tau} & (s = k + 1, \ldots, 2k), \end{cases}$$
$$y_{2k+1}^0 = y_{2k+2}^0 = 0, \tag{1.11}$$

The constants, like the respective λ_s, can be considered as a complex conjugate without violating generality,

$$\alpha_s = r_s e^{i\psi_s}, \quad \alpha_{k+s} = \bar{\alpha}_s = r_s e^{-i\psi_s},$$
$$r_s = |\alpha_s| = |\alpha_{k+s}| \quad (s = 1, \ldots, k). \tag{1.12}$$

In addtion, due to the autonomous character of system (1.8), again without violating generality, it can be assumed that

$$\alpha_k = \alpha_{2k} = r_k = \alpha, \quad \psi_k = 0. \tag{1.13}$$

Characteristic exponents $\lambda_s = \pm i$ of system (1.10) are critical, according to the terminology in § 3, chap. 10, and form a special leading group. The other two exponents $\lambda_{2k+1} = i\lambda_x$ and $\lambda_{2k+2} = i\lambda_x$ are also critical ones, but at $\lambda_x \neq n/2$, where n is an integer, they are not special, or special leading, exponents.

After simple calculations by expressions (3.5) and (3.6), chap. 10, and also expressions (1.4), (1.6), (1.7), (1.9), and (1.11), the following expressions for function P_s^* result:

$$\mu P_s^* = -i\left[\chi_s\alpha_s + \frac{p_s}{\lambda_x^2 - 1}\sum_{j=1}^{k} q_j\alpha_j + i\alpha_s(1 - \nu_s\alpha_s\alpha_{k+s})\,\alpha_s\right],$$
$$\mu P_{k+s}^* = i\left[\chi_s\alpha_{k+s} + \frac{p_s}{\lambda_x^2 - 1}\sum_{j=1}^{k} q_j\alpha_{k+j} - i\alpha_s(1 - \nu_s\alpha_s\alpha_{k+s})\,\alpha_{k+s}\right]$$
$$(s = 1, \ldots, k) \tag{1.14}$$

Another result is the following *basic equations* for determining the parameters α_s of the fundamental solution:*

* The number of critical roots of the leading group is denoted here as $2k$, while in § 3, chap. 10 this number was denoted as k.

$$\mu P_s \equiv - \alpha_{2k}\mu P_s^* - \alpha_s \mu P_{2k}^* \equiv$$

$$\equiv - \alpha_{2k}\left[- i\left(\chi_s\alpha_s + \frac{p_s}{\lambda_x^2 - 1}\sum_{j=1}^{k} q_j\alpha_j\right) + a_s\left(1 - \nu_s\alpha_s\alpha_{k+s}\right)\alpha_s\right] -$$

$$- \alpha_s\left[i\left(\chi_k\alpha_{2k} + \frac{p_k}{\lambda_x^2 - 1}\sum_{j=1}^{k} q_j\alpha_{k+j}\right) + a_k\left(1 - \nu_k\alpha_k\alpha_{2k}\right)\alpha_{2k}\right] = 0,$$

$$\mu P_{k+s} \equiv - \alpha_{2k}\mu P_{k+s}^* + \alpha_{k+s}\mu P_{2k}^* \equiv \tag{1.15}$$

$$\equiv \alpha_{2k}\left[i\left(\chi\alpha_{k+s} + \frac{p_s}{\lambda_x^2 - 1}\sum_{j=1}^{k} q_j\alpha_j + a_s\left(1 - \nu_s\alpha_s\alpha_{k+s}\right)\alpha_{k+s}\right.\right] -$$

$$+ \alpha_{k+s}\left[i\left(\chi_k\alpha_{2k} - \frac{p_k}{\lambda_x^2 - 1}\sum_{j=1}^{k} q_j\alpha_{k+j}\right) + a_k\left(1 - \nu_k\alpha_k\alpha_{2k}\right)\alpha_{2k}\right] = 0$$

$$(s = 1, \ldots, k).$$

Here

$$b_{sj} = - \frac{p_s q_j}{\lambda_x^2 - 1}. \tag{1.16}$$

Due to (1.12) and (1.13), $P_k^* = -P_{2k}^* = \overline{P}_{2k}^*$. at $s = k$ can be obtained from (1.14). Thus, it follows from (1.15), $P_{k+s} = \overline{P}_s$. To determine $2k - 1$ complex constants $\alpha_1, \ldots, \alpha_{k-1}, \alpha_{k+1} = \overline{\alpha}_1, \ldots, \alpha_{2k-1} = \overline{\alpha}_{k-1}, \alpha = \alpha_{2k} = \alpha_k$, which can be replaced, according to (1.12), with $2k - 1$ real constants $r_1, \ldots, r_k$, $\psi_1, \ldots, \psi_{k-1}$, it is sufficient to consider $2k - 1$ equations obtained after equating to zero of real and imaginary terms in the left-hand parts of the first k equations (1.15). The last equation $\operatorname{Im} P_k = -\alpha \operatorname{Im}(P_k^* + P_{2k}^*) = 0$ is excluded, since it is an identity due to $P_k^* = -P_{2k}^*$.

For synchronous vibrations of the considered type to exist, it is necessary that these $2k - 1$ equations have solutions in which ψ_s are real and r_s are positive.

According to the theorem in §3, chap. 10, sufficient conditions of stability of the synchronous motions corresponding to each such solution are, first of all, negative signs of the real parts of the roots χ of the algebraic equation of $(2k - 1)$-th order.

$$\left| \frac{\partial (\mu P_s)}{\partial \alpha_j} + \delta_{sj}\alpha\varkappa \right| = 0 \qquad (s, j = 1, \ldots, 2k - 1) \tag{1.17}$$

and, secondly, compliance with

$$\operatorname{Re}\left[\left\langle\left(\frac{\partial F_{2k+1}}{\partial y_{2k+1}}\right)\right\rangle - i\lambda_x \frac{P_{2k}^*}{2\alpha_{2k}\lambda_{2k}}\right] < 0, \tag{1.18}$$

The latter, after transformation considering (1.4), (1.7), (1.9), and (1.12) to (1.14), becomes

$$2n_x - \frac{1}{r_k}\frac{p_k\lambda_x}{\lambda_x^2 - 1}\sum_{j=1}^{k} q_j r_j \sin\psi_j - \lambda_x a_k\left(1 - \nu_k r_k^2\right) > 0. \tag{1.19}$$

A further analysis of the problem in its general formulation is difficult due to a need to solve the system of transcendent equations (1.15). Thus, the analysis which follows deals only with the case of quasi-identical clocks.

§ 2) Case of quasi-identical clocks.

Let the parameters of the clocks be identical up to terms of higher order than μ. Then, according to (1.3), (1.14), (1.6),

$$a_s = a, \quad \chi_s = \chi, \quad \nu_s = \nu, \quad p_s = p, \quad q_s = q, \tag{2.1}$$

After transformation from variables α_s to r_s and ψ_s, equations (1.15), considering (1.12) and (1.13), become

$$\begin{gathered}\mu P_s = -\pi r_k \left[ib \sum_{j=1}^{k} r_j e^{i(\psi_j - \psi_k)} + a\left(1 - \nu r_s^2\right) r_s e^{i(\psi_s - \psi_k)} - \right. \\ \left. - ib \sum_{j=1}^{k} r_j e^{i(\psi_s - \psi_j)} - a\left(1 - \nu r_k^2\right) r_k e^{i(\psi_s - \psi_k)} \right] = 0 \qquad (s = 1, \ldots, k),\end{gathered} \tag{2.2}$$

Here

$$b = -pq/(\lambda_\chi^2 - 1). \tag{2.3}$$

Solutions of equations (2.2) are considered, corresponding to oscillations of all pendulums with identical amplitudes

$$r_1 = \ldots = r_k = 1/\sqrt{\nu}, \tag{2.4}$$

The latter coincide with amplitude values when there is no coupling between the clocks, i.e., the platform is stationary ($b = 0$). For such a case, separating the real and imaginary terms in (2.2), we obtain $2k$ expressions for k constants $\psi_1, \ldots, \psi_k$ which should satisfy:

$$\begin{gathered}\sum_{j=1}^{k} \cos \psi_j = \sum_{j=1}^{k} \cos(\psi_s - \psi_j), \quad \sum_{j=1}^{k} \sin \psi_j = \sum_{j=1}^{k} \sin(\psi_s - \psi_j), \\ \psi_k = 0 \qquad (s = 1, \ldots, k).\end{gathered} \tag{2.5}$$

Equality $\psi_k = 0$ is added to these equations in accordance with (1.13).

Equations (2.5) are satisfied, for example, if

$$\psi_1 = \ldots = \psi_k = 0, \tag{2.6}$$

This corresponds to oscillations of all the pendulums with identical phases. However, equations (2.5) also have many solutions for which

$$\sum_{j=1}^{k} \cos\psi_j = 0, \quad \sum_{j=1}^{k} \sin\psi_j = 0, \quad \psi_k = 0. \tag{2.7}$$

These conditions are equivalent to the requirement that k unitary vectors e_s from which one (e_k) is directed along axis Ox of plane xOy, form on this plane a closed polygon (see Fig. 8), in which now $|e_s| = 1$. For $k = 2$ and $k = 3$ there are only unique polygons, but for $k > 3$ there is an infinite number of polygons. This means that for a number of clocks larger than 3, phases ψ_s are not uniquely determined from (2.2). To find the phases in accordance with the note 8) in § 7, chap. 10, the following approximations should be considered.

Amongst solutions complying with conditions (2.7), the following solutions should be recognized (for an arbitrary k):

$$\psi_s = \frac{2\pi}{k} j \quad (j = 1, \dots, k-1), \quad \psi_k = 0, \tag{2.8}$$

These solutions are correlated with regular k-polygons, i.e., the phases are uniformly distributed. Another special group of solutions (for even k) is

$$\psi_1 = \dots = \psi_{k/2} = \pi, \ \psi_{k/2+1} = \dots = \psi_k = 0, \tag{2.9}$$

which describe oscillations of one half of all the clocks in anti-phase in relation to the other half.

Thus, the synchronization of the clocks is always possible in principle, if their corresponding parameters differ only slightly. Of course, stability of the respective solutions should be proved. As is shown below, stability is assured at least for the case of two clocks. For solutions like (2.4), (2.6), and (2.7) and additional stability condition (1.19) synchronization can be achieved for any number of quasi-identical clocks, since it is assured by the positive sign of the resistance coefficient n_x.

The case of *two clocks* is considered in more detail. In this case, equations (2.2) have two solutions:

$$\begin{aligned} r_1 = r_2 = \sqrt{\nu}, \quad \psi_1 = 0, \quad \psi_2 = 0, \\ r_1 = r_2 = \sqrt{\nu}, \quad \psi_1 = \pi, \quad \psi_2 = 0, \end{aligned} \tag{2.10}$$

These correspond to synphaseous and antiphaseous motions of the pendulums, respectively. Analysis of the stability of this solution, considering (1.15), (2.1), and (2.10) results, for $k = 2$, in the following:

$$\begin{aligned} &\partial(\mu P_1)/\partial\alpha_1 = \left(a + ibe^{-i\psi_1}\right)/\sqrt{\nu}, &&\partial(\mu P_1)/\partial\alpha_2 = \left(ae^{i\psi_1} - ib\right)/\sqrt{\nu}, \\ &\partial(\mu P_1)/\partial\alpha_3 = \left(ae^{2i\psi_1} + ibe^{i\psi_1}\right)/\sqrt{\nu}, &&\partial(\mu P_2)/\partial\alpha_1 = -ib/\sqrt{\nu}, \\ &\partial(\mu P_2)/\partial\alpha_2 = \left(2a + ibe^{-i\psi_1}\right)/\sqrt{\nu}, &&\partial(\mu P_2)/\partial\alpha_3 = ib/\sqrt{\nu}, \\ &\partial(\mu P_3)/\partial\alpha_1 = ae^{-2i\psi_1}/\sqrt{\nu}, &&\partial(\mu P_3)/\partial\alpha_2 = -ae^{-i\psi_1}/\sqrt{\nu}, \\ &\partial(\mu P_3)/\partial\alpha_3 = \left(a - 2ibe^{-i\psi_1}\right)/\sqrt{\nu}. \end{aligned} \tag{2.11}$$

Deriving now equation (1.17), it is easy to find that its roots are the same for both solutions (2.10):

$$\varkappa_1 = -2\mu a, \quad \varkappa_{2,3} = \mu[-a \pm \sqrt{a^2 - 4b^2}] \tag{2.12}$$

They have negative real parts if only $a > 0$, a usual assumption.

Thus, in this case, regardless of ratio between the pendulum frequency $\tilde{\omega}$ and the natural frequency $p_x = \sqrt{c_x/M}$ of the platform, both synphaseous and antiphaseous motions of the pendulums are stable. In the case of the antiphaseous oscillations, the platform is practically stationary (a paradox of the nonworking connections, § 3, the introduction, and subsection 7, § 14, chap. 3).

It was shown in (60) that the same result is correct when the platform has three essential DOFs instead of one. It was noted in the introduction that antiphaseous synchronous oscillations of nominally identical clocks were observed by Huygens in his experiments, and both syn- and antiphaseous oscillations were observed in experiments in the "Mekhanobr" Institute.

§ 3) Some features of self-synchronization of pendulum clocks vs. features of self-synchronization of unbalanced rotors.

Some features of the synchronization of devices like pendulum clocks are the same as the features of the synchronization of mechanical vibro-exciters (see § 14, chap. 3) and other dynamic systems (§ 2, chap. 6). Thus, pendulums with identical or sufficiently close parameters are usually self-synchronizing. At the appropriate conditions, the effect of averaging the partial frequencies takes place. For pendulums, as with vibro-exciters, the paradox of nonworking connections occurs.

At the same time, many important features of the self-synchronization of clocks are substantially different. For example, as was pointed in § 2, the character of stable phase relations of pendulums might weakly depend or not depend at all on the number of DOFs of the connection systems (i.e., the platform to which the clocks are attached). The most substantial difference is that usually *only clocks with sufficiently close partial frequencies can self-synchronize, while in the case of rotating rotors the effect can occur also at very different partial angular velocities* (see footnote on p. 148). This is partially related to another practical and important difference: In the case of the self-synchronization of oscillating objects, like pendulum clocks, much less power can be transmitted between the objects than in the case of rotors (vibro-exciters). This power has an order of $N'_{max} = m\varepsilon A_1\omega^3\varphi_0$, where m is mass; ε is the length of the pendulum; A_1 is the amplitude of the platform vibrations; φ_0 is the angular amplitude of the pendulum oscillations; ω is the synchronous frequency of vibrations. For rotors, $N'_{max} = m\varepsilon A_1\omega^3$ (see § 6, chap. 3). If parameters m, ε, A_1, ω are assumed to be the same in both cases, then $N'_{max}/N_{max} = \varphi_0$. Thus, if $\varphi_0 = 0.1 \approx 5.7°$, then $N'_{max}/N_{max} = .01$, and for the parameters in the example in subsection 1, § 6, chap. 3, instead of $N_{max} = 800$ kW, it would be only $N'_{max} = 80$ kW. This fact does not now allow us to directly use self-synchronization of pendulums in crude, energy-intensive engineering systems with the same effectiveness as synchronization of the rotors (§ 15, chap. 3). However, there certainly would be cases when the limited energy transfer between the oscillating objects would be adequate. It might also happen that the energy transfer would be enhanced by some further developments.

§ 4) Synchronization of turbine blade vibrations.

It was experimentally observed that, at certain conditions, stable synchronous vibrations of the turbine blades develop and cause undesirable effects. This is another case of the adverse consequences of the synchronization phenomenon.

Self-synchronization of turbine blades was studied in (96 and 272) which considered the blades as quasi-linear oscillators (generators) forming a mechanical system with a cyclical symmetry. Vibrations are generated by the energy of the incoming flow; thus these are aeroelastic effects. Coupling between blade vibrations, especially significant for adjacent blades, is also mainly due to their interaction via the flow. Coupling via the disc to which the blades are attached is less important. The study has confirmed that self-synchronization always occurs if partial frequencies of the blades are only slightly different. Thus, an effective method of abatement of synchronous self-excited vibrations is an intentional detuning of the fundamental natural frequencies of the adjacent blades. In these systems, external synchronization of the blades due to circumferential nonuniformity of the flow is also possible.

Chapter 6

SYNCHRONIZATION IN ELECTROTECHNICS, RADIOTECHNICS, AND RADIOPHYSICS

§ 1) Synchronization at the parallel connection of electrical machines.

A stability problem for the steady regimes of the parallel operation of several synchronous electrical machines is one of the most important specific problems of the theory of synchronization of dynamic systems.* This problem is addressed in many publications, such as (101), (126), (157), (322). In the general formulation, the problem is so complicated that some important aspects are still not studied. As an example, cases in which stability analysis is associated with the analysis of a system of differential equations with periodic coefficients are hardly addressed. A survey of this problem is given in (136–138, 253).

First, a formulation of the problem of the parallel operation of some number k of generators $G_1, \ldots, G_k$ driving the same load. The condition of the s-th generator is characterized by one rotational coordinate, angular position φ_s of rotor in relation to stator, and also by the set of vibratory phase coordinates $x_1^{(s)}, \ldots, x_{r_s}^{(s)}$, which can be both electrical and mechanical parameters.

Each generator initially is driving loads $R_1, \ldots, R_k$, whose conditions are characterized by phase coordinates $u_1^{(s)}, \ldots, u_{\nu_s}^{(s)}$ (Fig. 41a). Then there are k independent autonomous systems, in which the following conditions are developing

$$\begin{gathered}\varphi_s = \omega_s t + \psi_s(\omega_s t), \quad x_j^{(s)} = x_j^{(s)}(\omega_s t), \quad u_\rho^{(s)} = u_\rho^{(s)}(\omega_s t), \\ (j = 1, \ldots, r_s; \quad \rho = 1, \ldots, \nu_s; \quad s = 1, \ldots, k),\end{gathered} \tag{1.1}$$

Here ψ_s, $x_j^{(s)}$, and $u_\rho^{(s)}$ are periodic functions of t with period $2\pi/\omega$, and each ω_s is a constant which can be called a partial frequency of the generator at a given load R_s. Due to the differences in loads R_s, manufacturing inaccuracies, and nonideality of the controllers, frequencies ω_s for different generators would be, generally speaking, different.

The next case is when all the generators are connected in parallel to drive the common load R whose condition is characterized by the phase coordinates $u_1, \ldots, u_\nu$ (Fig. 41b). The necessary condition for the normal operation of the system is the development in it of an operational regime with a certain common frequency ω, a synchronous frequency. A "gift of nature," frequently not appreciated, is the fact that such a synchronous regime turns out to be stable, at least locally, by itself, due to the internal features of the system's "generators-load." Thus, here also there is a tendency to synchronization. It is due to the same reasons which are shown in § 9, chap. 10, and in §§ 4 and 7, chap. 12, namely, the periodic character of the dependence of the interaction moments between the rotors on relative angles $\varphi_s - \varphi_j$. The conditions of the actual development of a synchronous regime of generators driving the common load are usually small differences in the partial frequencies ω_s and the

* In electrotechnics, "synchronization" is usually a system of procedures for initiating a synchronous regime (which is stable in a limited part of the phase space—see below) of the synchronous machine being switched on. However, when speaking about the stability of the parallel operation of several electrical machines, "synchronization" is usually understood in the same sense as in this book.

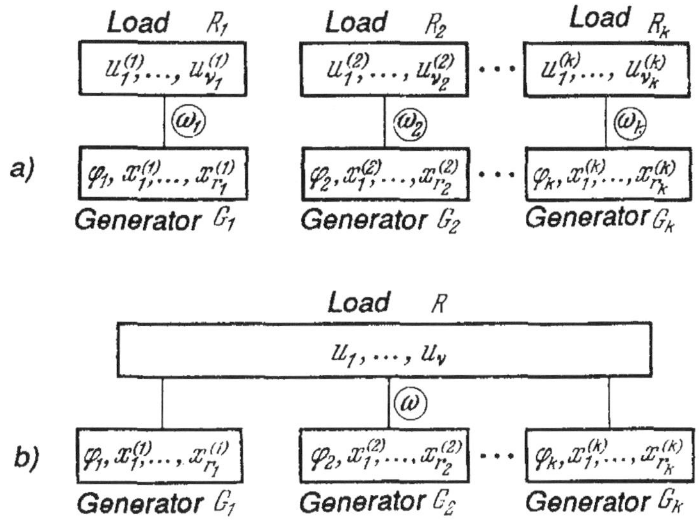

FIG. 41.

"correctness" of bringing the generators into synchronism. The first condition is assured by accuracy of the controllers. The second is associated with the fact that the "domain of attraction" of the synchronous regime in the phase space of the system is usually limited. This means that the regime might be stable locally but not globally. There are known practical occurrences of the generators dropping out of synchronism due to some perturbations in the system. In such cases, undesirable beats in the voltage, or even serious failures of aggregates, had developed.

Thus, a principal problem for parallel operation of the generators is in establishing conditions at which a regime with a common synchronous frequency ω develops in the unified system, notwithstanding the possible difference of partial frequencies ω_s. In other words, it is the basic problem of the synchronization of the theory-finding conditions of the existence and stability of the motions in the unified system, which can be described as

$$\varphi_s = \omega t + \psi_s(\omega t), \quad x_j^{(s)} = x_j^{(s)}(\omega t), \quad u_\rho = u_\rho(\omega t)$$
$$(j = 1, \dots, r_s; \quad \rho = 1, \dots, \nu; \quad s = 1, \dots, k), \tag{1.2}$$

Here ψ_s, $x_j^{(s)}$, and u_ρ are periodic time functions with the common period $2\pi/\omega$, where ω is a constant which is not accurately known beforehand.

In a similar way, the problem of the normal operation of the energy systems, each containing several synchronous machines, can be formulated. Again, the conditions for the existence and stability of motions like (1.2) should be analyzed in a case where partial frequencies ω_s of the individual systems are generally different.

A special class of synchronization problems is developed in conjunction with the electrotechnic methods of assuring identical rotation angles of several shafts, from which each or some are driven by some (not necessarily electric) motors. Such "electric shaft" systems are well known from the literature [e.g., (316)].

For a correct and cost-effective parameter selection for the electric shaft system, a similar problem of the synchronization of several coupled objects has to be solved. Dynamic, not just static, features of the objects connected to each shaft might, naturally, significantly affect the values of the system parameters providing synchronous and synphaseous rotation of shafts. Mechanical vibro-exciters are representative objects for which the solution of the synchronization/phasing problem by means of the electric shaft is of interest (see subsection 3, §11, chap. 3).

As a general rule, equations of motion for systems with electric machines can be written as Lagrange–Maxwell equations, which are generalizations of Lagrange equations for electromechanical problems. These differential equations are not given here. In the general case, they are so complex that even their derivation is a nontrivial problem, which is still not finally solved.

Using variables ψ_s, $x_j^{(s)}$, u_ρ and $\tau = \omega t$, the problem of parallel operation of electric synchronous machines, like all the synchronization problems previously considered, is reduced to an analysis of the existence and stability of the periodic (period 2π) solutions of some system of differential equations whose right-hand sides are also periodic functions of $\tau = \omega t$ with period 2π. However, the majority of the systems already studied are of such a nature that these equations can, after some assumptions, be transformed into variables for which the right-hand sides do not explicitly contain time. Steady regimes correspond to constant values of new variables.* Naturally, in such circumstances, the stability analysis is reduced to the consideration of a system of differential equations with constant coefficients.

Significant simplifications can be achieved due to the fact that differential equations of parallel operating electric machines and energy systems tolerate an effective introduction of small parameters. This is not only because of the character of the connections between objects being synchronized, but also because of a significant difference in orders of magnitude between the time constants characterizing the processes in the system (137, 286).

Besides the preceding formulated principal problem, a significant interest for the theory of the parallel operation of synchronous electrical machines is the study of the conditions of entraining into a synchronous regime a synchronous machine being connected to the operating network, as well as an accompanying transient process. These problems, which are quite difficult in general formulation, are not yet adequately studied. They are connected to consideration of the system "as a whole."

The synchronization problem for electrical generators is frequently formulated and solved as an external synchronization problem. It assumes that one of several generators in the system is much more powerful than another; thus their motion is prescribed and does not depend on the behavior of other objects. In such a case, synchronous frequency ω does, of course, coincide with the frequency generated by the "powerful" generators.

§ 2) Synchronization of vacuum-tube and transistor generators.

1. Brief survey of development and state of the art. A vacuum-tube generator is a self-exciting vibratory system including an electrical oscillator and a tube which transforms energy from a DC source into energy of constant-amplitude electrical vibrations. Recently, vacuum tubes were replaced by transistors. However, there is no principal difference between these devices, so in this paragraph a "generator" means either one of two types: generators of quasi-harmonic oscillations, whose behavior is described by quasi-linear differential equations (*Thomson*

* This is explained by the fact that such regimes of operation are considered which accurately enough can be reduced to so-called symmetric ones.

generators) and *relaxation generators*, whose oscillations are quite different from harmonic, and can even have discontinuities.

The synchronization problem for coupled generators is important for application to electronics and biophysics; see subsection 4.

There is an ample number of publications, started by (6, 305, 334, 335), where the problem was analyzed in the most simplistic formulation, as a problem of external synchronization (entrainment) of one generator by a given periodic excitation. A bibliography, together with the original results, can be found in (10, 112, 171, 181).

Such a simplification by the reduction of the problem to the degenerated problem of entrainment is not always possible, especially if the interacting generators have comparable power. Thus, more attention is paid to a mutual synchronization of several generators. For example, in (98, 319) two coupled Thomson generators with soft vibration excitation are considered.

Synchronization of quasi-identical Thomson generators is considered in (264). It was shown that with a changing degree of detuning, jumps of amplitude and frequency are possible; this was observed in (124). Frequency-phase diagrams were used for the same problems in (2).

Mutual synchronization of two quasi-linear and essentially nonlinear oscillators is addressed in (154) and (331).

Mutual synchronization of two inductively coupled reflective klystrons is analyzed in (289). Multiple synchronization in a two-contour generator is studied in (268). A stroboscopic method (see § 5, chap. 11) is used in (194) for the problem of mutual synchronization of two quasi-harmonic generators; a similar problem was analyzed in (26) using the Poincare method.

Mutual synchronization of three weakly coupled generators was analyzed in detail in [226, 227], using the van-der-Pol method, as well as approximate method from [288]. Cases of comparable and very diverse power of generators are considered. Domains of existence and stability of synchronous regimes are constructed as functions of degree of detuning and other parameters; experimental validation is also performed.

The systematic treatment of some problems of entrainment and the mutual synchronization of several generators are given in [17].

Mutual synchronization of two coupled relaxation generators (multivibrators) was considered in (73). Multiple synchronization regimes and demonstrated vibration generation in the coupled system in the case when separate generators are not excited were discovered. Since so-called parasitic DOFs were neglected, vibrations had to be considered as discontinuous. Some additional assumptions about system behavior were adopted, besides the known "jump hypothesis." These assumptions (without correlation with the synchronization problem) were justified in (244) by considering a system with "parasitic capacitors."

The method of point transformation was used for some problems with Thomson and relaxation generators in (128, 129) and (257, 258).

Problems of the mutual synchronization of generators were considered in (186, 187, 248) using the asymptotic method. Specifically addressed was the influence of time lag in coupling elements between generator contours on the synchronization process. Time lag is frequently an idealization (for simplicity) of action of wave connections between generators.

Some publications [see survey in (240) and subsection 2, chap. 9] consider an interesting problem of the interaction and synchronization of a large number of mutually coupled generators in connection with an alternative "order (e.g., synchronization) stochastic behavior."

Entrainment and mutual synchronization of generators when fluctuations are present are considered in (171, 181, 261).

An interest in the mutual synchronization of multiple generators is stimulated by biophysics and biology research, e.g., see chap. 8.

A brief characterization of the problems, some basic features, and the main application areas for the synchronization of the generators are given below.

2. Thomson generators [57] (van-der-Pol quasi-linear autogenerators). The most widely used model of quasi-harmonic generators is the so-called *van-der-Pol autogenerator*, described by a differential equation (*van-der-Pol equation*)

$$\ddot{x} + \tilde{\omega}^2 x = \mu a'(1 - \nu x^2)\dot{x}, \tag{2.1}$$

Here ω, a', ν and μ are positive constants, with μ assumed to be a small parameter. With large μ, the self-excited vibrations described by (2.1) attain the character of relaxation vibrations. Equation (2.1) corresponds to a primitive self-excited system. For this equation, the equilibrium position $x = 0$ is unstable, as is the equilibrium position for a linear equation [equation] approaching (2.1) at small x. At large x, equation (2.1) describes decaying vibrations (the term $\mu a' \langle(1 - x^2)\rangle$ is negative). The self-excited vibrations (a limiting cycle) are represented by a unique periodic solution of (2.1), which is characterized by some "intermediate" values of x:

$$x = (2/\sqrt{\nu})\cos\tau + O(\mu). \tag{2.2}$$

If there is a certain number k of generators with different values of parameters $\tilde{\omega}$, a', ν and μ, then they would be generating vibrations of different frequencies. If the generators are connected into a unified system, e.g., described by equations like

$$\ddot{x}_s + \tilde{\omega}_s^2 x_s = \left[a_s\tilde{\omega}\left(1 - \nu_s x_s^2\right)\dot{x}_s + \sum_{j=1}^{k}\left(b_{sj}^{(2)}\ddot{x}_j + b_{sj}^{(1)}\tilde{\omega}\dot{x}_j + b_{sj}^{(0)}\tilde{\omega}^2 x_j\right)\right]$$
$$(s = 1, \ldots, k), \tag{2.3}$$

where $b_{sj}^{(q)}, \tilde{\omega}$ are some constants described below, then synchronization could develop in such a system. This means vibrations with the same frequency ω or with frequencies $n_j\omega$, where n_j are integers. There is also possible self-synchronization on combination frequencies, see § 1, chap. 1. Equations (2.3) obviously correspond to the assumption of linear connections between generators. Terms with $b_{sj}^{(2)}$ represent inductive coupling; with $b_{sj}^{(1)}$ resistance coupling; and with $b_{sj}^{(0)}$ capacitance coupling.

Synchronization is considered for the purely inductive case when $b_{sj}^{(1)} = 0$, $b_{sj}^{(0)} = 0$, $b_{sj}^{(2)} = b_{sj}$. This case is somewhat degenerated, but a general case also would not cause specific difficulties. Considering a simple (nonmultiple) synchronization, it is assumed that frequencies of small free vibrations $\tilde{\omega}$ (*partial frequencies*) are about identical, i.e.,

$$\tilde{\omega}_s^2 = \tilde{\omega}^2(1 - \chi_s) \tag{2.4}$$

This is similar to analysis of the Huygens' problem [see expression (1.3), chap. 5 and its explanation].

Also, generators are considered to be quasi-linear and weakly coupled. Then, after the introduction of nondimensionless time $\tau = \bar{\omega} t$, system (2.3) can be written as

$$\frac{d^2 x_s}{d\tau^2} + x_s = \mu\Phi_s\left(\frac{dx_s}{d\tau},\ x_s;\ \frac{d^2x_1}{d\tau^2},\ \ldots,\ \frac{d^2x_k}{d\tau^2}\right) \quad (s = 1,\ \ldots,\ k), \tag{2.5}$$

where

$$\mu\Phi_s = a_s\left(1 - x_s^2\right)\frac{dx_s}{d\tau} + \chi_s x_s + \sum_{j=1}^{k} b_{sj}\frac{d^2x_j}{d\tau^2}. \tag{2.6}$$

After changing variables,

$$x_s = y_s + y_{k+s}, \quad \dot{x}_s = i(y_s - y_{k+s}) \tag{2.7}$$

system (2.5) is transformed into a canonic format

$$\dot{y}_s = \lambda_s y_s + \mu F_s \qquad (s = 1,\ \ldots,\ 2k), \tag{2.8}$$

corresponding to the format of equations (3.1), chap. 10. Here it is denoted

$$\lambda_s = i, \quad \lambda_{k+s} = -i, \quad F_s = -\frac{i}{2}\Phi_s, \quad F_{k+s} = \frac{i}{2}\Phi_s \tag{2.9}$$
$$(s = 1,\ \ldots,\ k).$$

The theorem from § 3, chap. 10, can be applied to (2.8). For this it should be noted that the characteristic exponents of the fundamental system corresponding to equations (2.8),

$$\dot{y}_s^0 = \lambda_s y_s^0 \qquad (s = 1,\ \ldots,\ 2k), \tag{2.10}$$

are critical and form a leading special group. This system has a family of periodic (period 2π) solutions that depend on $2k$ arbitrary constants $\alpha_1, \ldots, \alpha_{2k}$

$$y_s^0 = \alpha_s e^{\lambda_s t} = \begin{cases} \alpha_s e^{it}, & s = 1,\ \ldots,\ k, \\ \alpha_s e^{-it}, & s = k+1,\ \ldots,\ 2k, \end{cases} \tag{2.11}$$

Expressions for fundamental functions $P_s(\alpha_1, \ldots, \alpha_{2k})$ and basic equations for parameters $\alpha_1, \ldots, \alpha_{2k}$ of the fundamental solution, written in accordance with formulas (3.5) and (3.6) of chap. 10 and expressions (2.6) to (2.11), are exactly the same as equations (1.15) of chap. 5 related to the Huygens' problem. An equation whose root structures determine the stability of the motion, coincides with equation (1.17), chap. 5. The only difference is the absence of the additional stability condition (1.19), chap. 5, since the connection system does not introduce additional DOFs. In the Huygens' equation, parameter b_{sj}, corresponding to the inductive coupling coefficient, is also more complicated.

Thus, all the discussion at the end of § 1 and in § 2, chap. 5, can be repeated here (with the substitution of "clocks" and "pendulums" for "generators").

One of the main difficulties of a specific analysis of the synchronization of Thomson generators (as in the clock problem) is the solution of equations like

(1.15), chap. 5, for the steady amplitude and phase values in synchronous regimes, determining the condition for the existence of respective solutions (i.e., the search for synchronization domains in the parameter space), and screening of solutions corresponding to stable motions. To overcome this difficulty, some approximate methods were developed, such as (288). However, these methods are applicable only with some limiting assumptions, otherwise they may cause large errors. Recently, computer-based methods for direct solution of differential equations became popular (see below and also § 2, chap. 8).

The synchronization of Thomson generators is characterized by some peculiar features described in (171, 181, 247, 248); see also § 2, chap. 8.

One of these features is the narrowing of the domains of existence and stability of the generators with increasing number k of generators. At the same time, under certain conditions, synchronization is distinctly developing even for very large numbers of weakly interacting generators or nonlinear oscillatorlike objects in general. This was noticed during the numerical modeling of oscillators with an initial disturbance [see § 3, chap. VII of (266)]. The system behavior did not correspond at all to a uniform distribution of energy between DOFs. As was said in (266), in such systems "too much mixing" was not noticed. As could be seen from the description of the experiments, the motion of the oscillators was accompanied by synchronization phenomena.

Another example is treated in (304), also described in (240). In this paper, there is a numerical experiment with a large number of weakly coupled generators with randomly distributed partial frequency ω_s. Density of distribution is in accordance with the Lorentz distribution,

$$f(\omega) = \frac{g}{\pi\left[g^2 + (\omega - \omega_0)^2\right]}.$$

The system of interacting generators is described by equations

$$\dot{A}_s = \gamma_s A_s\left(A_{s0}^2 - A_s^2\right) + \alpha_s(A_1, \ldots, A_k;\ \varphi_1, \ldots, \varphi_k),$$

$$\dot{\varphi}_s = \omega_s + \beta_s(A_1, \ldots, A_k;\ \varphi_1, \ldots, \varphi_k) \qquad (s = 1, \ldots, k).$$

These are regular approximate equations for amplitudes A_s and phases φ_s which are obtained from initial equations like (2.3) by using, for example, van-der-Pol method of slowly changing amplitudes. It was denoted

$$\gamma_s = A_{s0} = 1, \quad \alpha_s = \frac{1}{k}\sum_{j=1}^{k}\left[A_j\cos(\varphi_j - \varphi_s) - A_s\right],$$

$$\beta_s = \frac{0,2}{k}\sum_{j=1}^{k} A_j\sin(\varphi_j - \varphi_s),$$

This corresponds to linear coupling between generators.

The experimental results are shown in Fig. 42, where $\psi = \varphi = \omega_0 t$ denote phases, e are squares of generator amplitudes, and T is period $2\pi/\omega_0$. It can be seen that with a significant initial scatter of amplitudes and phases, a synchronous regime of generator vibrations is developed after only 80 periods.

Thomson generator models are widely used for the study of synchronization not only in electronics and radiophysics, but also in mechanics and biology (see chap. 8). Essentially the same model was used for the analysis of synchronization of the pendulum clocks and turbine blades.

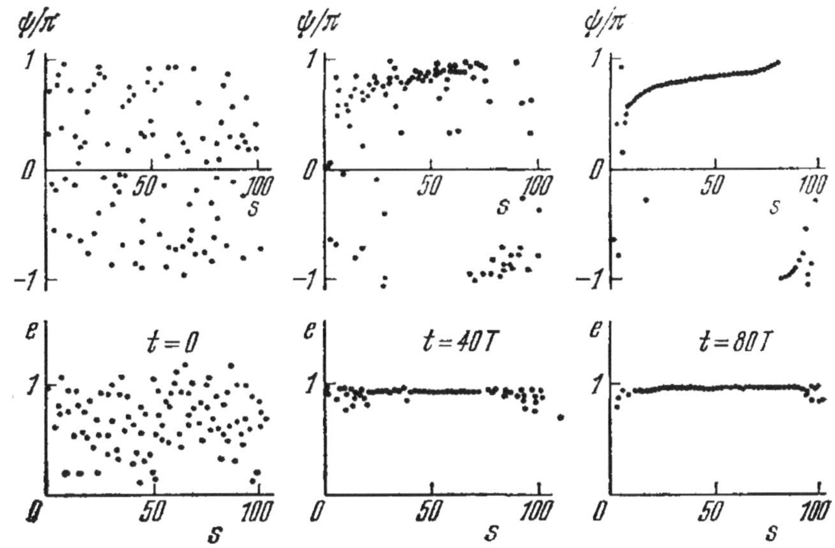

FIG. 42.

3. Relaxation generators. Synchronization of relaxation generators is of interest for various applications both from radiophysics and electronics and from biophysics and biology. Their analysis is much more difficult than that of Thomson generators, since it is based on considering differential equations which are nonlinear in the fundamental approximation. Mathematical models of relaxation generators are van-der-Pol equations (2.1) with "nonsmall" values of the coefficient $a = \mu a'$, Huygen's equations (see § 2, chap. 8), and also other types of equations and systems of equations.

For analysis of relaxation vibrations, it is frequently useful to consider equations with a small parameter at higher order derivatives, so-called singularly disturbed equations. This corresponds to neglecting, in the fundamental approximation, some DOF and usually leads to the development of discontinuous vibrations. Thus, analysis of the synchronization of the relaxation generators requires special methods and techniques.

A brief survey of the publications on the synchronization of relaxation generators, which treat problems of radiophysics and electronics, was given in subsection 1. Publications related to applications to chemistry and biology are discussed in § 2, chap. 8, where the basic features of the synchronization of the relaxation generators are described. Some of the principal features are widening of the synchronization domains during transition from quasi-linear vibrations to relaxation vibrations due to changing generator parameters, e.g., with increasing value of parameter $a = \mu a'$ in equations like (2.1). The latter is related to "clutching" of the generator vibrations by higher harmonics.

4. Applications of synchronization of generators. Systems of phase-based self-tuning. Self-synchronization and entrainment of generator effects found some important practical applications, e.g., see (112).

One of them is stabilization of the oscillation frequency of a powerful generator with an unstable frequency by a low-power but highly stable generator. In this case, for simple (nonmultiple) synchronization, the synchronized generator can be considered a power amplifier. It so happens that very high amplification coefficients (gains) are achievable. In the case of multiple synchronization, the generator being synchronized may resemble a frequency divider or multiplier.

The possibilities of using self-synchronization of multiple generators for generating precision frequency vibrations were mentioned in subsection 2. In some systems, several self-synchronizing generators are used just for obtaining the high power of the vibrations being generated by adding up powers of the individual generators.

While it is frequently possible to use self-synchronization of generators driving a common load, it is sometimes necessary to turn to forced synchronization. One of the widely used means for such synchronization is the system for phase-based self-tuning of frequency, e.g., as described in (298) and (271). In such systems, the phases of signals from the generators being synchronized are compared in the phase detector. After this, the error signal is used for controlling vibration frequency. Such systems have many positive features, but they are, of course, more complex than ones based on the use of self-synchronization.

§ 3) Synchronization of quantum generators.

The development of quantum generators in radio frequencies (masers) and optical frequencies (lasers) is one of the most important achievements of modern technical physics (144). Such devices found applications in many research and practical problems. In the future, they might be critical for improvements in transmission technology both for energy and information; in chemical technology; in the study and use of biological events; and in the study and use of controlled thermonuclear reactions.

Quantum electronics is based on the simultaneously induced discharge of energy by excited atoms, resulting in intensive, highly coherent, radiation. This effect was forecast by A. Einstein in 1917; in the 1930s some thoughts of its utilization were expressed by V. Fabricant. In 1954–55 Ch. Towns, and N. Basov and A. Prokhorov independently published a theory of masers and its implementation in radio-spectroscopes. The first laser was developed in 1960.

For some devices using quantum generators, the problem of mutual synchronization of several generators coupled by various mechanisms is important. Here is a formulation of this problem for the specific case of two interacting molecular generators (185). If x and x_2 are intensities of the electric field in resonators of the first and second generators, respectively, then, with some approximations, the processes in generators are described as

$$\begin{aligned} \ddot{x}_1 + \frac{\omega_1}{Q_1}\dot{x}_1 + \omega_1^2 x_1 + 4\pi \ddot{P}_1(x_1) + \alpha_{12}x_2 = 0, \\ \ddot{x}_2 + \frac{\omega_2}{Q_2}\dot{x}_2 + \omega_2^2 x_2 + 4\pi \ddot{P}_2(x_2) + \alpha_{21}x_1 = 0. \end{aligned} \tag{3.1}$$

Here $\omega_1, \omega_2, Q_1, Q_2$ are, respectively, frequencies and quality factors of the first and second resonators; α_{12}, α_{21} are coupling coefficients between generators, assumed to be small; $P_1(x_1)$ and $P_2(x_2)$ are, averaged along the lengths of resonators, polarizations of active media which are quite complex functions of their arguments.

The main problem is again to clarify the conditions at which, notwithstanding the differences in the generator parameters, vibrations with the common period $2\pi/\omega$ are generated. This is equivalent to the conditions for the existence and stability of periodic ($T = 2\pi/\omega$) solutions of equations (3.1). As before, this period is not known in advance and has to be determined during the solution process. Thus, in such a simplified formulation, the problem of mutual synchronization of molecular generators is in many ways analogous to the corresponding problem for vacuum-tube generators considered in § 2. The difference is only in the specifics of the structure of nonlinear functions in equations of motion. These specifics determine the operational specifics of coupled molecular generators, discussed for various cases in (21, 144, 185, 186). Equations of oscillations in generators are discussed in (177, 273).

The possibilities of forced synchronization of quantum generators by external sources are important for applications, e.g., see (72, 193).

§ 4) Synchronization in distributed systems and its applications.

Besides synchronization effects in systems which can be idealized as systems with a finite number of DOFs, as described in the previous sections, cases of synchronizations in systems whose characteristic dimensions have the same order of magnitude as the wave length of running or standing waves are quite important for applications. Such systems have to be considered as continuous systems and are described by partial differential equations.

In real systems, there are various interactions between waves of various lengths (between vibratory modes), which can be considered as weak in the sense of § 2, chap. 1. If wave frequencies are close or close to being commeasurable, modal synchronization and certain phase correlations can develop due to these interactions. Such, and even more complex, effects of synchronization are actually observed both for mechanical and for electromagnetic vibrations. They can (and do) find useful applications; on the other hand, they could create interferences in the operation of some devices.

The most fully explored now is the mutual modal synchronization in lasers (92, 145, 170).

An important application of a peculiar modal synchronization effect is the generation of ultrashort light pulses of high intensity. Such radiation can be used, for example, for the supercompression of matter and for reaching high temperatures for thermonuclear synthesis (238).

In "spectral language," pulse generation is the simultaneous generation of a large number of modes with an equidistant space–time specter. The latter can be developed if all the modes are spreading with the same speed, or are synchronized in this sense. Velocity synchronization of a large number of modes can be achieved either by the nonlinear interaction between them which is developing in the internal medium of the laser (internal synchronization) (92, 170), or by the external parametric action on a frequency which is close to the difference between the frequencies of two adjacent modes (external synchronization). Usually such action is realized either by the modulation of losses in the laser resonator (314) or by modulation of amplification of the active medium (279). Mathematically, the problem of forced modal synchronization can be reduced to a problem of synchronization of a large number of generators on combined frequencies (171).

An example of the adverse effect of synchronization in continuous systems is the synchronization of waves running in opposite directions in a laser gyroscope.

The laser gyroscope is an instrument for measuring the angular velocity of its supporting structure (92, 274). It can be used for measuring a component of the angular velocity of the Earth for orientation purposes. The main part of the gyroscope is a ring-shaped gas laser consisting of a gas-discharge tube inserted in one branch of the ring-shaped three- or four-mirror optical resonator. Fig. 43 shows a three-mirror resonator with mirrors 3_1, 3_2, 3_3. In such a laser, two waves are transmitted in opposite directions into each other, with wave frequencies being generally different. For example, if the ring-shaped laser is rotating about an axis perpendicular to its plane, then its "equivalent parameter" is different in the direction along and opposite to rotation. It causes "splitting" of the natural frequencies of the resonator by an increment proportional to the angular velocity,

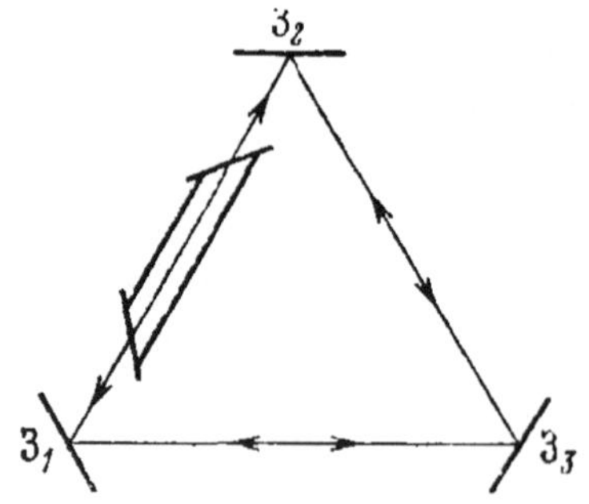

FIG. 43.

$$\Delta\omega_{1,2} = 8\pi n S/(\lambda L). \tag{4.1}$$

Here S is the surface area inside the laser outline, L is the perimeter of the outline, n is the number of revolutions in a unit for time, and λ is the average wavelength of the generated radiation.

Expression (4.1) is the basic one for the instrument. However, due to mutual synchronization of opposite-moving waves, expression (4.1) can be violated at low angular velocities (27, 169, 274). This effect significantly limits the sensitivity of the gyroscope and makes necessary special means for its alleviation.

In a single-mode approximation, the problem of the synchronization of oppositely moving waves in a ring-shape laser is reduced to a problem of the mutual synchronization of two generators described by equations similar to (3.1).

Chapter 7
SYNCHRONIZATION DURING MOTION OF CELESTIAL BODIES

§ 1) Remarkable integer ratios between revolution and spinning periods of celestial bodies and their explanation by synchronization theory.

Observations of remarkable integer ratios ("resonances") between the average angular velocities of spinning and revolutions (orbital motions) of celestial bodies were mentioned in the Introduction. Examples included the motions of the Moon (Cassini laws) and the combined motion of the Earth, Venus, and Mercury. Many such correlations in the solar system are known. Computations show that they are more numerous than can be explained from probability, e.g., see (202, 326). Thus, there is an opinion that such ratios are not accidental, but can be explained by the stability (at least local) of the motions (22–24, 88, 100, 202), i.e., by the tendency to self-synchronization. In some cases it can be proved by a more or less approximate analysis. An example is the establishment of the existence and stability of such solutions of differential equations, which correspond to Cassini laws for the Moon's motions. Some interesting results are published in (23, 100, 106, 111, 281).

When these ratios occur, there are very definite phase correlations between the motions. Thus, reunions of Saturn's satellites Enceladus and Dione always occur near Enceladus's pericenter (see § 3). Such phase correlations are typical for synchronization phenomena.

A hypothesis of full resonance (synchronization) between the revolutions of the large planets of the solar system was supported by the conclusion that their periods (periods of average motions) satisfy with a relatively high accuracy some linear homogeneous correlations with small integer coefficients (202).

As a general explanation of the tendency to synchronization of the orbital motions of celestial bodies, a purely heuristic "principle of minimum interaction" was suggested (324). It states that the motions of n bodies occur in such a way that the time-averaged value of the forcing function along the orbital trajectories is minimal. The correctness of a similar heuristic principle for a different system—a solid body moving along an elliptical trajectory around an attracting center—was confirmed by numerical integration in (22), and then a theoretical background for the search of similar principles was established in (25); see also in § 8, chap. 10.

It will be shown below that the solution of such problems does not require special heuristic principles. If the principle actually exists, it would directly follow from the integral stability criterion (extremal property) of synchronous motions. It was derived for systems with quasi-uniform rotations in a reasonably rigorous way from general differential equations of motion for such systems (40–42). Later this criterion, as noted in § 8, chap. 10, was generalized for many other systems (see also § 6, chap. 12). The integral criterion naturally leads to the expression for a function (potential function D) which, if it exists, is minimized on stable synchronous motions, and also leads to the conclusion that there is a tendency for synchronization in the considered orbital systems in the sense of § 1, chap. 1. Function D can be close to ones heuristically proposed in (22, 324); see below.

In the end of this chapter it will be shown that conclusions from use of integral stability criterion are in good correlation with some astronomic observations and calculations, and that they support the hypothesis (202).

§ 2) Potential function and integral stability criterion (extremal property) of synchronous motions for celestial-mechanical orbital systems.

The free orbital systems described in subsection 1, § 6, chap. 12, correspond to the systems considered in celestial mechanics, consisting of some number k of free solid bodies $B_1, \ldots, B_k$ (in the simplest case—point masses). The bodies are moving along closed trajectories around a certain central body B_0 (in the simplest case—also point mass) that has a much larger mass (see Fig. 44). The rotational coordinates $\varphi_s^0(t + \alpha_s)$ of the bodies when there is no interaction between them (in a more general case, when there is no interaction between the corresponding degrees of freedom) are determined by their autonomous character, up to arbitrary initial phases α_s [see expressions (6.1), chap. 12].

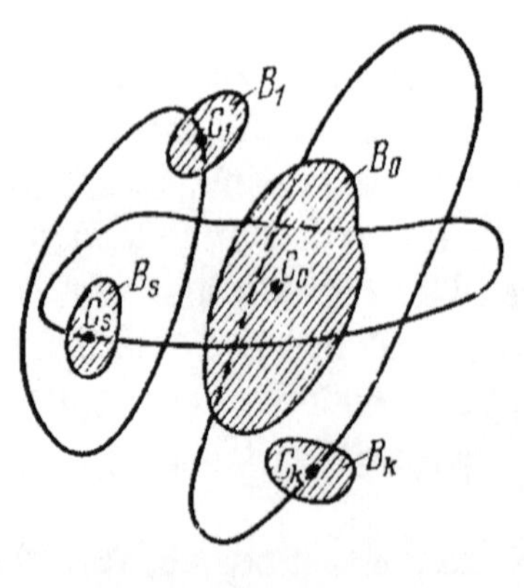

FIG. 44.

The basic equations [(4.6) or (5.8) of chap. 12], from which the phase values in feasible synchronous motions are determined, are as follows:

$$P_s(\alpha_1, \ldots, \alpha_k) = \frac{1}{e_s}\left(\frac{\partial \Lambda_0}{\partial \alpha_s} + A_s\right) = 0 \qquad (s = 1, \ldots, k) \tag{2.1}$$

As in § 6, chap. 12, the case is considered further when $\varphi_s^0 = \varphi_s^{0*}$ is an orbital coordinate whose number is equal to the number of bodies k. Here Λ_0 is averaged along a period $T = 2\pi/\omega$ of the synchronous motion, Lagrange function of the system of constraints (connections), calculated on the fundamental trajectories; A_s are averaged nonconservative generalized forces which are dissipative forces in this case and e_s is the so-called slope of the frequency characteristic.

It is assumed that there exists a potential of averaged nonconservative forces B, i.e., function $B = B(\alpha_1, \ldots, \alpha_k)$, complying with

$$\frac{\partial B}{\partial \alpha_s} = A_s. \tag{2.2}$$

Then, as noted in § 5, chap. 12, potential function D can be expressed as

$$D = -\sigma(\Lambda_0 + B), \tag{2.3}$$

Here $\sigma = \operatorname{sgn} e$ is the parameter determined by character of the anisochronicity of the objects, which is assumed to be identical for all s. According to the integral criterion, stable synchronous motions could correspond to points of crude (rough) minima of function D relative to phase differences $\alpha_s - \alpha_k$.

For the considered orbital system, the kinetic energy of supporting connections $T^{(\mathrm{II})}$ and the potential energy of supporting connections $\Pi^{(\mathrm{I})}$ are zeros, and potential energy of supported connections $\Pi^{(\mathrm{II})} = -U^{(\mathrm{II})}$, where $U^{(\mathrm{II})}$ is the potential of the gravity forces. Thus, according to § 5, chap. 12, the small vibrations of the supporting body expression (2.3) for potential function would be written as

$$D = [\langle\langle T^{(I)}\rangle\rangle - \langle\langle U^{(II)}\rangle\rangle - B]\sigma, \tag{2.4}$$

where $T^{(I)}$ is the kinetic energy of the central (supporting) body B_0.

If all the bodies are free point masses without gravity, then [e.g., see (121)]

$$U^{(II)} = \frac{1}{2} f \sum_{s=1}^{k} \sum_{j=1}^{k}{}' \frac{m_s m_j}{\Delta_{sj}},$$
$$\Delta_{sj} = \sqrt{(x_s - x_j)^2 + (y_s - y_j)^2 + (z_s - z_j)^2}; \tag{2.5}$$

$$h_s = -\frac{1}{2} m_s (f m_0 \omega_s)^{2/3}, \quad e_s = \frac{1}{\omega_s}\frac{dh_s}{d\omega_s} = -\frac{m_s}{3}(f m_0)^{2/3}\omega_s^{-4/3},$$
$$\sigma_s = \operatorname{sgn} e_s = -1. \tag{2.6}$$

Here f is the gravity constant; Δ_{sj} is the distance between masses m_s and m_j; h_s is the total energy of mass m_s in its elliptic Keplerlike motion around the stationary attracting central body, which is also considered a point mass; ω_s is the frequency of the revolution of mass m_s in the said motion (this frequency is called the ***average motion*** in celestial mechanics and is denoted as n_s);the ′ at the summation sign indicates omission of the term corresponding to $j = s$. Since in this case $\sigma_s = -1$, and the objects are softly anisochronous, it follows from (2.4)

$$D = \langle\langle U^{(II)}\rangle\rangle - \langle\langle T^{(I)}\rangle\rangle + B. \tag{2.7}$$

This function should minimize (relative to phases $\alpha_s - \alpha_j$) the stable synchronous motions of bodies m_s.

If $\langle\langle T^{(I)}\rangle\rangle$ and B are neglected, corresponding to an assumption of a much larger mass m_0 of the central body in comparison with masses $m_1, \ldots, m_k$, and also to an assumption of smallness of dissipative forces, then

$$D \cong \langle\langle U^{(II)}\rangle\rangle, \tag{2.8}$$

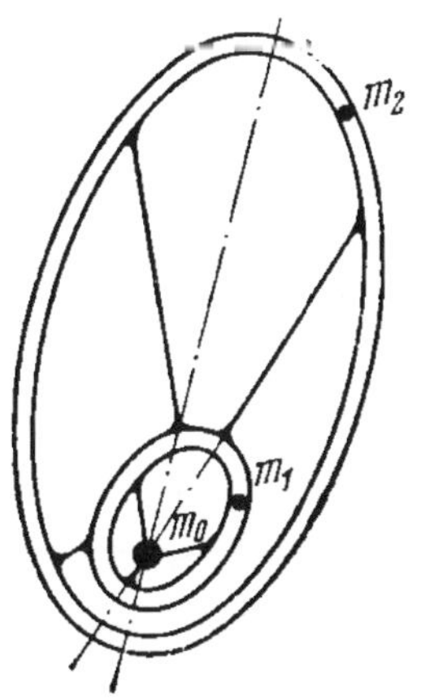

FIG. 45.

It is the conclusion on the "approximate minimality" of the average potential of the mutual attraction of masses in stable synchronous motions. Thus, the heuristic principle proposed in (324) [see also (106)] follows, at the appropriate assumptions, from the earlier integral stability criterion of synchronous motions established in (40–42).

It should be remembered that for the considered free orbital system, the stability conditions expressed by the integral criterion are the only necessary ones. They would also be sufficient if a corresponding nonfree system was considered instead of the free system. The former system can be one in which masses m_s are moving, for example, along rigid elliptical guide rods (see Fig. 45). Of course, such a nonfree system is quite a crude model of a planetary system, at least because the reduction in DOF of each mass m_s down to one causes a change in the character of the anisochronicity of the object. Indeed, in the case of point mass m_s which is moving in the gravitational field of the stationary central mass m_0 along the given curve, the total energy

$$h_s = \frac{m_s V_{s0}^2}{2} - \frac{f m_0 m_s}{r_{s0}},$$

Here V_{s0} and r_{s0} are, respectively, magnitude of velocity and distance between masses m_s and m_0 in a certain initial moment of time. Since now average motion ω_s does not depend on r_{s0}, but is determined only by initial velocity V_{s0}, then

$$e_s = \frac{1}{\omega_s}\frac{dh_s}{d\omega_s} = \frac{1}{\omega_s} m_s V_{s0} \frac{dV_{s0}}{d\omega_s}.$$

It is easy to show that $d\omega_s/dV_{s0} > 0$, or that ω_s is increasing with V_{s0}. Therefore, contrary to (2.6), $e_s > 0$ and $\sigma_s = \text{sgn } e_s = 1$, or the objects are rigidly anisochronous. Also rigidly anisochronous are solid bodies spinning around certain fixed axes ($h_s = {}^1\!/_2 I_s\omega_s^2$, $e_s = I_s > 0$, $\sigma_s = 1$).

As a result, for the nonfree (guided) system, at stable synchronous motions, an approximate minimization would occur not for the average potential of gravity forces, $\langle(U^{(II)})\rangle$, but for the average potential energy

$$\langle(\Pi^{(II)})\rangle = -\langle(U^{(II)})\rangle,$$

This means that the average potential $\langle(U^{(II)})\rangle$ for such motions would be (also approximately) not minimum, but maximum.

Using A. Lyapunov's terminology, the stability of the described nonfree orbital system corresponds to the so-called conditional stability of the respective free orbital system (178).

§ 3) Case of planar orbital system with point masses. Comparison with some astronomical observations and computations.

Detailed expressions for potential function $D \approx \langle(U^{(II)})\rangle$ are derived below for several cases when the orbits of all the bodies $m_1, \ldots, m_k$ considered as point masses are coplanar. This condition is sufficiently satisfied for the large planets of the solar system and many of their satellites. Then, by using the integral criterion (extremal property), feasible values of the relative phases of motion of two bodies along the orbits in stable synchronous (resonance) motions will be found and compared with data from astronomical observations and computations.

For prescribed elliptical orbits and the motions of the bodies along these orbits corresponding to the fundamental solution (the so-called *undisturbed Kepler motion*), parameters and notation usual for celestial mechanics will be used [e.g., see (121)]. When this notation coincides with the designation for different parameters elsewhere in this book, it will be marked with an "*".

The stationary attracting center (central body m_0) is located at some point O, (see Fig. 46); ON is some fixed direction ("*line of nodes*"). Directions $O\xi_s$ toward pericenters of orbits ("*lines of apses*") are defined by angles ω_s^* ("*angular distances between pericenters and nodes*"). The elliptical orbit of body m_s is defined by three parameters (elements): angle ω_s^*, major axis a_s and eccentricity e_s. The position of a body on the orbit is determined by polar coordinates r_s and v_s, with polar angle v_s ("*true anomaly*") counted from line of apses $O\xi_s$.

The following expansions by powers of eccentricity e_s can be written (121):

$$r_s = a_s\left[1 - e_s \cos n_s^*(t - \tau_s) + \ldots\right],$$
$$v_s = n_s^*(t - \tau_s) + 2e_s \sin n_s^*(t - \tau_s) + \ldots, \tag{3.1}$$

Here n_s^* are the average angular velocities of the revolutions of the bodies along the orbits, called "*average motions*" in celestial mechanics. Elsewhere in this book,

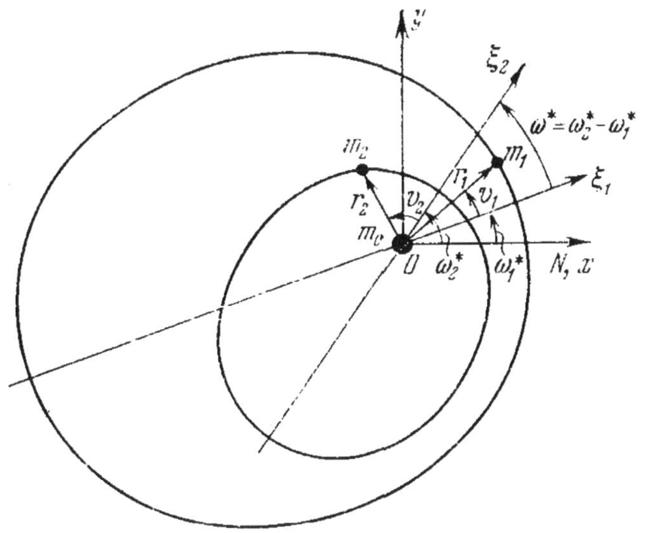

FIG. 46.

these parameters, which are the frequencies of motion of isolated objects, were denoted as ω_s or $\sigma_s\,\omega_s$. Following is a correlation between n_s^* and a_s:

$$2\pi/T_s = n_s^* = (fm_0)^{1/2} a_s^{-3/2}, \tag{3.2}$$

This represents the third Kepler law (T_s are the periods of the revolutions of the bodies). Parameter τ_s in (3.1) is the *moment of the passing of body m_s through the pericenter*; at $t = \tau_s$, $v_s = 0$.

The angular positions of bodies m_s in relation to the stationary direction ON are determined by angles (latitude argument)

$$u_s = \omega_s^* + v_s = \omega_s^* + n_s^*(t - \tau_s) + 2e_s \sin n_s^*(t - \tau_s) + \ldots, \tag{3.3}$$

Thus, the fundamental solution in this problem can be described by

$$r_s = a_s\left[1 - e_s \cos n_s^*(t - \tau_s) + \ldots\right],$$
$$u_s = \omega_s^* + n_s^*(t - \tau_s) + 2e_s \sin n_s^*(t - \tau_s) + \ldots\Big] \tag{3.4}$$
$$(s = 1, \ldots, k).$$

This solution contains $4k$ arbitrary constants: $3k$ constants a_s, e_s, ω_s^* determining elliptical orbits of bodies, and k constants $\omega_s^* - n_s^*\tau_s$, playing the role of initial phases α_s in the expressions of the general theory. The latter constants (or connected to them constants τ_s) have to be determined using the integral stability criterion. Constants n_s^* in (3.4) (average motions) are not independent, but are expressed via a by expressions (3.2).

Since the cartesian coordinates of bodies m_s are determined by $x_s = r_s \cos u_s$ and $y_s = r_s \sin u_s$ (Fig. 46), the expression for function $\langle(U^{(II)})\rangle$ can be written, according to (2.5), as

$$\langle(U^{(II)})\rangle = \frac{1}{2} f \sum_{s=1}^{k} \sum_{j=1}^{k}{}' \left\langle \frac{m_s m_j}{\sqrt{r_s^2 + r_j^2 - 2r_s r_j \cos(u_s - u_j)}} \right\rangle \tag{3.5}$$

or, considering expansions (3.4), as

$$\langle(U^{(II)})\rangle = \frac{1}{2} f \sum_{s=1}^{k} \sum_{j=1}^{k}{}' m_s m_j \left\{ \left\langle \frac{1}{\{a_s^2 + a_j^2 - 2a_s a_j \cos[(n_s^* - n_j^*)t + \alpha_{sj}]\}^{1/2}} \right\rangle + \right.$$

$$\left. + 2e_s \left\langle \frac{A_{sj}(t)}{\{a_s^2 + a_j^2 - 2a_s a_j \cos[(n_s^* - n_j)t + \alpha_{sj}]\}^{3/2}} \right\rangle \right\}, \tag{3.6}$$

It is denoted here

$$\alpha_{sj} = -\alpha_{js} = \omega_s^* - \omega_j^* - n_s^* \tau_s + n_j^* \tau_j, \tag{3.7}$$

$$A_{sj}(t) = a_s^2 \cos n_s^* (t - \tau_s) - a_s a_j \{\cos[n_j^* (t - \tau_j) + \omega_j^* - \omega_s^*] +$$

$$+ \sin n_s^* (t - \tau_s) \sin[(n_s^* - n_j^*) t + \alpha_{sj}]\},$$

The omitted terms are at least second or higher order relative to eccentricity e_s.

When you perform averaging, you arrive at

$$D \approx \langle(U^{(II)})\rangle =$$

$$= \frac{1}{2\sqrt{2}} \frac{f}{a} \sum_{\substack{s=1 \\ n_s^* = n_j^*}}^{k} \sum_{j=1}^{k}{}' \frac{m_s m_j}{\sqrt{1 - \cos\alpha_{sj}}} + \frac{f}{4} \sum_{\substack{s=1 \\ n_s^*/n_j^* = l/(l\pm1)}}^{k} \sum_{j=1}^{k}{}' \frac{e_s}{a_s} \{2b_3^{(p_{sj})}(z_{js}) -$$

$$- z_{js}[3b_3^{(p_{js})}(z_{js}) - b_3^{(q_{sj})}(z_{js})]\} \cos(n_s^* \tau_s + p_{sj}\alpha_{sj}) + \ldots, \tag{3.8}$$

Here

$$z_{js} = a_j/a_s = (n_s^*/n_j^*)^{2/3}, \quad p_{sj} = n_s/(n_s - n_j), \quad q_{sj} =$$

$$= (2n_s^* - n_j^*)/(n_s^* - n_j^*), \tag{3.9}$$

and $b_3^{(r)}(z) = b_3^{(-r)}(z)$ denote coefficients in expansion

$$(1 + z^2 - 2z\cos\lambda)^{-3/2} = \frac{1}{2} \sum_{r=-\infty}^{\infty} b_3^{(r)}(z) \cos r\lambda, \tag{3.10}$$

These are called Laplace coefficients and are expressed via complete elliptic integrals (259). Nonessentials for further derivations of the term, which does not depend on $\omega_s^* - n_s^* \tau$, are omitted in (3.8).

Correlation $n_s^* = n_j^*$, written under the first double summation sign, shows that only terms corresponding to all the bodies with the same average motions n_s^* are

are considered [thus, according to (3.2), also with the same semiaxes a_s]. Equally $n_s^*/n_j^* = l/(l \pm 1)$, written under the second double summation sign, shows that only those terms are considered which correspond to all pairs of bodies whose average motions n_s^* and n_j^* are related as consequent integers (l is an integer, $l \neq 0$, $l \pm 1 \neq 0$).

It follows from (3.8) that, in this analysis, in the case of purely circular planar motion only simple (nonmultiple) synchronization is feasible. For elliptical orbits with small eccentricities e_s, consideration of the terms of the first order relative to e_s in the expansion of the averaged potential $\langle(U^{(II)})\rangle$ allows us to account for the interactions which could cause synchronization of the bodies whose average motions relate as consecutive whole numbers ("resonances of the type $n_s^*/n_j^* = l/(l \pm 1)$"). If higher order terms relative to e_s in this expansion are considered, then interactions which may cause synchronization with $n_s^*/n_j^* = n/(n+q)$, where $q = 2, 3, \ldots$, would be accounted for. Of course, at small eccentricities these interactions would be weakened as compared to $n_s^* = n_j^*$ and $n_s^*/n_j^* = l/(l \pm 1)$ since the respective terms in the $\langle(U^{(II)})\rangle$ expansion would be of a higher order of smallness relative to e_s.

Let's consider in more detail the simplest orbital system consisting of only two revolving bodies ($k = 2$).

Initially, let the orbits be circular and $n_1^* = n_2^* = n^*$. Then $a_1 = a_2 = a$, and according to (3.8)

$$D \approx \langle(U^{(II)})\rangle = \frac{1}{2\sqrt{2}}\frac{f}{a}\frac{m_1 m_2}{\sqrt{1-\cos\alpha_{12}}} \tag{3.11}$$

The minimum of the potential function would be at

$$\alpha_{12} = \omega_1^* - \omega_2^* + n^*(\tau_2 - \tau_1) = \pi.$$

It can be assumed in this case that $\omega_1^* = 0$ and $\omega_2^* = 0$, since any point of the circular orbit can be considered as its pericenter. Thus, the stability condition following from the integral criterion is satisfied for the "antiphase" motion, at which bodies are located at the ends of the same diameter (see Fig. 47).

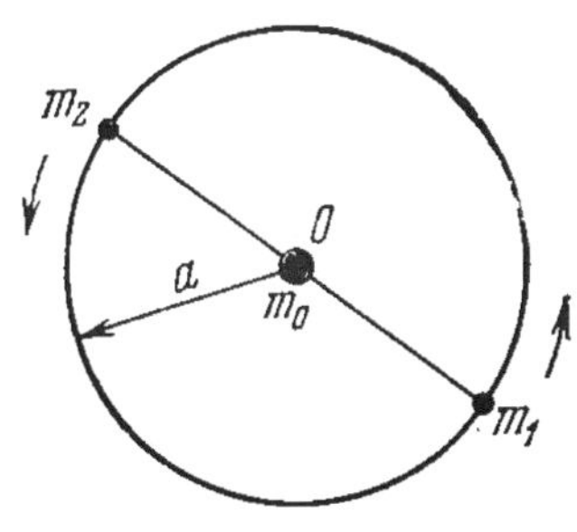

FIG. 47.

For the case of two bodies moving in the same direction along elliptical orbits with small eccentricities, it will be assumed that the average motion of the second body is twice as fast as the average motion of the first one. Then, according to (3.2), (3.7) and (3.9)

$$n_1^* = n^*, \quad n_2^* = 2n^*,$$

$$z_{12} = a_1/a_2 = (n_2^*/n_1^*)^{2/3} = 2^{2/3} = 1{,}59, \quad z_{21} = 1/z_{12} = 0{,}629;$$

$$\alpha_{12} = \omega_1^* - \omega_2^* - n^*\tau_1 + 2n^*\tau_2, \quad \alpha_{21} = \omega_2^* - \omega_1^* - 2n^*\tau_2 + n^*\tau_1;$$

$$p_{12} = \frac{n_1^*}{n_1^* - n_2^*} = -1, \quad p_{21} = \frac{n_2^*}{n_2^* - n_1^*} = 2, \quad q_{12} = \frac{2n_1^* - n_2^*}{n_1^* - n_2^*} = 0,$$

$$q_{21} = \frac{2n_2^* - n_1^*}{n_2^* - n_1^*} = 3, \tag{3.12}$$

Expression (3.8) would be transformed into

$$\langle (U^{(II)}) \rangle =$$

$$= \frac{f}{4}\frac{e_1}{a_2} z_{21} \{2b_3^{(1)}(z_{21}) - z_{21}[3b_3^{(2)}(z_{21}) - b_3^{(0)}(z_{21})]\} \cos(2n^*\tau - \omega^*) +$$

$$+ \frac{f}{4}\frac{e_2}{a_2} \{2b_3^{(2)}(z_{12}) - z_{12}[3b_3^{(1)}(z_{12}) - b_3^{(3)}(z_{12})]\} \cos 2(n^*\tau - \omega^*).$$

(3.13)

Here it is denoted

$$\tau^* = \overset{*}{\tau}_2 - \overset{*}{\tau}_1, \quad \omega^* = \overset{*}{\omega}_2 - \overset{*}{\omega}_1. \tag{3.14}$$

At $z_{12} = 1.59$ and $z_{21} = 0.629$, there will be the following values for Laplace coefficients: $b_3^{(1)}(z_{12}) = 1.23$; $b_3^{(2)}(z_{12}) = 0.913$; $b_3^{(3)}(z_{12}) = 0.649$; $b_3^{(0)}(z_{21}) = 6.03$; $b_3^{(1)}(z_{21}) = 4.82$; $b_3^{(2)}(z_{21}) = 3.66$. Substituting them into (3.13), finally

$$D \approx \langle (U^{(II)}) \rangle =$$

$$= \frac{f}{4a_2}[4{,}11e_1 \cos(2n^*\tau - \omega^*) - 3{,}00e_2 \cos 2(n^*\tau - \omega^*)] =$$

$$= \frac{f}{4a_2}\mathcal{M} \cos(2n^*\tau - \delta), \tag{3.15}$$

where

$$\mathcal{M} = \sqrt{16{,}9e_1^2 + 9{,}00e_2^2 - 24{,}7e_1e_2 \cos\omega^*},$$

$$\mathcal{M} \cos\delta = 4{,}11e_1 \cos\omega^* - 3{,}00e_2 \cos 2\omega^*, \tag{3.16}$$

$$\mathcal{M} \sin\delta = 4{,}11e_1 \sin\omega^* - 3{,}00e_2 \sin 2\omega^*.$$

From this, in accordance with the integral stability criterion, it follows that a stable synchronous motion is correlated with the following value of the time difference between bodies passing through the pericenters

$$\tau = \tau_2 - \tau_1 = (\delta + \pi)/2n^*, \tag{3.17}$$

The value $\tau = \delta/2n^*$ corresponds to an unstable synchronous motion.

Let's consider two more cases. In the first one, it is assumed that $e_1 = 0$, i.e., "external" (the farthest from the attracting center) body m_1, has a circular orbit in an undisturbed motion (Fig. 48a). Without losing generality, it can be assumed that $\omega^* = \omega_2^* - \omega_1^* = 0$, since any point of the circle can be considered as a pericenter. According to (2.15),

$$D \approx \langle (U^{(II)}) \rangle = -f\,\frac{3{,}00}{4}\frac{e_2}{a_2} \cos 2n^*\tau$$

and the stable motion would occur at

$$\tau = \tau_2 - \tau_1 = 0, \tag{3.18}$$

This means that in the stable synchronous motion, the bodies are configured as in Fig. 48a. This case is described in astronomy and celestial mechanics as: ***coincidences and oppositions of the bodies (while observed from the central body) occur at the internal body passing its pericenter.***

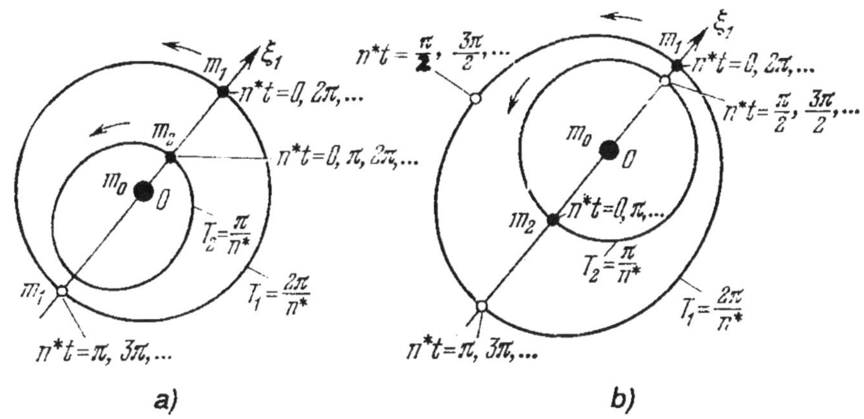

FIG. 48.

The assumption in the second case is $e_2 = 0$, i.e., the circle is an undisturbed orbit of the internal body m_2. Assuming, as before, $\omega^* = 0$, from (3.15)

$$D \approx \langle (U^{(\mathrm{II})}) \rangle = f\, \frac{4{,}11}{4} \frac{e_1}{a_2} \cos 2n^*\tau$$

For a stable synchronous motion

$$\tau = \tau_2 - \tau_1 = \pi/(2n^*). \tag{3.19}$$

The configuration for this motion is shown in Fig. 48b. In this motion, the coincidence of the bodies ***occurs when the external body is passing through its apocenter, and the opposition occurs when it is passing through its pericenter.***

These results can be compared with astronomical observation/computation data.

It is stated in (100) that Saturn's satellites Dione and Enceladus have periods of revolution $T_1 = 2.73691$ days and $T_2 = 1.37028$ days, respectively; thus the ratio of their average motions is $n_1^* / n_2^* = T_1 / T_2 = 1.99734$, or very close to two. Orbits of both satellites are coplanar, and their excentricities are, respectively, $e_1 = 0.0021$ and $e_2 = 0.0045$. If it is assumed, that $e_1 \approx 0$ (a possible error is discussed below), it becomes the first of the considered cases, when coincidence of bodies in a stable motion occurs while the internal body is passing through its pericenter. This result correlates with observations, that coincidences of Dione and Enceladus always occur near Enceladus's pericenter [e.g., see (100)]. In (100), are given qualitative thoughts in support of the stability of the motion of satellites like Enceladus and Dione.

The average motions of two solar system planets, Neptune and Uranus, also have a ratio of about 2:1 ($n_2^* / n_1^* = 0.0117/0.0060$). Table 4 lists the values of the elements characterizing the positions and motions of these planets, from (12).

TABLE 4. Elements of the orbits of Neptune and Uranus (for January 7, 1979).

Planet	Element					
	Average motion n_s^*	Eccentricity e_s	Average epoch anomaly $\mathscr{M}_{0s}$	Longitude of ascending mode Ω_s	Longitude of pericenter $\bar{\pi}_s$	Inclination i_s
Neptune ($s = 1$)	0,0060	0,00984	196,89°	131,53°	61,63°	1,772°
Uranus ($s = 2$)	0,0117	0,0491	52,70°	74,01°	170,12°	0,771°

Since the relative inclination of orbits $i = i_2 - i_1$ is small, they may be assumed to be coplanar. Also, due to the absence other large planets in the solar system whose average motions relate to those of Neptune and Uranus as $l/(l \pm 1)$,* the influence of other planets on synchronization in the Sun–Neptune–Uranus system can be neglected. Let's check to see if the results obtained before for the case of two bodies with $n_1^* / n_2^* = 2{:}1$ are applicable.

Following are the correlations between $\mathscr{M}_{0s}$, Ω_s, π_s and the elements ω_s^*, τ_s used before [e.g., see (121)]

$$\omega_s^* = \pi_s - \Omega_s, \quad \mathscr{M}_{0s} = n_s^* (t_0 - \tau_s), \tag{3.20}$$

where t_0 is an initial moment of time t. Thus, considering the data from Table 4,

$$\omega^* = \omega_2^* - \omega_1^* = \pi_2 - \Omega_2 - (\pi_1 - \Omega_1) = \\ = 170{,}12° - 74{,}01° - (61{,}63° - 131{,}53°) \approx 166°, \tag{3.21}$$

$$n^* (\tau_2 - \tau_1) = n^*\tau = \\ = -\left(\frac{\mathscr{M}_{02}n^*}{n_2^*} - \frac{\mathscr{M}_{01}n^*}{n_1^*}\right) = -\left(\frac{\mathscr{M}_{02}}{2} - \mathscr{M}_{01}\right) = -\left(\frac{52{,}70°}{2} - 196{,}89°\right) \approx \\ \approx 171°. \tag{3.22}$$

If it is assumed, as in the previous example, that $e_1 = 0$, or that Neptune's orbit is circular, then from a minimum condition for (3.15) considering (3.21) $n^*\tau = \omega^* = 166°$, which is only 5° less than in (3.22). However, due to the availability of more accurate data, $n^*\tau$ can also be found from the more accurate formula (3.17). From Table 4

$$\mathscr{M} \cos \delta = 4{,}11 \cdot 0{,}00984 \ \cos 166° - 3{,}00 \cdot 0{,}0491 \ \cos 332° = -0{,}169;$$
$$\mathscr{M} \sin \delta = 4{,}11 \cdot 0{,}00984 \ \sin 166° - 3{,}00 \cdot 0{,}0491 \ \sin 332° = 0{,}0789;$$

From these data, $\delta = 155°$, and according to (3.17), $n\tau^* = 0.5(155° + 180°) = 167.5°$, which is even closer to $n^*\tau = 171°$ from (3.22).

Thus, the correlation of the results obtained from the integral stability criterion and astronomical data is quite reasonable.

*See § 5; one exception is Pluto, but its mass is more than 7000 times smaller than the mass of Uranus or Neptune.

§ 4) Tendency toward synchronization in the considered systems. Role of dissipative forces.

Due to the periodicity of function Λ_0 by α_s (or, in a special case, of function $<(U^{(II)})>$ by $\omega_s^* - n_s^*\tau_s$), there is a tendency toward synchronization in the considered systems, at least when averaged dissipative generalized forces $|A_s|$ are sufficiently small as compared with generalized potential forces $|\partial\Lambda_0/\partial\alpha_s|$ or, in a special case, with forces $|\partial<(U^{(II)})>/\partial(\omega_s^* - n_s^*\tau_s)|$; see also § 9, chap. 10, and § 7, chap. 12.

However, a tendency toward synchronization can be suppressed by dissipative factors. Thus, dissipative forces have a dual influence on the synchronization in the considered system. On one hand, they assure asymptotic stability along the corresponding coordinates; on the other hand, they could cause violation of the conditions of existence of synchronous motions, i.e., to the suppression of the tendency toward synchronization. At some conditions, dissipative forces can cause a loss of motion stability.

Considering dissipation in the celestial-mechanics orbital systems, at a formal approach [analysis of (2.1)] causes the disappearance of steady, specifically periodic or synchronous, motions with a nonchanging synchronous frequency. However, due to the smallness of the dissipative forces, deviations from periodicity and changes in synchronous frequency are noticeable only at very large time intervals compared with the specific time lengths on which the motions are analyzed here.

The role of dissipative forces in the development of synchronisms (resonances) during the process of the evolution of the orbital and spinning motions of celestial bodies was thoroughly analyzed in (23, 100, 202, 281). Basically, this role is as follows. Dissipation causes changes in the ratios between the average motions and the spinning of bodies during the process of evolution. Entrainment into a stable synchronous motion and a lengthy "trapping" occurs (or, better, may occur) in cases when these ratios become close to the appropriate critical values.

§ 5) A. Molchanov's hypothesis.

Understanding the general character of the synchronization effects induced several interesting ideas about the synchronizationlike character of some phenomena in both animate and inanimate nature. One important idea is the already mentioned hypothesis by A. Molchanov about the full resonance (synchronous) character of the orbital motion of large planets in the solar system (202). He discovered, that the average motions of these nine planets $n_1^*, \ldots, n_9^*$ are approximately correlated by nine linear homogeneous expressions with integer coefficients $k_1^{(j)}, \ldots, k_9^{(j)}$.

$$k_1^{(j)}n_1^* + \ldots + k_9^{(j)}n_9^* = 0 \qquad (j = 1, \ldots, 9) \tag{5.1}$$

The important fact is (see subsection 3, § 3, of the Introduction) that numbers $k_s^{(j)}$ are small. It is obvious that if the values of $k_s^{(j)}$ were not limited, then using the format of (5.1), correlation between any values of n_s^* could not be presented with any desirable accuracy.

Table 5 lists values $k_s^{(j)}$ from (202) leading to a very accurate compliance with (5.1) with a small "correction" of the observed values of average motions n_s^*. These values in relation to the average motion of Jupiter, their corrected values $\tilde{n}_s^*$ and the "relative corrections" $\Delta n_s^*/n_s^* = (n_s^* - \tilde{n}_s^*)/n_s^*$, are also listed. The corrections do not

TABLE 5. Resonance correlations in the solar system.

j	Planet	n_s^*	$\tilde{n}_s^*$	$\Delta n_s^* / n_s^*$	$k_1^{(j)}$	$k_2^{(j)}$	$k_3^{(j)}$	$k_4^{(j)}$	$k_5^{(j)}$	$k_6^{(j)}$	$k_7^{(j)}$	$k_8^{(j)}$	$k_9^{(j)}$
1	Mercury	49,22	49,20	0,0004	1	—1	—2	—1	0	0	0	0	0
2	Venus	19,29	19,26	0,0015	0	1	0	—3	0	—1	0	0	0
3	Earth	11,862	11,828	0,0031	0	0	1	—2	1	—1	1	0	0
4	Mars	6,306	6,287	0,0031	0	0	0	1	—6	0	—2	0	0
5	Jupiter	1,000	1,000	0,0000	0	0	0	0	2	—5	0	0	0
6	Saturn	0,4027	0,400	0,0068	0	0	0	0	1	0	—7	0	0
7	Uranus	0,14119	0,14286	—0,0118	0	0	0	0	0	0	1	—2	0
8	Neptune	0,07197	0,07143	0,0075	0	0	0	0	0	0	1	0	—3
9	Pluto	0,04750	0,04762	—0,0025	0	0	0	0	0	1	0	—5	1

exceed 1.2%. A similar table has been compiled in (202) for the satellites of Jupiter, Saturn, and Uranus.

These tables create the strong impression that the principal bodies of the solar system are in a synchronous regime or are very close to one. However, a convincing dynamic analysis has still not been performed, thus there still are some doubts. An interesting discussion is given in (23). Some thoughts based on the results presented in that book are as follows.

It seems that tendency toward synchronization in orbital systems established here is a convincing argument in support of the Molchanov hypothesis. Our results also show a possible effective direction for more detailed studies of resonance motions in the solar system—to use the integral stability criterion in conjunction with astronomic observation data. Some simple examples of this approach were given in § 3. It would be interesting to perform such a study, using computers, for the whole system of large planets in the solar system, as well as for satellite systems. It would allow us to find out, on the one hand, if phase relations at real motions of the respective bodies are in correlation with the phases calculated from minimum condition for function $D = \langle (U^{(\mathrm{II})}) \rangle$ and, on the other hand, how far it is from the minimum, and the evolutionary trend, for the value of $\langle (U^{(\mathrm{II})}) \rangle$ as determined from astronomical observations. Such a study should be performed considering thoughts from § 3 about the relative intensity of the interactions between the bodies with the various ratios of their average motions n_s^*.

A study of the evolution of orbital systems by a numerical experiment would also be very interesting— i.e., the computer integration of the appropriate system of differential equations. Such results are given in (315). It was shown that with a random initial distribution, planets are evolving relatively rapidly (on an order of 10^4 years) to quasi-steady orbits for which one could explicitly observe a commensurability of the average motions, i.e., synchronization [see also (106)].

Chapter 8

SYNCHRONIZATION PHENOMENA IN CHEMISTRY, BIOLOGY, AND MEDICINE

§ 1) Rhythms and cyclicity in nature.

Oscillation phenomena are widely spread both in inanimate and animate nature. As was stated in (313), "it is said that nature does not stand emptiness but does like very much rhythm and cyclicity."

Oscillations in the course of chemical reactions were discovered in the last century. Now many such reactions are known, specifically those occurring at the metal-solution-boundary and causing such terms as "iron nerve" and "mercury heart"(125). However, the most important are oscillatory processes and various rhythms in biochemical and biological systems, where they are distinctly traced starting from live micromolecules and cells right up to macroscopic systems of live organisms and communities of organisms. Specifically, various oscillations and rhythms in the human organism are of extreme importance.

These processes attract ever growing attention from chemists, physicists, biologists, physicians, control specialists and mathematicians. The state of the art in this area can be assessed from (28, 29, 84, 125, 149, 150, 229, 236, 247).

An especially important role in chemical, biochemical, and biological oscillatory systems belongs to self-excited vibrations phenomena. Their study started from a classical publication (336), in which heart action is modeled by coupled relation generators triggered in succession. Up to now, van-der-Pol generators, both quasi-harmonic and relaxation, as well as "predator-prey" models (317, 337) have been the basis for mathematical modeling of oscillatory processes in such systems.

With all its successes, the study of oscillations in biochemical and biological systems is just emerging. For example, the mechanism of so-called *biological clock* still has not been satisfactorily explained. This "clock" develops a sense of time in many biological objects, from primitive to the most sophisticated. Its accuracy is amazing and sometimes is used by animals for orientation purposes; in some cases it reaches single minutes or even seconds a day (28). There are suggestions that the solution of the problem of malignant tumors might be connected with studies of the mechanism of the biological clock, whose regimes are different in normal and cancerous cells [see § 3 and (125, 247)].

There is still no convincing answer for the basic quasi-philosophical question of the internal causes of the broadest occurrences of the oscillatory phenomena in nature. There are just some thoughts and hypotheses.

Thus, one thought (200) is that oscillations are characteristic for any biological "species." "Balanced" species merge with the medium during the evolution process, while the "unstable" ones are destroyed. As is said in (200), "a view of an individual as a system which, on the one hand, retains its structure and, on the other hand, is capable of internal motion, is in the best way represented by oscillatory systems."

Another view, advocated for example in (125), is that the vibratory (including self-excited vibrations) character of many processes in biochemical systems is due to the presence in them of appropriate (usually, nonlinear) interactions between the variables.

The third view, not excluding the previous one, and the one which the author prefers, is based on the perception that vibratory systems and processes are characterized by some degree of optimality. Therefore, it seems there exists some general principle due to which nature specifically "prefers" vibratory processes. It would be important and interesting to formulate such a principle.

§ 2) Synchronization phenomena in biochemical and biological systems.

Since synchronization phenomena can be encountered practically everywhere there are vibrations, they are also important for the subject systems.

In many cases, the synchronization of biological objects can be considered as entrainment by a prescribed rhythm which is generated by one powerful object, with feedback influence being insignificant.

The day–night cycle caused by the spinning of the Earth plays an important role in the lives of humans and animals, as does the period of the year, which is due to the revolution of the Earth on its orbit around the Sun. These rhythms are characteristic of many physiological processes and also of the mechanical phenomena and chemical reactions occurring on Earth.

Rhythms related to the cyclical activity of the Sun with periods of eleven and twenty-two years are also distinctly traced. It was shown that Sun spots are related to such processes as temperature variations and intensity variations of streams in the world's oceans; reproduction and migration of fish, locusts, and some other species; epidemics and pandemics of flu, plague, flare-ups of rabbit-fever, foot-and-mouth disease and some other diseases; see (296, 297).

Besides "heliology," which studies the influence of solar processes on live organisms, "selenobiology," which studies the influences from the Moon, is now emerging (84).

An important application of entrainment phenomena in medicine is pacemakers, which send electric pulses into specific areas of the heart, thus dictating to the muscles a necessary contraction frequency (usually 60–70 per min) (13).

Recently, the role of mutual synchronization in chemical, biochemical, and biological systems has been more widely accepted (125, 149, 150, 236, 247). As is stressed in (125, 247), any occurrence of self-excited vibrations in such distributed systems as is observed in macrovolumes, is possible only because of synchronization of the vibrations occurring in various parts of these macrovolumes.

Several mathematical models of synchronization in distributed systems, each of whose "elementary volume" is a self-excited generator of some type, are studied in (247). These models are used to describe "wavelike" chemical reactions, contractions of muscle fibers, transmission of nerve excitations, etc. Models such as van-der-Pol generators (quasi-harmonic and relaxation), the modified "predator-prey" model by Lottki-Volterra, and a recent Higgins model are used for elementary self-oscillatory systems.

The most simplistic synchronization model in the unidirectional distributed systems ("pipes") studied in (247) is described by a system of partial differential equations

$$\frac{\partial x}{\partial t} = y + D_x \frac{\partial^2 x}{\partial r^2},$$
$$\frac{\partial y}{\partial t} = a\,[1 - \nu(r)\,x^2]\,y - \omega^2(r)\,x + D_y \frac{\partial^2 y}{\partial r^2}. \tag{2.1}$$

Here the unknown functions $x(r, t)$ and $y(r, t)$ are deviations of the concentration of substances X and Y from some average values; r is a space coordinate; t is time;

D_x, D_y, and a are constants with D_x and D_y representing diffusion coefficients; $\nu(r)$ and $\omega(r)$ are some functions of r. At $D_x = D_y = 0$, system (2.1) is reduced to one second-order equation

$$\frac{\partial^2 x}{\partial t^2} + \omega^2(r)\, x = a\,[1 - \nu(r)\, x^2]\, \frac{\partial x}{\partial t}. \tag{2.2}$$

This corresponds to the van-der-Pol equation in each point r = const [see (2.1) in chap. 6]. In other words, it is postulated that the cells or elementary volumes of the chemical systems ("reactors") are similar to van-der-Pol generators. At $D_x \neq 0$ and $D_y \neq 0$, system (2.1) describes reactors connected by diffusion processes. Coefficients $\omega(r)$ and $\nu(r)$ determining, respectively, partial frequencies and amplitudes of steady vibrations of uncoupled reactors, are considered, in general, different at different points of the medium.

For solution of the problem, (2.1) is replaced by a system of ordinary differential equations that describe discrete generators. The system represents "pipe" as a simple number k of separate volumes (reactors) with full internal mixing, with each reactor being connected by two adjacent ones by means of diffusion,

$$\begin{aligned}
&\dot{x}_1 = y_1 + \beta\,(x_2 - x_1),\\
&\dot{y}_1 = a\left(1 - \nu_1 x_1^2\right) y_1 - \omega_1^2 x_1 + \alpha\,(y_2 - y_1);\\
&\dot{x}_s = y_s + \beta\,(x_{s+1} - x_{s-1} - 2x_s),\\
&\dot{y}_s = a\left(1 - \nu_s x_s^2\right) y_s - \omega_s^2 x_s + \alpha\,(y_{s+1} + y_{s-1} - 2y_s)\\
&\qquad\qquad (s = 1, \ldots, k-1);\\
&\dot{x}_k = y_k + \beta\,(x_{k-1} - x_k),\\
&\dot{y}_k = a\left(1 - \nu_k x_k^2\right) y_k - \omega_k^2 x_k + \alpha\,(y_{k-1} - y_k).
\end{aligned} \tag{2.3}$$

To derive these equations, the conditions of the impermeability of the "pipe" ends were considered. Coupling coefficients α and β between the reactors (generators) are expressed via D_x and D_y by formulas

$$\alpha = D_y(k/l_0)^2, \quad \beta = D_x(k/l_0)^2,$$

where l_0 is the length of "pipe"; $\omega_s = \omega(r_s)$, $\nu_s = \nu(r_s)$, r_s is the coordinate of the center of the s-th elementary volume.

For the case of small a, α and β system (2.3) was analyzed by asymptotic methods and the method of (288). For nonsmall a, α and β, i.e., for the case of strong nonlinearity and strong coupling, it was analyzed by analog modeling.

Some results of the study are:

1) The synchronous regime is characterized by definite correlations between the phases of the generators (reactors), being significantly dependent on their detuning. At an unlimited increase in the number k of generators, the distribution of phase shifts develops a continuous "standing phase wave," in which the maximum value of the phase difference of the generators does not exceed π. A discrete analog of such wave takes place at finite k. This refers specifically to phase shifts or the starting phases of vibrations. The vibrations themselves of the unidirectional continuum or chain of generators look like a running wave; see below.

2) The synchronous frequency ω at small a, α and β is, in the first approximation,

$$\omega^2 = \sum_{s=1}^{k} A_{0s}^2\omega_s^2 \Big/ \sum_{s=1}^{k} A_{0s}^2, \tag{2.4}$$

Here $A_{0s} = 2/\sqrt{\nu_s}$ are the amplitudes of the self-excited vibrations of uncoupled generators. Thus, ω does not exceed the largest and is not less than the smallest of partial frequencies ω_s.

3) Under other equal conditions, the "leading" generators which entrain all the others into a synchronous regime are the ones with the highest partial frequencies.

4) The generator which is detuned in relation to all other generators, all having identical partial frequencies, is more easily synchronized the closer it is to the center of the chain. The largest jump in the phase difference of the vibrations in the synchronous regime is observed between the detuned generators and those adjacent to them.

5) Simple (nonmultiple) synchronization is the easiest to achieve.

6) The synchronization range gets wider with increasing parameter a/ω, which characterizes the degree of nonlinearity of the generators, i.e., with transition from the quasi-harmonic to the relaxation regime. This is especially true for synchronization bands with multiple frequencies. Synchronization bands are also widened, increasing coupling parameters $2\alpha/a$ and $2\beta/a$.

Plots in Figs. 49 and 50 illustrate conclusions 5) and 6). Figure 49 shows the width of simple synchronization Δ for a chain of six generators vs. parameters $2\alpha/a$ and $2\beta/a$. It is assumed that detuning is introduced in the first generator in the chain, and that all other generators have the same partial frequencies. Figure 50 shows, for the same chain, domains of simple and multiple synchronization in the plane of parameters $2\alpha/a$, ω_1/ω at $\beta = 0$.

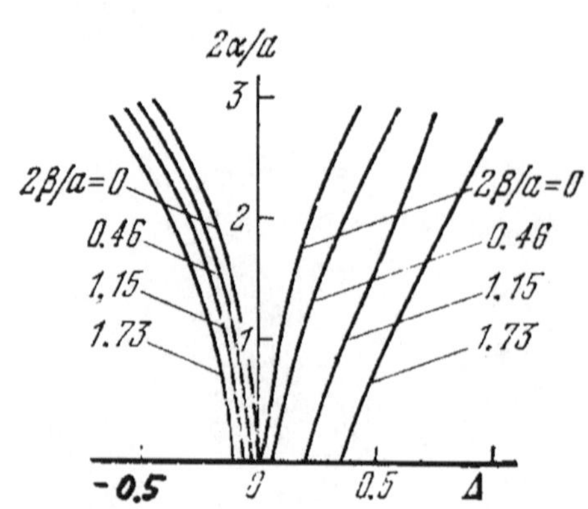

FIG. 49.

7) For slight nonlinearity and weak coupling, the width of the synchronization band is cut in half by increasing the number of generators in the chain from $k = 2$ to $k = \infty$.

This system represents one of the models of so-called *active media*, in which amplification of the self-excitation of oscillations is possible. Such models attracted attention recently since they describe some distributed chemical systems, working bodies of lasers, nerve fibers, cell communities, distributed populations, etc. One of the main features of the active media is the feasibility of the development in them of wave processes whose parameters do not depend on the initial, and sometimes on boundary, conditions and geometric dimensions of the system.

It is easy to see that these results allow us to explain the development and transmission of a running wave which was observed, for example, in long tubes in which an oxidation reaction with self-excited oscillations takes place. As was stated in 1), in a steady synchronous regime, the distribution of the vibratory phases of the generators configures a certain "standing wave." As a result, certain concentra-

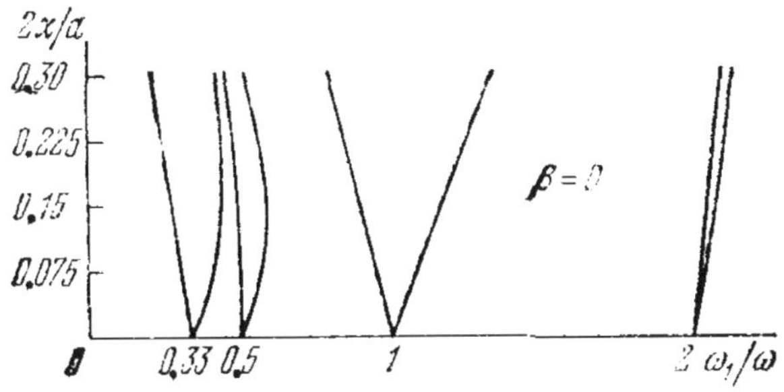

FIG. 50.

tion levels, such as $x_s = x^*$, would be achieved in adjacent reactors at sequential moments of time, i.e., a wave of concentration change would be transmitted along the tube. This wave in some cases may be observed as a color wave (e.g., if at $x_s < x^*$ a solution is transparent and at $x_s > x^*$ it is colored).

Some features of the synchronization of a chain of diffusion-connected relaxation generators of sawlike oscillations, as well as short pulse oscillations, are studied in (247). Such self-excited oscillations are especially important for biological systems, since they adequately model, for example, processes of nerve excitation.

As a mathematical model for a generator of the short pulse relaxation oscillations, the following equations can be used which were introduced by Higgins to describe glycolitic self-excited oscillations (at $m = 0$),

$$\dot{x} = \alpha\,[1 - (1 - m)\,xy - mx], \quad \dot{y} = xy - \frac{q+1}{q+y}\,y \qquad (2.5)$$

Here α, m, q are constants.

Besides self-synchronization of a chain of diffusion-connected self-exciting objects, it is very important for biologists to study self-synchronization of such objects which are diffusion-connected via a common medium. with which the objects are connected through the exchange of some substances. Individual cells, cell communities, and even individual organisms could be such objects. Several examples of such self-synchronization are known. One represents synchronous oscillations of nucleus sizes of malignant tumor cells immersed in one solution (197). The discovery of this effect started some interesting hypotheses on the mechanism of cancerous growth and possible ways of its abatement (see § 3). There are some indications of cycle synchronization in organisms of animals coupled by a common habitat [e.g., see (247)].

Of the simplest models of self-exciting "reactors" coupled via a common medium, (247) is based on the assumption of the van-der-Pol equation. It is assumed that there are k such reactors in which a reaction with substances X and Y can take place. The reaction has the property of self-exciting of oscillations. The reactors are immersed in a neutral medium (solution) with which they can exchange molecules of the substances X, Y. These molecules can spread in the medium by diffusion. It is assumed in this primitive case that there occurs a full internal mixing in the medium; thus concentrations of substances are instantaneously equalized in the whole volume. Let x_s, y_s be, as before, deviations from some fixed values of concentrations of substances X, Y in reactors, and let x, y be the respective deviations of concentrations in the medium. Then the differential equations of the system become

$$\dot{x}_s = y_s + \beta_+ (x - x_s),$$
$$\dot{y}_s = a\left(1 - \nu x_s^2\right) y_s - \omega_s^2 x_s + \alpha_+ (y - y_s) \tag{2.6}$$
$$(s = 1, \ldots, k);$$

$$\dot{x} = \beta_- \sum_{j=1}^{k} (x_j - x), \quad \dot{y} = \alpha_- \sum_{j=1}^{k} (y_j - y). \tag{2.7}$$

Here β_+ and α_+ are the diffusion coefficients of substances X and Y out of the reactors which are identical for all reactors; β_- and α_- are the coefficients of diffusion from the medium into the reactors. If permeability of the reactor walls for molecules X and Y is the same in both directions, then $\beta_- = \beta_+ v/v_0$, $\alpha_- = \alpha_+ v/v_0$, where v, v_0 are the volume of the reactor and the medium, respectively. Constants ω_s, a and ν have the same meaning as in (2.3). Expressions (2.6) are equations of the objects and (2.7) are equations for the connection system.

Results of the study of reactor self-synchronization for $\beta_+ \ll a \ll \omega_s$, $\beta_+ \ll a\nu \ll \omega_s$, are given in (247). For example, expressions are found for the bandwidth of synchronization for reactors differing only by partial frequencies ω_s. It is shown that in the case of the uniform distribution of frequencies, the band of synchronization is somewhat smaller than in the case when all but two of the reactors have the same frequencies $\omega_s = \omega_0$, and two reactors have frequencies $\omega_0 + \Delta\omega$ and $\omega_0 - \Delta\omega$. The band is somewhat larger if half of the reactors have one and the other half have another partial frequency. It is proved that the relaxation character of self-excited vibrations leads to about the same increase in the bandwidth of synchronization compared with a quasi-harmonic regime, as in the case of the chain of reactors.

Qualitatively new results are obtained for the synchronization of the reactors in a medium with a finite diffusion. This problem is analyzed in (247) with the assumption that the medium is boundless and unidirectional. A "cartridge" of k reactors is connected with each point $r = r_0$ of this medium. The reactors thus are in a condition equivalent to placement into a medium with full mixing. However, for reactors in different cartridges, and thus clustered in different points r of the medium, concentrations of substances X and Y in the medium are, generally speaking, different and are determined by the diffusion processes in it.

The principal feature of the synchronization in such a system is the fact that such reactors are sufficiently coupled through the medium (in relation to synchronization), and are located at distances $s = \sqrt{2D_y/\omega}$ from each other. Here D_y is the diffusion coefficient for molecules of substance Y (correspondingly D_x is assumed to be zero), and ω is the synchronous frequency. As a result, the whole medium is separated into "*clusters*"—areas with dimensions of the order of s inside which the synchronization of reactors with certain frequencies $\omega^{(i)}$ takes place. In general, the synchronous frequency is different for different clusters. As in the case of the synchronization of a chain of reactors, the maximum phase shift inside each cluster does not exceed π.

The development of the synchronous clusters is also characteristic of chains, and also for planar and 3-dimensional "networks" of diffusion-coupled reactors (generators) in the presence of noise. Such networks are modeling, respectively, one-, two- and three-dimensional space, at each point of which a reaction with self-exciting oscillations occurs. Without noise and detuning, synchronous regimes in these networks are established across the whole space, i.e., dimensions of the synchronous cluster are not limited. With the presence of noise, the dimensions of the clusters in one- and two-dimensional cases become limited and shrink with the increasing intensity of noise. However, in the 3-dimensional case with noise intensity not exceeding a certain level, the cluster can be of unlimited size. It is noted that these results were experimentally confirmed.

Mutual synchronization is also observed in the behavior of human collectives and animal groups. It is known that in a large auditorium (sometimes several thousands of people who try to applaud as loudly as possible), random applause changes relatively rapidly to synchronized and synphaseous applause. Another example is a group of people walking "in step" with a band or by the count of a commander. This is similar to what was called forced synchronization in the Introduction.

Similar phenomena sometimes occur in flocks of birds and in schools of fish: the motions of the wings and of the tails are synchronous.

N. Winer suggested that the rhythmical flicking of light emanating from fireflies located in the range of "mutual visibility" is not an optical illusion, but is an example of mutual synchronization and phasing.

§ 3) Some unsolved problems and related hypotheses (telepathy and cancer problem).

A leading role of synchronization phenomena in the considered cases looks indisputable. In addition, hypotheses of the synchronization nature of the mechanisms of some now unsolved biological phenomena were recently suggested.

It is known that electrodes placed on the scalp will pick up a signal voltage varying in time. Its record represents a characteristic electroencephalogram for the condition of the brain.

An example of spectral analysis of electric signals of the brain is sketched in Fig. 51a. The abscissa represents the frequency in hertz, and S the ordinate-spectral power density. The most prominent activity of biopotentials is near 10 Hz; this feature is called an *alpha-rhythm*. A finer structure of the alpha peak (81) is shown in Fig. 51b. It can be seen that there are two valleys on each side of the peak, and the spectrum looks like there is a "frequency attraction."

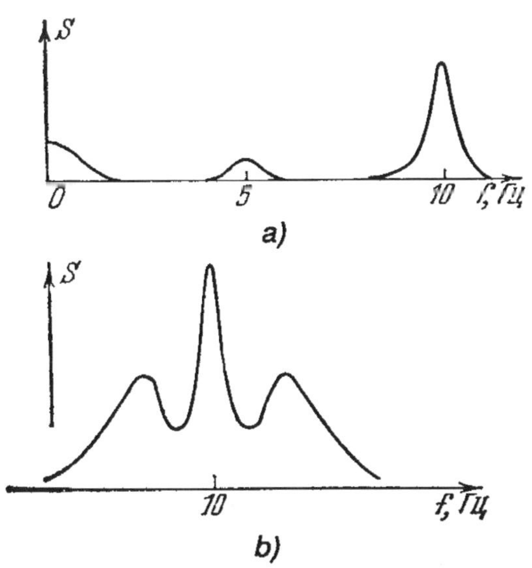

FIG. 51.

The explanation by N. Winer (88, 89) suggests that the brain contains autogenerators* of sorts whose partial frequencies are close to 10 Hz.

Due to interaction of these autogenerators a mutual "frequency attraction" occurs, i.e., self-synchronization with the frequency about 10 Hz.

N. Winer mentioned some, at least indirect, confirmations of this hypothesis. In one experiment, a man was placed in a room to the ceiling of which a sheet of steel was suspended.

*Before N. Winer, in 1923 a hypothesis about the oscillatory nature of the structural components of nerve fabric was suggested in (134).

An AC voltage with amplitude of about 400 V was connected to this sheet. As a result, electrostatic induction developed in each object in this room. It was observed that at a frequency about 10 Hz the man was feeling "definitely unpleasant perceptions."

Supporting the suggestion of the presence in the brain of autogeneratorlike structures with a frequency of about 10 Hz is the fact that after subjecting the brain to an unexpected electric pulse, there is a distinctly observable relaxation with a frequency of about 10 oscillations per second. Besides, an entrainment of the alpha-rhythm frequency by an external excitation source (flashes of light) with a frequency of about 10 Hz was observed.

It can be noted that the test described by Winer (with a steel sheet) shows that a signal can be sent into the human brain via not necessarily conventional senses.

Maybe this test prompted Winer to an attempt to find a scientific explanation for the effects of the exchange of information between people located at such distances or in such conditions when an ordinary sensual perception is impossible. Such effects, whose veracity is doubted by many (not without reason), are called telepathic (83, 90). These alleged effects were studied by biologists, psychologists, and physicists.

In the end of the last and the beginning of this century an hypothesis of the electromagnetic nature of communications between nervous systems of humans was popular. It was rejected, though, by computations showing that the intensity of the electromagnetic fields created by the brain biopotentials is far below the threshold level which, as physiologists claim, can be perceived by the nervous system, e.g., see (11).

A recent publication (180) describes the measurements of the spectrum of biopotentials of some biological objects in 1–150 Hz range and the direct detection of electromagnetic radiation at frequencies of 3 and 150 kHz . It concluded that electromagnetic radiation is too weak to be an information carrier in a biological connection, at least on frequencies below 150 Hz. But the authors do not exclude the possibility of the involvement of higher frequency electromagnetic radiation in biological communications. Such a high-frequency radiation could be, in their opinion, a resonance for radiation of some protein complex.

This can play the role of carrier frequency, with information transmitted from the biopotentials by means of the modulation of the carrier frequency.

The synchronization theory shows that with both electromagnetic and another "vibratory" mechanism of interaction, the sensitivity threshold should be considered differently. One of the principal features of the synchronization of the various autogenerators (see § 3 of the Introduction) is the absence of a synchronization threshold. However weak the coupling between such objects, synchronization would occur if the difference between the partial frequencies of the objects is small enough.* Synchronization is also accompanied by definitive phase relations between the oscillations, while without synchronization the phases are random.

Thus, with the partial frequencies sufficiently close, even an infinitesimally weak coupling might cause a qualitative change in the motion's character, thus a corresponding weak signal can be perceived. If so, then there is no universal physiological threshold of sensitivity. It would be different for different parts of people. Of course, the intensity of the transmitted signal plays a certain role. However, even a very weak signal can be received if it has a specific spectrum which plays a key role.

*An important issue of the time necessary for establishing a synchronous regime is not addressed here [e.g., see (171, 247)]. Of course, the time is increasing and approaching infinity with the weakening of the connections.

It is known that a necessary condition for the success of telepathic tests is the conformity of participants. They have to present a so-called telepathic pair. If the reality of telepathic phenomena is assumed, then it can be guessed that people comprising such a pair have close partial frequencies of generated electromagnetic waves.

The conclusion of the absence of the "synchronization threshold" relates to deterministic systems. In the presence of noise, such a threshold does exist, e.g., see (181, 247). However, it is determined differently and can be, under certain conditions, much lower than the "physiological threshold" mentioned before.

A bold and far-reaching hypothesis about synchronization in biology was suggested by N. Winer while he was studying possible mechanisms of self-organizing and self-reproduction of biological objects (89). He noted that one of the basic problems in biology is the study of a method by which the main substances comprising genes or viruses or possible specific carcinogenic substances are self-reproduced from materials not having that specificity, e.g., from a mixture of amino-acids and nucleic acids. It is said that one macromolecule of these substances acts as a template by which smaller molecules from the mixture form into an analogous macromolecule. Still, hardly anything is known about the mechanism of such reproduction.

Winer suggested that the active carrier of specific information is in the frequency content of its molecular radiation, mainly in infrared electromagnetic waves or even lower. Winer writes: "It might happen that some of the constitutive substances of the virus, in some circumstances, emanate infrared oscillations which have a capacity to assist in formation of other molecules of the virus from an amorphous magma of amino acids and nucleic acids. It is possible that such a phenomenon can be viewed as a kind of mutual attractive interaction between frequencies."

Of course, it is not the only feasible mechanism and thus it requires a thorough validation. One way to do it, suggested by Winer, is to study the absorption and radiation spectra of the virus substance, and then analyze the action of light at these frequencies on the rate of reproduction of the viruses in the appropriate mixture of nutrients.

Another hypothesis of the reproduction of cancerous cells, also based on synchronization effects, is suggested in (247) (pp. 278–279). It is based on the results in (197) where synchronous pulsations of the size of the nucleus in cells of Erlich carcinoma were discovered. Later, oscillations of the dimensions of the cells themselves were discovered, as were periodic variations of the permeability of plasma membranes (85). It was suggested in (247) that the intensive growth of a cancerous tumor is due to synchronous oscillatory processes. If this concept is confirmed, new methods for fighting cancer can be developed, by abating or changing the characteristics of synchronous oscillations. One case described in the literature is of the successful curing of a child hopelessly ill with leukemia, which can be explained as the result of influencing the synchronization of the life cycles of cells.

Interesting thoughts on a possible correlation between the cancer problem and the phenomenon of the biological clock and the synchronization phenomena are given in (125).

§ 4) Conclusion.

Examples of the spread of the synchronization phenomena in chemical, biochemical, and biological systems can be added to. The systematic description of facts and results now known deserve a separate book, not smaller than this one. We only wanted to show the importance of the synchronization problems in these areas, stress the main direction of development, and list some publications known to us.

Chapter 9
SYNCHRONIZATION AND SOME PHYSICAL PROBLEMS

§ 1) Stochastics and synchronization—two polar types of behavior in dynamic systems.

The experimental data and theoretical results presented in the preceding chapters have shown that the tendency toward synchronization is characteristic of a broad class of various natural and engineering objects.

As was noted in the Introduction, this tendency is one form of the self-organization of matter and one type of tendency of dynamic systems toward orderly behavior. An opposite type of behavior of a dynamic system is its stochastic, unregulated behavior, when a regular dynamic description becomes useless. Experiments show that one tendency prevails is some cases, while the second tendency prevails in others. Such a dialectic situation—a struggle between "order" and "disorder"—is typical not only for temporal but also for spatial and temporal–spatial organization of material forms.*

Still, very recently it was accepted that a tendency toward disorderly, stochastic behavior is characteristic only of systems consisting of a very large number of similar elements, e.g., a gas volume containing a large number of molecules. Recently it was shown [see (77, 203, 219, 240)] that under certain conditions and also in quite wide areas of space of parameters, systems with a 3-dimensional phase space (such as a pendulum with a vibrating support; a ball bouncing above a vibrating plane; some types of autogenerators; etc.) behave essentially stochastically.

On the other hand, results in previous chapters testify about the feasibility of an opposite situation: systems containing sufficiently large numbers of similar objects—Thomson and relaxation generators, vibro-exciters, electrical generators and motors—can under reasonably broad conditions perform stable synchronous motions, i.e., behave in a highly orderly way.

A view of synchronization as one manifestation of the tendency toward self-organization and order, which is opposite to the tendency toward randomness and disorder, was expressed by the author in (57) and is now shared by many researchers. Both in (57) and this book, results are given that allow us to judge when the tendency toward synchronization, i.e., to order, prevails. On the other hand, there are now some interesting results which allow us to predict conditions of stochastic behavior of a dynamic system [see surveys (203, 240)]. However, the problem of the clarification of the general conditions determining when one or another of these two opposite tendency prevails still remains unsolved.

§ 2) A possible role of synchronization phenomena in microcosm.

The universality of the synchronization phenomena in general, and specific features of the synchronization of orbital systems and systems of various vibration generators, bring the thought that synchronization might play a significant role in the microcosm, in which vibratory and rotational motions of various objects are

---

*An interesting popular book (141) is dedicated to "order and disorder in the atom world."

quite common (63). It seems that this thought deserves attention from theoretical physicists. In this light, some quotations from mechanics and specialists in nonlinear vibrations might be of interest.

N. Chetaev said that " . . . Stability, which is principally a general phenomenon, should demonstrate itself somehow in basic laws of nature" (295). Following this idea, he came to the hypothesis that only selected (elite) trajectories of dynamic systems can be stable. Later, in [202], A. Molchanov complemented the ideas of Chetaev from other positions, by expressing the assumption that the discrete character of the orbits of "evolutionarily mature" dynamic systems are associated with "resonances," or in other words with synchronization.

It seems that the results of the study of synchronization of the orbital systems presented in §§ 6 and 7, chap. 12, which specifically established the tendency of such systems toward synchronization, testify in support of the mentioned hypotheses. That is, it can be assumed from these results that the quantum character of the microcosm processes, as well as the peculiar quantum character of the orbits in celestial systems (see chap. 7), is a manifestation of the general principle—i.e., the tendency toward synchronization, or stability, of the corresponding "elite" motions. It is possible that using this principle would help to explain and describe quantum–mechanical phenomena in a nonformal way.

PART THREE

MATHEMATICAL APPARATUS AND PRINCIPAL RESULTS OF THE GENERAL THEORY OF SYNCHRONIZATION OF WEAKLY COUPLED DYNAMIC OBJECTS

Chapter 10

THEOREMS OF EXISTENCE AND STABILITY OF PERIODIC AND SYNCHRONOUS SOLUTIONS FOR DIFFERENTIAL EQUATIONS CONTAINING A SMALL PARAMETER (POINCARE AND LYAPUNOV METHODS)

§ 1) Preliminary comments. Poincare and Lyapunov methods in the theory of periodic solutions.

In the first segment of this chapter the principal ideas of the Poincare and Lyapunov methods are described. Later several theorems related to the existence and stability of the periodic and synchronous solutions of certain systems of differential equations containing a small parameter are given (without proofs). These theorems further develop and complement in certain directions the classical apparatus of the Poincare and Lyapunov theories of periodic solutions for systems containing a small parameter.

The principal feature of the subject equations, which is characteristic for synchronization problems, is the existence of a family of periodic (or synchronous) solutions for this family, depending on a certain number of arbitrary parameters $\alpha_1, \ldots, \alpha_k$, for the fundamental system. The principal feature of the theorems is the possibility of expressing the principal result by means of the so-called generating (or fundamental) functions $P_s(\alpha_1, \ldots, \alpha_k)$ and the integral stability criterion. These features of the mathematical apparatus of the synchronization theory were discussed in detail in chap. 2.

A majority of the described results were developed recently specifically under the influence of the need for developing the synchronization theory and its immediate applications. These results represent a basis for the mathematical apparatus which is used in this book for the solution of specific problems.

First, the basic ideas of the perturbation method by Poincare and Lyapunov will be briefly described.

The mathematical basis for the perturbation method in the application to the theory of periodic solutions for differential equations had been developed in classical publications by H. Poincare at the end of nineteenth century (325). A major role in using this method is played by the theory of motion stability and the theory of periodic solutions of differential equations developed at about the same time by A. Lyapunov (178).

The Poincare and Lyapunov methods were recently developed further. They were successfully used for many important application-related problems and are now some of the most effective and universal methods in the theory of nonlinear vibrations.

Let's consider a system of differential equations:

$$\dot{x}_s = X_s(x_1, \ldots, x_n, t) + \mu F_s(x_1, \ldots, x_n, t, \mu) \quad (s = 1, \ldots, n), \tag{1.1}$$

Here, X_s and F_s are analytical functions of $x_1, \ldots, x_n$ in a closed domain G, to which all the solutions considered below belong. These functions are also continuous T-periodic functions of t (time) in every point of G. It is assumed that F_s are also analytical functions of a small parameter $\mu \geq 0$ at $\mu \leq \mu_0$, where $\mu_0 > 0$. In the majority of application problems listed above, rigid requirements related to the smoothness of the right-hand sides of the equations, which make the analysis much easier, are sufficient;* these requirements could be significantly relaxed (see § 7). The system (1.1) is not autonomous; it is assumed that the right-hand sides of the equations contain time explicitly. The study of an autonomous system has some specifics (e.g., see §§ 3 and 5).

Finding the periodic solutions of the system

$$\dot{x}_s^0 = X_s(x_1^0, \ldots, x_n^0, t) \qquad (s = 1, \ldots, n), \tag{1.2}$$

which is obtained from (1.1) at $\mu = 0$ and called *a fundamental system*, might be easier than finding the periodic solutions of (1.1). Let's assume that T-periodic solutions of the fundamental system (*fundamental solutions*) exist and are known. There is a question of whether these solutions can be "trusted" in the sense that they uniquely correspond to the nearest periodic solutions of the original system, i.e., to the solutions of the original system becoming, at $\mu = 0$, solutions of the fundamental system. Sometimes it is also necessary to find solutions of the original system with a higher accuracy, or by accounting for the terms containing μ.

The Poincare theory gives the principal solution of these problems. Specifically, he has shown that correspondence between the solutions of (1.1) and (1.2) does not always occur. Depending on the character of the right-hand side parts of equations (1.1), it might occur that a periodic solution of the fundamental system (1.2) does not correspond to a periodic solution of the original system (1.1). On the other hand, there can be cases when one solution of the fundamental system corresponds to several or even an infinite number of periodic solutions of the original system.

All these special cases present the highest interest for the theory of nonlinear vibrations in general and for the synchronization theory specifically.

In the practical application of the Poincare method, a periodic solution of system (1.1) is sought as a series

$$x_s(t) = x_s^0(t) + \mu x_s^{(1)}(t) + \mu^2 x_s^{(2)}(t) + \ldots \tag{1.3}$$

with T-periodic coefficients of whole positive powers of parameter μ. However, in some cases the expansion can be done by fractional powers of μ. After substituting expressions (1.3) into (1.1) and expanding their right-hand sides by powers of μ, expressions of coefficients at the same powers of μ in both sides are equated. Then the following systems of equations would result for determining approximations for $x_s^{(1)}, x_s^{(2)}, \ldots$

*The case of the so-called singular perturbations when the equations contain a small parameter at derivatives $\dot{x}_s$ is not considered here.

$$\dot{x}_s^{(1)} = \sum_{j=1}^{n} p_{sj}(t)\, x_j^{(1)} + F_s(x_1^0, \ \ldots, x_n^0, t, 0) \qquad (s = 1, \ldots, n), \tag{1.4}$$

$$\dot{x}_s^{(q)} = \sum_{j=1}^{n} p_{sj}(t)\, x_j^{(q)} + \Phi_s^{(q)}(x_1^0, \ldots, x_n^0; \ldots; x_1^{(q-1)}, \ldots, x_n^{(q-1)}, t),$$

$$(q = 2, 3, \ldots; \ s = 1, \ldots, n). \tag{1.5}$$

There is a notation in (1.4) and (1.5) of

$$p_{sj}(t) = \left(\frac{\partial X_s}{\partial x_j}\right), \tag{1.6}$$

Here round brackets mean that the derivatives enclosed in them are calculated for the fundamental solutions $x_s = x_s^0$, i.e., for the T-periodic solution of the system (1.2). If the fundamental solution is known, the recurrent systems of equations (1.4) and (1.5) could be successively used to find functions $x_s^{(1)}$, $x_s^{(2)}$, ... Each of these systems is a system of linear nonhomogeneous part

$$\dot{z}_s = p_{s1}(t) z_1 + \ldots + p_{sn}(t) z_n, \tag{1.7}$$

whose coefficients are T-periodic time functions due to the T-periodicity of solution $x_s^0(t)$.

Thus, the answer to the basic questions formulated above depends on the possibility of finding unique T-periodic solutions $x_s^{(q)}(t)$ for equations (1.4) and (1.5), and also on the convergence of series (1.3), at least at sufficiently small μ. The solutions of these questions depend significantly on the character of system (1.7), which is most critical for further analysis. According to Poincare, these equations are called *equations in variations for the fundamental system, derived for the fundamental solution.*

It is established in the theory of linear differential equations that if system (1.7) does not show T-periodic solutions, then T-periodic solutions of equations (1.4) and (1.5) certainly exist and are unique. Studies show that in this simplistic case, the periodic solution $x_s^0(t)$ of the fundamental system (1.2) corresponds, at least with sufficiently small μ, to a single analytical solution (1.3) of the original system (1.1), which becomes $x_s^0(t)$ at $\mu = 0$.

The case when the system in variations (1.7) has T-periodic solutions is much more complex. Such a case is encountered, for example, when the fundamental system (1.2) allows a family of T-periodic solutions

$$x_s^0 = x_s^0(t, \alpha_1, \ldots, \alpha_k) \qquad (s = 1, \ldots, k), \tag{1.8}$$

This family depends on some number $k \le n$ of arbitrary parameters $\alpha_1, \ldots, \alpha_k$. Just such a case is of the highest interest for synchronization theory. It is assumed that parameters α_j enter expressions (1.8) independently, which means that the rank of matrix $\|\partial x_s^0/\partial \alpha_j\|$ is k. Then, according to the Poincare theorem (184, 325), the system in variations certainly has k periodic (with period T) solutions

$$z_{sj}(t) = \frac{\partial x_s^0}{\partial \alpha_j} \qquad (s = 1, \ldots, n; \ j = 1, \ldots, k). \tag{1.9}$$

Here and below, the first subscript of z shows the function number, and the second subscript shows the number of the solution.

The following system, which is called *conjugate* to (1.7), in this case also has at least k periodic solutions $z^*_{sj}(t)$

$$\dot{z}^*_s + p_{1s}(t)\, z^*_1 + \ldots + p_{ns}(t)\, z^*_n = 0, \qquad (1.10)$$

The analysis of the stability of the periodic solutions found by the Poincare perturbation method has some specifics. According to the Lyapunov theory (178), stability analysis depends on the character of the solutions of the so-called *system of equations in variations for equations (1.1) and solution (1.3)*

$$\dot{y}_s = \sum_{j=1}^{n} \left[\left(\frac{\partial X_s}{\partial x_j} \right)_* + \mu \left(\frac{\partial F_s}{\partial x_j} \right)_* \right] y_j \qquad (s = 1, \ldots, n), \qquad (1.11)$$

This is a linear homogeneous system with T-periodic coefficients. Round brackets with subscript "*", in which the derivatives are enclosed, show that the latter were computed for this solution. Equations (1.11) describe the behavior in time of the small deviations of the motion from the considered periodic motion. If all the solutions of system (1.11) approach zero at $t \longrightarrow \infty$, then the considered motion is asymptotically stable. The existence of at least one unlimited, increasing solution corresponds to instability, and cases when there are nontrivial periodic solutions would require, as a rule, special analysis.

One important specific of the analysis of the character of the solutions of the system in variations (1.11) is the presence in it of the small parameter. If system (1.7), obtained from (1.11) at $\mu = 0$, has only decaying at solutions $t \longrightarrow \infty$, then the studied motion is also asymptotically stable at sufficiently small μ. If system (1.7) has at least one unlimitedly increasing at $t \longrightarrow \infty$ solution, then the studied motion is unstable at sufficiently small μ. When system (1.7) has periodic solutions to analyze stability even at sufficiently small μ, the terms containing μ in equations (1.11) should be considered.

The last "critical" case is of the greatest interest for synchronization theory, since it corresponds to the case when the fundamental system has a family of T-periodic solutions. Theorems about the existence and stability of the periodic solutions refer to just such cases.

An important case of the fundamental solution for synchronization theory, depending on an arbitrary number of parameters $\alpha_1, \ldots, \alpha_k$, was initially considered by I. Malkin in (184); Poincare studied only the case of one parameter. Malkin obtained, for various cases, equations like $P_s(\alpha_1, \ldots, \alpha_k) = 0$ $(s = 1, \ldots, k)$ to determine these parameters ("basic equations"). Algebraic equations like $|\partial P_s/\partial\alpha_j - \delta_{sj}\chi| = 0$ $(s, j = 1, \ldots, k)$, from whose roots the statement of stability depends (see § 2, chap. 2 and also below), were obtained by the author for quasi-linear systems (33, 34), and then by others for more complex systems. An important generalization is made by R. Nagaev for quite a general case of a quasi-periodic family of fundamental solutions (216). The results of these and other publications on the development of the Poincare method, which are of interest for synchronization theory, are described below.

The systematic treatment of the Poincare and Lyapunov methods in nonlinear vibration theory can be found in (57, 74, 184, 188, 237, 260, 280, 312), and a survey on the development of these methods in (54, 57, 291).

§ 2) Quasi-linear nonautonomous systems.

Let's consider a system of equations

$$\dot{y}_s = \lambda_s y_s + f_s(\omega t) + \mu F_s(y_1, \ldots, y_l, \mu, \omega t) \quad (s = 1, \ldots, l), \tag{2.1}$$

Here λ_s are constants; F_s are analytical functions of $y_1, \ldots, y_l$ in some close domain G, to which all the solutions of the equations considered below are assumed to belong, and also of parameter $\mu \geq 0$ at its sufficiently small values. Functions F_s and f_s are periodic (period $T = 2\pi/\omega$) and continuous by time t, with functions F_s satisfying these conditions at any point in the domain*).

Parameters λ_s are *characteristic exponents* of the fundamental system corresponding to system (2.1),

$$\dot{y}_s^0 = \lambda_s y_s^0 + f_s(\omega t) \qquad (s = 1, \ldots, l), \tag{2.2}$$

These could be real or complex. In the latter case, it is assumed that λ_s appear as complex conjugate pairs, and that complex conjugate functions f_s and F_s correspond to the complex conjugate λ_s. Just such a situation is typical for systems like (2.1) which are obtained by an appropriate transformation of the original differential equations containing only real parameters.

For application to the synchronization theory, it is assumed that there might be any number of zeros and purely imaginary, possibly equal, values among the values of λ_s. Such characteristic exponents will be called *critical*. The real parts of all other exponents, being, in general, complex numbers, will be assumed to be positive.

Let's divide all critical exponents into groups, each of which includes multiple exponents and exponents differing from each other by $in\omega$, where n is any positive or negative integer or zero. From these groups, two *special groups* are selected which include exponents of the type $\lambda = in\omega/2$, where n is the positive or negative integer or zero. The *first special group (leading special group)* would include exponents $\lambda = in\omega$, and the *second special group*, exponents $\lambda = i(2n+1)\omega/2$. The imaginary parts ν_s of all the other critical exponents are different from the numbers of the format $n\omega/2$. The groups including such exponents will be called *nonspecial*.

A characteristic exponent from a nonspecial group r_s is denoted as $\lambda_s^{(r)} = i\nu_s^{(r)}$. According to the aforesaid,

$$\lambda_s = \begin{cases} in_s\omega, & s = 1, \ldots, k, \\ \frac{1}{2} i(2n_s+1)\omega, & s = k+1, \ldots, k+m, \\ i\nu_s^{(r)}, & s = k+m+m_1+\ldots+m_{r-1}+1, \ldots \\ & \ldots, k+m+m_1+\ldots+m_r \\ & (r = 1, \ldots, p), \\ -u_s + i\nu_s, & s = k+m+m_1+\ldots+m_p+1, \ldots, l \\ & (u_s > 0). \end{cases} \tag{2.3}$$

Here k is the number of exponents in the leading special group; m is the number of

*The listed restrictions could be significantly relaxed (see § 7). This also refers to the theorems in §§ 3–6.

exponents in the second special group; p is the overall number of nonspecial groups; and, finally, according to the algorithm of division into groups,

$$|\nu_s^{(r)} - \nu_j^{(r)}| = n\omega; \quad |\nu_s^{(r)} - \nu_j^{(q)}| \neq n\omega, \quad r \neq q. \tag{2.4}$$

Let's explain the purpose of the exclusion into special groups of characteristic exponents $\lambda = in\omega/2$. Since it is assumed that all complex (including pure imaginary) characteristic exponents appear only as complex conjugated pairs, the characteristic exponents of special groups $\lambda = in\omega/2$ get put, in accordance with (2.4), into the same special group as their complex conjugates $\bar{\lambda} = -in\omega/2$. However, complex conjugate exponents $\lambda = i\nu\omega$ are put into different nonspecial groups. In other words, *special groups contain pairs of complex conjugate exponents, while nonspecial groups do not.*

For each nonspecial group another nonspecial group can be found which contains characteristic exponents which are complex conjugates for the first group. Thus, the total number of nonspecial groups p is always even, and it can be written that $p = 2p'$, where p' is an integer. It will be further assumed, for specificity, that a nonspecial group number $p' + r$ contains exponents which are complex conjugate relative to exponents from group number r. In other words, in addition to (2.3), it is assumed that

$$\nu_s^{(r)} = -\nu_{s+m_1+\dots+m_{p'}}^{(p'+r)}, \quad m_r = m_{p'+r}$$
$$(r = 1, \dots, p'; \quad s = k + m + m_1 + \dots + m_{r-1} + 1, \dots \tag{2.5}$$
$$\dots, k + m + m_1 + \dots + m_r).$$

Together with the characteristic exponents λ, the *roots of the characteristic equation* ρ are also used, and are related to λ as $\rho = e^{\lambda T} = e^{2\pi\lambda/\omega}$. The structure of the characteristic exponents defined by (2.3) and the structure of the differential equations (2.1) correspond to a k-times root of the characteristic equation $\rho = 1$, m-times root $\rho = -1$; m_r-times ($r = 1, \dots, p$) roots different from ± 1 but having unitary magnitudes; and, finally, $l - (k + m + m_1 + m_p)$ roots whose magnitudes are less than one. Linear elemental dividers* correspond to all multiple roots.

Finally, the natural assumption is made that the fundamental system (2.2) tolerates the periodic solutions of the same period $T = 2\pi/\omega$ as the period of the right-hand sides of equations (2.1). This requirement leads to the following conditions on functions $f_s(\omega t)$ corresponding to variables from the leading group:

$$\int_0^T f_s(\omega t)\, e^{-in_s\omega t} dt = 0 \qquad (s = 1, \dots, k). \tag{2.6}$$

If conditions (2.6) are satisfied, then fundamental system (2.2) does not allow a single solution, but a family of T-periodic solutions, depending on k arbitrary constants α_s:

$$y^0(\omega t) = \begin{cases} \alpha_s e^{in_s\omega t} + u_s(\omega t), & s = 1, \dots, k, \\ u_s(\omega t), & s = k + 1, \dots, l. \end{cases} \tag{2.7}$$

Here $u_s(\omega t)$ denotes $2\pi/\omega$-periodic time functions defined by

*Some definitions from the theory of systems of linear differential equations with periodic (a special case—with constant) coefficients are not given here. If needed, they can be found in (294, 295, 57).

$$u_s(\omega t) = \begin{cases} \int_0^t f_s(\omega t_1)\, e^{in_s\omega(t-t_1)} dt_1, \quad s = 1, \ldots, k, \\ \frac{1}{1 - e^{2\pi\lambda_s/\omega}} \left[\int_t^{2\pi/\omega} f_s(\omega t_1)\, e^{(2\pi - \omega t - \omega t_1)} dt_1 + \right. \\ \left. + \int_0^t f_s(t_1)\, e^{\lambda_s(t-t_1)} dt_1 \right], \quad s = k+1, \ldots, l. \end{cases} \tag{2.8}$$

The following theorem is true with the formulated assumptions (33, 57).

Theorem. *Periodic solutions of the nonautonomous set of equations (2.1), which become periodic solution (2.7) of the fundamental system (2.2) at* $\mu = 0$ *can correspond only to such values of constants* $\alpha_1, \ldots, \alpha_k$ *which satisfy equations*

$$P_s(\alpha_1, \ldots, \alpha_k) \equiv \langle F_s(y_1^0, \ldots, y_l^0, 0, \omega t)\, e^{-in_s\omega t} \rangle = 0 \\ (s = 1, \ldots, k). \tag{2.9}$$

If for a certain set of constants $\alpha_1 = \alpha_1^*, \ldots, \alpha_k = \alpha_k^*$, *complying with (2.9), the real parts of all the roots of algebraic equations**

$$\left| \frac{\partial P_s}{\partial \alpha_j} - \delta_{sj}\varkappa \right| = 0 \qquad (s, j = 1, \ldots, k), \tag{2.10}$$

$$\left| \left\langle \frac{\partial F_s(y_1^0, \ldots, y_l^0, 0, \omega t)}{\partial y_j^0}\, e^{i(n_j - n_s)\omega t} \right\rangle - \delta_{sj}\varkappa \right| = 0 \\ (s, j = k+1, \ldots, k+m), \tag{2.11}$$

$$\left| \left\langle \frac{\partial F_s(y_1^0, \ldots, y_l^0, 0, \omega t)}{\partial y_j^0}\, e^{i(\nu_j^{(r)} - \nu_s^{(r)})t} \right\rangle - \delta_{sj}\varkappa \right| = 0 \\ (s, j = k + m + m_1 + \ldots + m_{r-1} + 1, \ldots \\ \ldots, k + m + m_1 + \ldots + m_r;\quad r = 1, \ldots, p' = p/2) \tag{2.12}$$

are negative, then with sufficiently small μ, *this set of constants would indeed correlate to a unique, analytical in relation to* μ, *and asymptotically stable periodic solution of equations (2.1). At* $\mu = 0$ *it transforms into fundamental solution (2.7).*

If the real part of at least one of the roots of equations (2.10)–(2.12) is positive, then the corresponding solution is unstable. If there are purely imaginary or zero roots, a special analysis is required.

Thus, the stability analysis for the periodic solutions of (2.1) is reduced to the regular Horwitz problem for algebraic equations whose coefficients are expressed through the right-hand sides of the first $k + m + m_1 + \ldots + m_{p'}$ equations (2.1).

---

*An abbreviated notation is used for the determinant of the q-th order, with components a_{sj} as $|a_{sj}|$ $(s, j = 1, \ldots, q)$, as well as the Kronecker symbol δ_{sj} designating a number equal to one at $s = j$ and to zero at $s \neq j$. Thus, (2.10) is a brief presentation of equation (2.2) of chap. 2. A reminder: angle brackets < > denote averaging of the enclosed expression by variable t from 0 to $2\pi/\omega$, or by variable $\tau = \omega t$ from 0 to 2π.

§ 3) Quasi-linear autonomous systems.

The system

$$\dot{y}_s = \lambda_s y_s - \mu F_s(y_1, \ldots, y_l, \mu) \qquad (s = 1, \ldots, l), \tag{3.1}$$

is now considered. It differs from (2.1) only by the fact that the right-hand side parts do not depend explicitly on t.

It is known (75, 147) that periodic solutions of system (3.1) with period $T_{q,\sigma}$, which is close to $T^0_{q,\sigma} = 2\pi q/\nu_\sigma$ (q is an integer), can correspond to any part of purely imaginary numbers $\lambda_s = \pm i\nu_\sigma$. Therefore, considering the existence and stability of solutions having period T close to some of these periods, it is assumed that at least of the numbers $\lambda_s = 2\pi i n_s/T = i\omega n_s$ from the leading group, ($1 \le s \le k$) is not zero. Otherwise, the period of the assumed solution would be indeterminate.

The fundamental system, corresponding to equations (3.1), is

$$\dot{y}^0_s = \lambda_s y^0 \qquad (s = 1, \ldots, l) \tag{3.2}$$

This allows a periodic solution with the period $T = 2\pi/\omega$

$$y^0_s = \begin{cases} \alpha_s e^{in_s\omega t}, & s = 1, \ldots, k, \\ 0, & s = k+1, \ldots, l, \end{cases} \tag{3.3}$$

which depends on k arbitrary parameters $\alpha_1, \ldots, \alpha_k$.

In this case of an autonomous system, the periodic solutions of equations (3.1), transforming at $\mu = 0$ into fundamental solutions (3.2), would have a period generally different from $T = 2\pi/\omega$ and depending on μ. Denoting this period as

$$T^* = T^*(\mu) = T[1 - \delta(\mu)], \tag{3.4}$$

where $\delta(\mu)$ is an unknown function ("*period correction*") to be determined in the solution process. It is obvious that $\delta(0) = 0$.

With the formulated conditions, the following theorem is true (34, 57).

Theorem. *Periodic (period $T^*(\mu) = T[1 - \delta(\mu)]$) solutions of the autonomous set of equations (3.1), becoming at $\mu = 0$ periodic (period T) solutions (3.3) of the fundamental system (3.2), can correspond only to such values of constants $\alpha_1, \ldots, \alpha_{k-2}$, $\alpha_{k-1} = \alpha_k = \alpha$, which satisfy equations*

$$P_s(\alpha_1, \ldots, \alpha_k) \equiv \alpha_k n_k P^*_s - \alpha_s n_s P^*_k = 0 \qquad (s = 1, \ldots, k-1), \tag{3.5}$$

where

$$P^*_s = P^*_s(\alpha_1, \ldots, \alpha_k) = \langle F_s(y^0_0, \ldots, y^0_l, 0)\, e^{-in_s\omega t}\rangle = $$
$$= \langle F_s(\alpha_1 e^{in_1\omega t}, \ldots, \alpha_k e^{in_k\omega t}, 0, \ldots, 0)\, e^{-in_s\omega t}\rangle \; (s = 1, \ldots, k). \tag{3.6}$$

*If for a certain set of constants $\alpha_1 = \alpha^*_1, \ldots, \alpha_{k-2} = \alpha^*_{k-2}, \alpha_{k-1} = \alpha_k = \alpha^*$ which satisfies equations (3.5), the real parts of all roots χ of the following algebraic equations**

$$\left| \frac{\partial P_s}{\partial \alpha_j} - \alpha_k n_k \delta_{sj} \varkappa \right| = 0 \qquad (s, j = 1, \ldots, k-1), \tag{3.7}$$

*For an explanation of the notation, see the previous footnote.

$$\left|\left\langle\left(\frac{\partial F_s}{\partial y_j}\right)e^{i(n_j-n_s)\omega t}\right\rangle-\delta_{sj}\left(\frac{1}{2}\frac{(2n_s+1)P_k^*}{n_s\alpha_k}+\varkappa\right)\right|=0$$
$$(s,\ j=k+1,\ \ldots,\ k+m), \tag{3.8}$$

$$\left|\left\langle\left(\frac{\partial F_s}{\partial y_j}\right)e^{i\left(\nu_j^{(r)}-\nu_s^{(r)}\right)t}\right\rangle-\delta_{sj}\left(\frac{\nu_s^{(r)}P_k^*}{n_k\omega\alpha_k}+\varkappa\right)\right|=0 \tag{3.9}$$

are negative, then for sufficiently small μ this set of constants would indeed correspond to a unique, analytical in relation to μ, stable (asymptotically orbitally stable), periodic solution of the set (3.1) with period $T^*(\mu) = T[1-\delta(\mu)]$. At $\mu = 0$, it becomes a periodic (period T) solution (3.3) of the fundamental system (3.2).

If the real part of at least one root of equations (3.7)–(3.9) is positive, then the respective solution is unstable. In the case of purely imaginary or zero roots, an additional analysis is generally required.

With accuracy up to terms of order μ, the period correction is determined by

$$\delta(\mu)\approx\mu\,\frac{P_k^*\left(\alpha_1^*,\ \ldots,\ \alpha_{k-2}^*,\ \alpha^*,\ \alpha^*\right)}{i\alpha^* n_k}. \tag{3.10}$$

Thus, as in the case of the autonomous quasi-linear system (3.1), the stability analysis of the periodic solutions is reduced to the Horwitz problem for algebraic equations whose coefficients are expressed directly via the right-hand side parts of the first $k + m + m_1 + \ldots + m_{p'}$ equations (3.1).

§ 4) Nonautonomous sets close to arbitrary nonlinear ones.

Let's consider a set of differential equations

$$\dot{x}_s = X_s(x_1,\ \ldots,\ x_l,\ \omega t)+\mu F_s(x_1,\ldots,\ x_l,\ \mu,\ \omega t)\qquad(s=1,\ldots,\ l), \tag{4.1}$$

Here X_s and F_s are the analytical functions of $x_1, \ldots, x_l$ in the close domain G which is assumed to contain all the periodic solutions considered below. Functions X_s and F_s are assumed in relation to the continuous periodic time function of period $T = 2\pi/\omega$. In addition, F_s are analytical functions of parameter $\mu \geq 0$ at its sufficiently small values.

The fundamental system corresponding to (4.1) is

$$\dot{x}_s^0 = X_s\left(x_1^0,\ \ldots,\ x_l^0,\ \omega t\right)\qquad(s=1,\ \ldots,\ l) \tag{4.2}$$

This allows for a family of T-periodic solutions

$$x_s^0 = x_s^0\left(\omega t,\ \alpha_1,\ \ldots,\ \alpha_k\right)\qquad(s=1,\ \ldots,\ l), \tag{4.3}$$

It depends on $k \leq l$ arbitrary parameters $\alpha_1, \ldots, \alpha_k$ and is analytical in relation to those parameters near the "fundamental" values $\alpha_1 = \alpha_1^*, \ldots, \alpha_k = \alpha_k^*$ analyzed below. It is assumed that parameters $\alpha_1, \ldots, \alpha_k$ enter into (4.3) independently, or at least that one of the k-th-order determinants contained in the matrix

$$\left\|\frac{\partial x_s^0}{\partial\alpha_j}\right\|\qquad(s=1,\ \ldots,\ l;\ j=1,\ \ldots,\ k), \tag{4.4}$$

does not become zero at the mentioned fundamental values of the parameters.*

*Here, as before, the determinant of the q-th order with elements α_{sj} is denoted as $|\alpha_{sj}|$ $(s, j = 1, \ldots, q)$; similarly, $\|\alpha_{sj}\|$ $(s = 1, \ldots, l; j = 1, \ldots, k)$ denotes a matrix with l rows and k columns.

Variations of the equations (see § 1) corresponding to equations (4.2) and solution (4.3) are

$$\dot{z}_s = p_{s1}z_1 + \ldots + p_{sl}z_l \qquad (s = 1, \ldots, l), \tag{4.5}$$

Here

$$p_{sj} = p_{sj}(\omega t) = \left(\frac{\partial X_s}{\partial x_j}\right) = \frac{\partial X_s(x_1^0, \ldots, x_l^0, \omega t)}{\partial x_j^0} \tag{4.6}$$

are T-periodic functions. The round brackets enclosing the derivatives from X_s show (as before) that the derivatives are computed for one of the fundamental solutions (4.3).

Due to the known property of the variations of the equation (see § 1), set (4.5) allows at least k periodic solutions like

$$z_{sj}(t) = \frac{\partial x_s^0}{\partial \alpha_j} \qquad (s = 1, \ldots, l;\ j = 1, \ldots, k), \tag{4.7}$$

which are obtained by differentiating equations (4.3) by the parameters. It is also known that at least k periodic solutions exist for the set of linear equations

$$\dot{z}_s^* + p_{1s}z_1^* + \ldots + p_{ls}z_l^* = 0 \qquad (s = 1, \ldots, l), \tag{4.8}$$

which is called conjugate to (4.5). It is assumed that the number of independent T-periodic solutions of systems (4.5) and (4.8) is exactly k; thus (4.5) does not have periodic solutions independent of (4.7). Then, the characteristic equation of the set of variations would certainly have k-times multiple root $\rho = 1$. The magnitudes of all other roots will be assumed to be less than one. The multiple "critical" root $\rho = 1$ corresponds to the linear elementary dividers, and it is always possible to choose the periodic solutions of the conjugate system in such a way that they satisfy correlations

$$\sum_{s=1}^{l} z_{si}z_{sj}^* = \delta_{ij} = \begin{cases} 1, & i = j \\ 0, & i \neq j. \end{cases} \tag{4.9}$$

With these assumptions, the following theorem [suggested in (184) and generalized in (57)] is true.

Theorem. *The periodic solutions of the nonautonomous set (4.1), which become periodic solutions (4.3) of the fundamental set (4.2) at $\mu = 0$, may correspond only to such values of constants $\alpha_1, \ldots, \alpha_k$ which satisfy*

$$P_s(\alpha_1, \ldots, \alpha_k) \equiv \left\langle \sum_{j=1}^{l} F_j(x_1^0, \ldots, x_l^0, 0, \omega t)\, z_{js}^*(t) \right\rangle = 0. \tag{4.10}$$

If for a certain set of constants $\alpha_1 = \alpha_1^*, \ldots, \alpha_k = \alpha_k^*$, which satisfy (4.10), real parts of all roots $\varkappa$ of the k-th-order algebraic equation

$$\left| \frac{\partial P_s}{\partial \alpha_j} - \delta_{sj}\varkappa \right| = 0 \qquad (s, j = 1, \ldots, k) \tag{4.11}$$

are negative, then at sufficiently small μ this set of constants is indeed correlated with the unique, analytical relative to μ, asymptotically stable periodic solution of (4.1). This solution changes into the fundamental solution (4.3) at $\mu = 0$.

If the real part of at least one root of equation (4.11) is positive, then the corresponding solution is unstable. With purely imaginary or zero roots, an additional analysis is required in the general case.

§ 5) Autonomous systems, close to arbitrary nonlinear ones.

Let's consider an autonomous system

$$\dot{x}_s = X_s(x_1, \ldots, x_l) + \mu F_s(x_1, \ldots, x_l, \mu) \qquad (s = 1, \ldots, l). \tag{5.1}$$

Here X_s and F_s are analytical functions of $x_1, \ldots, x_l$ in the close domain G, which also contains all periodic solutions analyzed below. Also F_s are analytical functions of parameter $\mu \geq 0$ at its sufficiently small values.

It is assumed that the fundamental system corresponding to (5.1),

$$\dot{x}_s^0 = X_s(x_1^0, \ldots, x_l^0) \qquad (s = 1, \ldots, l) \tag{5.2}$$

allows for a family of $T = 2\pi/\omega$-periodic solutions

$$x_s^0 = x_s^0(\omega t, \alpha_1, \ldots, \alpha_{k-1}), \tag{5.3}$$

It also depends, besides on constant α, which can always be added to t by the autonomous character of the system, on some number $k - 1 \leq l - 1$ of arbitrary parameters $\alpha_1, \ldots, \alpha_{k-1}$. Let's assume that solution (5.3) is analytical relative to these parameters near the "fundamental" values $\alpha_1 = \alpha_1^*, \ldots, \alpha_{k-1} = \alpha_{k-1}^*$ considered below. As in § 4, it is assumed that parameters $\alpha_1, \ldots, \alpha_{k-1}$ enter expressions (5.3) independently. This means that at the mentioned fundamental values of parameters $\alpha_1 = \alpha_1^*, \ldots, \alpha_{k-1} = \alpha_{k-1}^*$, at least one of the determinants of $(k-1)$th order contained in matrix (4.4) is not zero.

As in § 4, the set of equations in variations (4.5), which are due to the already mentioned property, allows k solutions like

$$z_{sj}(t) = \frac{\partial x_s^0}{\partial \alpha_j} \qquad (j = 1, \ldots, k-1), \tag{5.4}$$

$$z_{sk}(t) = \dot{x}_s^0, \tag{5.5}$$

These are now obtained by the differentiating solutions (5.3), respectively, by $\alpha_1, \ldots, \alpha_{k-1}$ and by t (reminder: one more constant α can always be added to t).

Contrary to the quasi-linear autonomous system (3.1), period T of the fundamental solution of (5.2) may depend on parameters $\alpha_1, \ldots, \alpha_{k-1}$. Here, though, a simpler case is considered, i.e., when T does not depend on any of these parameters. In such a case, it can be stated that derivatives (5.4) are T-periodic functions of t and equations in variations thus allow k periodic solutions of the same period. The same number of T-periodic solutions would also allow set (4.8), which is conjugate relative to the system in variations. These solutions are denoted as $z_{sj}^*(t)$ $(j = 1, \ldots, k)$, and it is assumed that the conjugate system does not have other periodic solutions. Then too, the system in variations does not have periodic solutions independent of (5.4) and (5.5). This means that this system corresponds to the k-times multiple root $\rho = 1$ of the characteristic equation. As in § 4, it will be assumed that all other roots of the characteristic equation have magnitudes less than one. Then, multiple root $\rho = 1$

would correspond to linear elementary dividers, and it can be concluded that expressions (4.9) are satisfied.

With these assumptions, the following theorem is true (57, 184, 207).

Theorem. *Periodic solutions with period $T^*(\mu) = T[1 - \delta(\mu)]$ of the original set of equations (5.1), which become periodic (period T is not dependent on parameters $\alpha_1, \ldots, \alpha_{k-1}$) solutions of family (5.3) of the fundamental set (5.2) at $\mu = 0$, may correspond only to such values of parameters $\alpha_1, \ldots, \alpha_{k-1}$ of said family, which satisfy equations*

$$P_s(\alpha_1, \ldots, \alpha_{k-1}) \equiv \left\langle \sum_{j=1}^{l} F_j(x_1^0, \ldots, x_l^0, 0)\, z_{js}^*(t) \right\rangle = 0 \qquad (5.6)$$

$$(s = 1, \ldots, k-1).$$

A unique, analytical relative to μ, and stable (asymptotically orbitally stable) periodic solution of (5.1) with period $T^*(\mu)$, becomes a fundamental one at $\mu = 0$, and indeed corresponds to a certain solution of (5.6) at sufficiently small μ, if all roots of algebraic equation of $(k-1)$th order

$$\left| \frac{\partial P_s}{\partial \alpha_j} - \delta_{sj}\varkappa \right| = 0 \qquad (s, j = 1, \ldots, k-1) \qquad (5.7)$$

have negative real parts. If there is at least one root of (5.7) which has a positive real part, then this solution is unstable. The case of zero or purely imaginary roots generally requires a special analysis.

Up to terms of the order μ, correction $\delta(\mu)$ to the period of the solution is determined by

$$\delta(\mu) \approx \mu P_k(\alpha_1, \ldots, \alpha_{k-1}) = \mu \left\langle \sum_{s=1}^{l} F_s(x_1^0, \ldots, x_l^0, 0)\, z_{sk}^*(t) \right\rangle. \qquad (5.8)$$

§ 6) Systems close to arbitrary nonlinear ones in case of quasi-periodic family of fundamental solutions (216).

Let's consider the same general nonlinear system as in § 4,

$$\dot{x}_s = X_s(x_1, \ldots, x_l, \omega t) + \mu F_s(x_1, \ldots, x_l, \mu, \omega t) \qquad (6.1)$$
$$(s = 1, \ldots, l),$$

However, it is assumed that the respective fundamental system

$$\dot{x}_s^0 = X_s(x_1^0, \ldots, x_l^0, \omega t) \qquad (s = 1, \ldots, l) \qquad (6.2)$$

allows for a family of quasi-periodic solutions

$$x_s^0 = x_s^0(\omega t; \psi_1, \ldots, \psi_k; h_1, \ldots, h_n). \qquad (6.3)$$

Here $k + n \leq l$,

$$\psi_s = \omega_s t + \alpha_s \qquad (s = 1, \ldots, k), \qquad (6.4)$$

Functions x_s^0 are 2π-periodic by each of arguments ψ_s; α_s and h_ν are arbitrary constants. Further, the first m frequencies ω_s are essentialy dependent on constant h_ν, and $k-m$ subsequent frequencies ω_s are equal to the frequency of the external excitation ω, or

$$\omega_s = \omega_s(h_1, \ldots, h_n) \qquad (s = 1, \ldots, m),$$
$$\omega_s = \omega \qquad (s = m+1, \ldots, k), \tag{6.5}$$

with $m \le k$. In other words, it will be assumed that the fundamental system allows for a family of quasi-periodic solutions depending on $k + n$ arbitrary constants α_s $(s = 1, \ldots, k)$ and h_ν $(\nu = 1, \ldots, n)$. Such cases occur in analyses of synchronization of quasi-conservative systems.

Family (6.3) contains a subfamily of $T = 2\pi/\omega$-periodic solutions described by

$$\omega_s(h_1, \ldots, h_n) = \omega \qquad (s = 1, \ldots, m). \tag{6.6}$$

This subfamily, which is considered below, thus contains only $k + n - m$ independent constants. Therefore, the system of equations in variations (4.5), which correspond to equations (6.2) and solution (6.3), in this case allows k independent T-periodic solutions

$$z_{sj}(t) = \frac{\partial x_s^0}{\partial \alpha_j} \qquad (s = 1, \ldots, l;\ j = 1, \ldots, k) \tag{6.7}$$

and n linearly ascending solutions

$$\tilde{z}_{s\nu}(t) = \frac{\partial' x_s^0}{\partial h_\nu} = \sum_{i=1}^{m} \frac{\partial x_s^0}{\partial \psi_i}\frac{\partial \omega_i}{\partial h_\nu} + \frac{\partial x_s^0}{\partial h_\nu} \qquad (s = 1, \ldots, l;\ \nu = 1, \ldots, n), \tag{6.8}$$

Here the prime denotes full partial differentiation. It is easy to compile such linear combinations of solutions $\tilde{z}_{s\nu}(t)$, so that the first m would still be linearly ascending, and the other $n - m$ would be periodic (period T). Such combinations, which are also solutions of equations in variations (4.5), would be

$$z_{s\nu}(t) = \frac{\partial x_s^0}{\partial \alpha_\nu} t + \sigma_{s\nu}(t) \qquad (\nu = 1, \ldots, m), \tag{6.9}$$

$$z_{s\nu}(t) = \frac{\partial x_s^0}{\partial h_\nu} - \sum_{i=1}^{m} \frac{\partial \omega_i}{\partial h_\nu}\sigma_{si}(t) \qquad (\nu = m+1, \ldots, n), \tag{6.10}$$

$$(s = 1, \ldots, l).$$

Here $\sigma_{s\nu}(t)$ denote T-periodic functions determined by

$$\sigma_{s\nu}(t) = \sum_{i=1}^{m} \chi_{i\nu} \frac{\partial x_s^0}{\partial h_i}, \tag{6.11}$$

where $\chi_{i\nu}$ are parameters which are determined from sets of linear differential equations

$$\sum_{i=1}^{m} \chi_{i\nu} \frac{\partial \omega_s}{\partial h_i} = \delta_{s\nu} = \begin{cases} 1, & s = \nu, \\ 0, & s \neq \nu \end{cases} \qquad (s, \nu = 1, \ldots, m). \tag{6.12}$$

It can be assumed that these equations have a single solution due to the assumption of an essential correlation between frequencies $\omega_1, \ldots, \omega_m$ and constants $h_1, \ldots, h_m$.

Functions x_s^0 in expressions (6.9) and (6.10) are assumed to be T-periodic, i.e., it is assumed that (6.5) and (6.6) are satisfied.

Thus, the system in variations (4.5) allows $k + n - m$ to have mutually independent T-periodic solutions (6.7) and (6.10). Let's assume that the system does not have other independent T-periodic solutions. Then, the conjugate system (4.8) would allow for the same number of independent T-periodic solutions $z_{s1}^*(t), \ldots, z_{s,k+n-m}^*(t)$. It will be assumed that they are chosen to satisfy the following conditions of orthogonality and normalization:

$$\begin{aligned} &\sum_{s=1}^{l} z_{sr}^*(t)\,\sigma_{si}(t) = \delta_{ir} && (i = 1, \ldots, m), \\ &\sum_{s=1}^{l} z_{sr}^*(t)\, z_{si}(t) = \delta_{ir} && (i = m+1, \ldots, k), \\ &\sum_{s=1}^{l} z_{sr}^*(t)\, z_{s\nu}(t) = \delta_{k-m+\nu,r} && (\nu = m+1, \ldots, n). \end{aligned} \tag{6.13}$$

In these assumptions, the following theorem is true (216).

Theorem. *$T = 2\pi/\omega$ periodic solutions of nonautonomous set (6.1) which become, at $\mu = 0$, solutions of the fundamental system (6.2), may correspond only to such values of constants $\alpha_1, \ldots, \alpha_k$ and $h_1, \ldots, h_n$, which are solutions of the system consisting of $k + n - m$ equations*

$$P_s(\alpha_1, \ldots, \alpha_k;\, h_1, \ldots, h_n) = \left\langle \sum_{j=1}^{l} F_j(x_1^0, \ldots, x_l^0, 0, \omega t)\, z_{js}^*(t) \right\rangle = 0 \tag{6.14}$$

$$(s = 1, \ldots, k + n - m),$$

and also in equations (6.6). A unique, analytical relative to μ, and asymptotically stable solution of this kind indeed corresponds to each solution $\alpha_1^*, \ldots, \alpha_k^*$; $h_1^*, \ldots, h_n^*$ at a sufficiently small μ if:

1) *All the roots of the algebraic m-order equation*

$$\left| \frac{\partial P_s}{\partial \alpha_j} - \delta_{sj}\lambda_1^2 \right| = 0 \qquad (s, j = 1, \ldots, m) \tag{6.15}$$

are real and negative, and for each root the following inequality is satisfied

$$\lambda_2 = \frac{1}{2\sum_{r=1}^{m} a_r a_r^*} \sum_{i=1}^{m}\sum_{r=1}^{m}\left[\sum_{s=1}^{m} \chi_{si}\frac{\partial P_r}{\partial h_s} + \left(\sum_{s=m+1}^{k}\frac{\partial P_r}{\partial \alpha_s}\frac{\partial P_s}{\partial \alpha_i} + \sum_{s=m+1}^{n}\frac{\partial^* P_r}{\partial h_s}\frac{\partial P_{s+k-m}}{\partial \alpha_i}\right) - p_{ri}\right] a_i a_r^* < 0. \tag{6.16}$$

2) *All the roots λ_3 of the algebraic equation of $(k + n - 2m)$th order which is derived by equating to zero the determinant of the set of linear homogeneous equations*

$$\sum_{i=1}^{k} \frac{\partial P_r}{\partial \alpha_i} a_i + \sum_{s=m+1}^{n} \frac{\partial^* P_r}{\partial h_s} b_s = \begin{cases} 0 & (r = 1, \ldots, m), \\ \lambda_3 a_r & (r = m+1, \ldots, k), \\ \lambda_3 b_{r-k+m} & (r = k+1, \ldots, k+n-m), \end{cases} \tag{6.17}$$

have negative real parts.

If there is at least one positive number among λ_1 and λ_2, or there is at least one complex or pure imaginary number among λ_1, and also if at least one λ_3 has a positive real part, then the corresponding solution is unstable. A case of zero λ_1, λ_2 or pure imaginary λ_3 requires a special analysis.

It is denoted in (6.15)–(6.17)

$$\frac{\partial^*}{\partial h_s} = \frac{\partial}{\partial h_s} - \sum_{i=1}^{m} \sum_{r=1}^{m} \chi_{ir} \frac{\partial \omega_r}{\partial h_s} \frac{\partial}{\partial h_i},$$

$$p_{ri} = \left\langle \sum_{s=1}^{l} z_{sr}^* (t) \frac{\partial x_s^{(1)}}{\partial \alpha_i} \right\rangle - \frac{\partial P_r}{\partial \alpha_i}, \tag{6.18}$$

where $x_s^{(1)} = x_s^{(1)}(t)$ is the first approximation to the periodic solution being sought, i.e., coefficients at μ in expansions $x_s(t) = x_s^0(t) + \mu x_s^{(1)}(t) + \ldots$, which are determined from

$$\dot{x}_s^{(1)} = \sum_{j=1}^{l} \left(\frac{\partial X_s}{\partial x_j} \right) x_j^{(1)} + F_s (x_1^0, \ldots, x_l^0, 0, \omega t) \qquad (s = 1, \ldots, l); \tag{6.19}$$

As before, the round brackets embracing the derivatives $\partial X_s / \partial x_j$ indicate that these derivatives are computed for the fundamental solution x_s^0 and are at $\mu = 0$.

As a_i and a_*^r are designated solutions of mutually conjugated sets of linear homogeneous algebraic equations

$$\sum_{i=1}^{m} \frac{\partial P_r}{\partial \alpha_i} a_i = \lambda_1^2 a_r \quad (r = 1, \ldots, m),$$

$$\sum_{r=1}^{m} \frac{\partial P_r}{\partial \alpha_i} a_r^* = \lambda_1^2 a_i^* \quad (i = 1, \ldots, m). \tag{6.20}$$

These equations have solutions determined with accuracy up to the constant multiplier if roots λ_1 of equation (6.15) are simple or multiple but with simple elementary dividers, which is assumed.

The stability conditions related by expressions (6.15) and (6.16) are due to the equation in variations having m solution groups

$$z_{sj}(t) = \frac{\partial x_s^0}{\partial \alpha_j}, \quad z_{sj}(t) = \frac{\partial x_s^0}{\partial \alpha_j} t + \sigma_{sj}(t) \qquad (j = 1, \ldots, m),$$

which correspond to zero (that is, critical) characteristic exponents with quadratic elementary dividers (57, 184). The exponents of the system in variations for equations (6.1) being transformed in the latter at $\mu \longrightarrow 0$ ones, are analytical relative to $\mu^{1/2}$. Corresponding to the periodic solutions of equations in variations

$$z_{sj}(t) = \frac{\partial x_s^0}{\partial \alpha_j} \qquad (j = m+1, \ldots, k),$$

$$z_{sj}(t) = \frac{\partial x_s^0}{\partial h_j} - \sum_{i=1}^{m} \frac{\partial \omega_i}{\partial h_j} \sigma_{si}(t) \qquad (j = m+1, \ldots, n)$$

are zero critical characteristic exponents with simple elementary dividers. Exponents of the system in variations for equations (6.1), being transformed in the latter at $\mu \longrightarrow 0$ are analytical in relation to μ. The stability conditions related by equations(6.17) are stipulated by the presence of such exponents.

In this case, contrary to what was considered before, the stability conditions depend not only on fundamental but also on the first approximation to the analyzed periodic solution.

The described result is easy to extrapolate to a case of vibrations with commeasute frequencies, when, instead of correlations (6.5) and (6.6), there are equalities

$$\omega_s = n_s \omega \qquad (s = 1, \ldots, k), \tag{6.21}$$

where n_s are integers.

§ 7) Additional remarks to theorems of §§ 2–6.

1) The left-hand parts of equations (2.9), (3.5), (4.10), (5.6), (6.14) are the fundamental functions themselves, and these equations are the same basic equations referred to in § 2, chap. 2.

2) Equations like (2.1) and (3.1), referred to in theorems §§ 2 and 3, are sometimes called canonic. The following general quasi-linear sets can be reduced to the same format.

$$\dot{x}_s = \sum_{j=1}^{l} a_{sj}(t) x_j + \varphi_s(t) + \mu \Phi_s(x_1, \ldots, x_l, \mu, t) \qquad (s = 1, \ldots, l), \tag{7.1}$$

Here a_{sj}, φ_s, and Φ_s are $T = 2\pi/\omega$-periodic time t functions if the linear elementary dividers correspond to all the multiple roots ρ of the characteristic equation of the linear part of the set. In the autonomous case, a_{sj} and φ_s are constant, and Φ_s does not explicitly depend on t. Transition from equations like (7.1) to equations like (2.1) and (3.1) is achieved by a nonsingular linear transformation of the variables. The actual finding of this transformation, although it requires in a general case a knowledge of the fundamental system of solutions for an appropriate linear set of differential equations with periodic coefficients, is feasible during the solution of many application problems due to the appropriate structure of matrix $\|a_{sj}(t)\|$.

3) The theorems in §§ 2 and 3 also remain true for a set of equations only partially reduced to the canonic format, i.e., only in relation to variables corresponding to critical characteristic exponents. Basically, to obtain equations (2.9), (2.10) and (3.5), (3.7), it is sufficient to bring the set to the canonic format only for variables $x_1, \ldots, x_k$ corresponding to the leading critical group of exponents. Moreover, equations associated with noncritical variables might not even be quasi-linear. It is only important that all noncritical characteristic exponents of the set of equations in the variations have negative real parts.

4) All theorems are directly applicable not only to the purely periodic, but also to the synchronous solutions of the corresponding systems [see expressions (1.1) in chap. 1].

5) In many practical, important problems leading to analyses of periodic or synchronous solutions, e.g., in problems of internal synchronization of dynamic systems, the period of the periodic solution is not known in advance.* In such cases it is natural to try to determine this period from the condition that it has to be as close as possible to the period $T^*(\mu)$ of the original system (3.1) or (5.1) and, specifically, from the condition that the period correction $\delta(\mu)$ with accuracy up to terms of the order μ were becoming zero.† Then, according to (3.10) and (5.8), instead of $k-1$ equations (3.5), there are k equations

$$P_s(\alpha_1, \ldots, \alpha_k) = 0 \qquad (s = 1, \ldots, k). \tag{7.2}$$

From these equations, $k-1$ constants $\alpha_1, \ldots, \alpha_{k-1}$ and period T can be determined.

Due to this circumstance, it is possible to formulate the resulting solution for some synchronization problems in a format suitable for cases of both external and internal synchronization. Moreover, at certain conditions it is possible not to consider the most complex problem of internal synchronization, and to get results by purely formal transformations from the solution of the more simple problem of external synchronization.

6) The conditions of the theorems in §§ 2–5 contained limitations related to the linear character of the elementary dividers corresponding to the multiple roots of the characteristic equation for the system in variations. However, in § 6, exponents with elementary quadratic dividers were also tolerated. The latter case can be encountered in some synchronization problems. Examples are synchronization problems for quasi-conservative objects, and generally problems which can be reduced to autonomous analyses of the fundamental approximation problems for which the period T of the fundamental solution depends on at least one of the constants α_s. The case of quadratic elementary dividers of multiple roots for some specific systems is discussed in (208–210, 213, 215, 283), partially described in (57).

7) The direct use of theorems allows us to analyze the issue of the presence or absence of the occurrence of synchronization, and also fully determine the fundamental periodic (or synchronous) solutions and analyze the stability of the solutions. Technically, this amounts to constructing and solving basic equations $P_s(\alpha_1, \ldots, \alpha_k) = 0$ and also to the analysis of the signs of the real parts of the algebraic equations roots, i.e., to the well-known Horwitz problem. In cases of quasi-linear sets there are usually no special problems. For systems which are close to nonlinear, difficulties are associated with a need to find a system of periodic solutions $\overset{*}{z}_{sj}$ of the set of equations conjugate to equations in variations, i.e., with the solution of linear differential equations with periodic coefficients. However, in some practical, important problems such a set of solutions can be found without particular difficulty. As it was noted in § 2, chap. 1, the difficulties are due not to the degree of complexity of the whole system, but due to degree of complexity of the individual isolated objects.

*The case of the unknown period of the fundamental approximation should be discriminated from cases when the period of synchronous motion is not known in the internal synchronization problems (see § 1, chap. 1).

†Such a technique was used for the solution of some problems of nonlinear vibration theory, e.g., see (31, 57, 74).

Usually, to solve application synchronization problems, it is enough to establish the fact of the existence of a stable synchronous motion, and to compute it in a fundamental approximation. When there is a need for more accurate solutions, they can be sought as expansions by the whole positive powers of μ due to the analytical character of this parameter. Issues related to the practical determination of the expansion coefficients are addressed in (57, 66, 184).

8) It is stated in the theorem formulations that if equations (2.10), (3.7), (4.11), (5.7), and (6.15) have a zero root, an additional analysis is needed. Most interesting for the synchronization theory are such cases when the existence of one or several zero roots of these equations is associated with the fact that the basic equations (2.9), (3.5), (4.10), (5.6) and (6.14) allow us to find only some unknowns α_s. In relation to other unknowns or their combinations they become identities. Equations(2.10), (3.7), (4.11), (5.7), and (6.15) allow us in such cases to obtain only necessary stability conditions. Such cases, which were mentioned in § 5, chap. 3, and in § 2, chap. 5, and which arise during the solution of some multiple synchronization problems, are discussed in (191) and (15, 16). It is shown in these publications that the parameters α_s of the fundamental solution, which are left indeterminate, may be found from existence and stability conditions for the higher approximations. The missing stability conditions can also be obtained from analyses of higher approximations.

9) The requirements for the right-hand side parts of the differential equations discussed in §§ 2–6 could be weakened considerably while retaining most of the results (147, 184). For example, in many cases function X_s in equations (2.1), (3.1), (4.1), (5.1), and (6.1) does not have to be analytical in respect to variables $x_1, \ldots, x_l$, but should only have continuous second-order partial derivatives. Functions F_s can be required just to have continuous partial derivatives of the first order. However, from the application standpoint, such generalizations are hardly very important, since in typical problems the right-hand sides of equations of motion are either analytical or have discontinuities (are "piecemeal analytical"). The latter cases, which occur, for example, in problems of the synchronization of mechanical objects in systems with dry friction and impacts, are addressed in (152, 215, 217, 218), which are partly summarized in (57).

A generalization of theorems like those given in §§ 2–4 on nonlinear quasi-periodic systems was performed in (184), and on quasi-linear quasi-periodic systems with time lags in (301, 302).

§ 8) Integral stability criterion of periodic and synchronous motions.

The basic contents of the theorems in this chapter can be, under certain conditions, reformatted to make it convenient both for the solution of specific application problems and for the study of the general features of the synchronization of mechanical objects.

Let's consider first the case of nonautonomous systems and remember that according to the theorems, periodic (or synchronous) motions comply with simple solutions

$$\alpha_1 = \alpha_1^*, \ldots, \alpha_k = \alpha_k^*$$

of some set of equations

$$P_s(\alpha_1, \ldots, \alpha_k) = 0 \qquad (s = 1, \ldots, k), \tag{8.1}$$

which are called *basic equations*.

The stability of each specific motion depends on the signs of the real parts of roots χ of a k-th order algebraic equation

$$\left| \frac{\partial P_s}{\partial \alpha_j} - \delta_{sj}\varkappa \right|_{\alpha_q = \alpha_q^*} = 0 \qquad (s, j = 1, \ldots, k). \tag{8.2}$$

Function $P_s(\alpha_1, \ldots, \alpha_k)$ is expressed via certain integrals computed during one motion period T and depends on the right-hand sides of the considered differential equations. These functions were called *fundamental*.

It is assumed that there exists a function $D(\alpha_1, \ldots, \alpha_k)$ of the parameters of the fundamental solution, which is continuous, has continuous partial derivatives up to and including second order, and is such that

$$\frac{\partial D}{\partial \alpha_s} = -P_s(\alpha_1, \ldots, \alpha_k) \qquad (s = 1, \ldots, k). \tag{8.3}$$

Another assumption is that functions P_s are real near the fundamental values of parameters $\alpha_1^*, \ldots, \alpha_k^*$, which are also assumed real.* Conditions for function D are

$$\frac{\partial D}{\partial \alpha_s} = 0 \qquad (s = 1, \ldots, k), \tag{8.4}$$

These conditions coincide with equations (8.1).

Function D will be called *potential function*. It is obvious that for its existence, satisfaction of the following expression is necessary and sufficient

$$\frac{\partial P_s}{\partial \alpha_j} = \frac{\partial P_j}{\partial \alpha_s} \qquad (s, j = 1, \ldots, k).$$

Expanding this function by the Taylor formula near fundamental values $\alpha_1 = \alpha_1^*, \ldots, \alpha_k = \alpha_k^*$ in accordance with (8.1) and (8.3), we obtain

$$D(\alpha_1, \ldots, \alpha_k) - D(\alpha_1^*, \ldots, \alpha_k^*) =$$
$$= \sum_{s=1}^{k}\sum_{j=1}^{k}\left(\frac{\partial^2 D}{\partial \alpha_s \partial \alpha_j}\right)_{\alpha_q = \alpha_q^*}(\alpha_s - \alpha_s^*)(\alpha_j - \alpha_j^*) + \ldots =$$
$$= -\sum_{s=1}^{k}\sum_{j=1}^{k}\left(\frac{\partial P_s}{\partial \alpha_j}\right)_{\alpha_q = \alpha_q^*}(\alpha_s - \alpha_s^*)(\alpha_j - \alpha_j^*) + \ldots, \tag{8.5}$$

Here terms higher than second order in respect to $(\alpha_q - \alpha_q^*)$ are not shown.

For function D to have a minimum at the point $(\alpha_1^*, \ldots, \alpha_k^*)$ it is sufficient that the quadratic form in (8.5) be nonzero (sign-definite) positive. In turn, it is necessary and sufficient for all roots of the k-th-order algebraic equation

$$\left| \frac{\partial^2 D}{\partial \alpha_s \partial \alpha_j} + \delta_{sj}\varkappa \right|_{\alpha_q = \alpha_q^*} = 0 \qquad (s, j = 1, \ldots, k) \tag{8.6}$$

*A reminder: The fundamental values of the parameters are the ones in compliance with (8.1).

to be negative. All roots of this equation are real due to the theorem of the characteristic values of the real symmetric matrix (97).

If at least one root of (8.6) is positive, then the quadratic form is not sign-definite, and there is no minimum. The case when there are zero roots is suspicious and, generally, requires analysis of higher order terms of expansion (8.5).

However, equation (8.6), with existing correlations (8.3), is equivalent to equation (8.2). Thus, the conditions for a rigorous minimum of function D in point $(\alpha_1^*, \ldots, \alpha_k^*)$, based on the analysis of the second-order terms in expansion (8.5), coincide in this case with the stability conditions of the periodic (or synchronous) solutions of the appropriate sets of equations.*

If equation (8.6) has zero roots with other negative roots, then there exist necessary conditions for a minimum (which, in this case, will be noncrude, might be nonrigorous, or be absent at all). Corresponding conditions would give only a part of the existence and stability conditions for periodic or synchronous motions. This means that these conditions will be only necessary and not sufficient [see also remark 8 of § 7 and also § 9].

Thus, in problems of the existence and stability of the periodic or synchronous motions, function D plays the same role as potential energy Π does in the problem of finding and analyzing the stability of equilibrium positions for a conservative system with holonomic stationary constraints. In other words, for periodic (or synchronous) solutions of such equations, theorems analogous to the theorems of Lagrange (Dirichlet) and Lyapunov are applicable (178, 183).

The formulated *stability criterion* may be called integral, since function D, in accordance with what was said before about fundamental functions P_s, depends on the motion characteristics averaged during the period, calculated for the fundamental solution.

If motion stability criteria amount only to the requirement that the real parts of equation (8.2) be negative, then the integral criterion formulates both necessary and sufficient stability criteria. When some other conditions are present, the conditions expressed by the integral criterion are just necessary. However, as was noted in § 2, chap. 2, even in such cases, in many applications, conditions formulated by equations like (8.2), or by the integral criterion, play the major role. Accordingly, they are called basic, and other conditions, if available, auxiliary.

With minor modifications, whose character is clear from remark 5), § 7, the integral stability criterion can also be formulated for an autonomous system.

The class of systems for which the integral stability criterion is applicable can be broadened significantly if it is noted that it requires only the existence of function D, which satisfies not conditions (8.3) but less rigid correlations

$$b_{s1}\frac{\partial D}{\partial \alpha_1} + b_{s2}\frac{\partial D}{\partial \alpha_2} + \ldots + b_{sk}\frac{\partial D}{\partial \alpha_k} = P_s(\alpha_1, \ldots, \alpha_k) \qquad (s = 1, \ldots, k), \tag{8.7}$$

Here $b_{sj} = b_{js}$ are any real constants only complying with

$$B(z, z) \equiv \sum_{s=1}^{k}\sum_{j=1}^{k} b_{sj}z_s z_j > 0, \tag{8.8}$$

---

*The minimum of a function F in point $\mathcal{M}_0$ is called rigorous if inequality $F(\mathcal{M}) > F(\mathcal{M}_0)$ stands for all points $\mathcal{M}$ from some ε-vicinity of point $\mathcal{M}_0$. If correlation $F(\mathcal{M}) \geq F(\mathcal{M}_0)$ stands, then the minimum is nonrigorous. A rigorous minimum of the function, located by analysis of second-order terms in its (8.4)-like expansion, will be called a *crude minimum*.

or a condition that corresponds to them in quadratic form is positive. This is proved in (57).

It also follows from the aforesaid that the integral criterion is correct if the function D exists which complies with

$$\frac{\partial D}{\partial \alpha_s} = -(P_1 c_{s1} + P_2 c_{s2} + \ldots + P_k c_{sk}) \qquad (s = 1, \ldots, k), \qquad (8.9)$$

Here $c_{sj} = c_{js}$ are any real numbers such as

$$C(z, z) = \sum_{s=1}^{k} \sum_{j=1}^{k} c_{sj} z_s z_j > 0. \qquad (8.10)$$

If the integral criterion applies, then stability analysis becomes much easier, since stability conditions can be explicitly written. Indeed, according to the Sylvester theorem (97), positiveness of all the major minors of its matrix is the necessary and sufficient condition of positiveness of a quadratic form. Thus, according to (8.6), the necessary and sufficient conditions for the minimum of function D in the point $\alpha_1 = \alpha_1^*, \ldots, \alpha_k = \alpha_k^*$ are the inequalities

$$\left| \frac{\partial^2 D}{\partial \alpha_i \partial \alpha_j} \right|_{\alpha_q = \alpha_q^*} > 0 \qquad (i, j = 1),$$

$$\left| \frac{\partial^2 D}{\partial \alpha_i \partial \alpha_j} \right|_{\alpha_q = \alpha_q^*} > 0 \qquad (i, j = 1, 2),$$

$$\cdots\cdots\cdots\cdots\cdots$$

$$\left| \frac{\partial^2 D}{\partial \alpha_i \partial \alpha_j} \right|_{\alpha_q = \alpha_q^*} > 0 \qquad (i, j = 1, \ldots, k).$$

The latter are especially convenient for the solution of a problem of the synthesis of a system of objects having synchronous motion with a prescribed combination of fundamental parameters $\alpha_1^*, \ldots, \alpha_k^*$.

In the case when analytical study is difficult or undesirable, the presence of the integral criterion allows us to use numerical algorithms to search the extrema of multivariable functions.

The importance of the integral criterion is also due to the fact that, in some cases, function D has a distinct physical meaning. It also facilitates the synthesis of synchronously operating objects. Thus, for the self-synchronization of debalanced vibro-exciters, the potential function turns out to be equal to, averaged over a period, the value of the Lagrange function for the system, which is taken with the opposite sign and computed for the fundamental solution (see § 7, and subsection 4, § 8, chap. 3). For this case the integral criterion was originally formulated from intuitive ideas, later validated by some examples, in (40), and then substantiated and generalized by the author in (41, 43). Further development was done in (25, 42, 52, 57, 81, 164, 165, 209, 211, 213, 215, 282, 283, 285, 286). What follows are some general results for canonic and similar systems from (25, 81).

Paper (81) analyzes a canonic system with Hamilton function $H(t; q_1, \ldots, q_n; p_1, \ldots, p_n; \mu)$, which is a quasi-periodic time function and can be twice differentiated by all arguments. It is assumed that fundamental system

$$\dot{q}_s = \frac{\partial H_0}{\partial p_s}, \quad \dot{p}_s = -\frac{\partial H_0}{\partial q_s} \qquad (s = 1, \ldots, n),$$

$$H_0 = H(t;\ q_1, \ldots, q_n;\ p_1, \ldots, p_n;\ 0) \tag{8.11}$$

has a quasi-periodic fundamental solution

$$q_s = q_{s0}(t;\ a_1, \ldots, a_n;\ b_1, \ldots, b_n),$$

$$p_s = p_{s0}(t;\ a_1, \ldots, a_n;\ b_1, \ldots, b_n), \tag{8.12}$$

Here a_j and b_j are the initial values of variables q_s and p_s at $t = 0$. Then, by some sufficiently general assumptions, it is proved that the stable solution corresponds to points of rigorous minima and maxima of the averaged Lagrange function of the system

$$\Lambda_0(a_1, \ldots, a_n;\ b_1, \ldots, b_n) =$$

$$= \lim_{T \to \infty} \frac{1}{T} \int_0^T \left[\sum_{s=1}^{n} p_{s0}\dot{q}_{s0} - H(t;\ q_{10}, \ldots, q_{n0};\ p_{10}, \ldots, p_{n0};\ 0) \right] dt =$$

$$= \lim_{T \to \infty} \frac{1}{T} \int_0^T (L)\, dt, \tag{8.13}$$

which is computed on fundamental solutions. With this, solutions (8.12) can also be just periodic and synchronous. In such a case, averaging in (8.13) is performed during some finite period $T = 2\pi/\omega$. As parameters a_j and b_j constants can be taken which do not coincide with the initial conditions of variables q_j and p_j. It is also obvious that the result also remains when there is no small parameter in the system, that is, in the case when the fundamental system coincides with the original one ($H \equiv H_0$).

In paper (25) a periodic system somewhat more general than the canonic one (a system for maintaining a phase volume) was considered. It was shown that for the occurrence of stable periodic or synchronous motions it is necessary and sufficient that a function of initial values of phase variables $K(x_0)$ exist, which has a rigorous maximum or minimum relative to the said initial values, and that this function be correlated with some function of the phase coordinates and time $\chi(x, t)$ by an integral expression like (8.13). However, the issue of the method for finding functions χ and K remains open. According to the abovementioned paper (81), in the case of canonic systems such functions are L and Λ, respectively.

Finally, the integral criterion allows us to establish a tendency toward synchronization for dynamic objects with quite general assumptions about their properties, and on connection system properties (see § 9 of this chapter, as well as § 7, and subsection 4, § 8, chap. 3; § 2, chap. 7; § 7, chap. 12).

If the equilibrium conditions of the system are considered to be degenerated periodic motions of an arbitrary period, then function D, for some mechanical systems, becomes under certain conditions equal to the potential energy of the system. Thus, there is not only an analogy but a direct connection between the presented results and the Lagrange (Dirichlet) theorem.

§ 9) Existence of stable periodic or synchronous motions (tendency toward synchronization) in the case of periodic potential function and in some other cases.

In many cases (see chaps. 3, 12), potential function D is periodic in relation to each argument $\alpha_1, \ldots, \alpha_k$ with a certain period T. Periodicity can be established based on general considerations, without actual construction of a specific expression for $D(\alpha_1, \ldots, \alpha_k)$. In such cases, with sufficiently general assumptions, function D has minima in some points. If these minima turn out to be coarse (which also takes place at relatively general assumptions), then the stable periodic or synchronous motions definitely exist. Thus, there is a tendency toward synchronization in the system in the sense of § 1, chap. 1.

For example, let's consider the simplest case when D is a function of just one variable α. This case represents the case of the internal synchronization of two objects. It is assumed that D, being T-periodic by α, does not have intervals of constancy. Since it is continuous, function D reaches its minimum value inside the interval $0 \leq \alpha \leq T$. If this value occurs at the internal point α^* of this interval, then at the adopted assumption D would have a rigorous minimum in this point. If $(\vartheta^2 D/\vartheta\alpha^2)_{\alpha=\alpha^*} \neq 0$ (it is satisfied, "as a rule"), then this minimum would be a coarse one, and stable periodic or synchronous motions would certainly exist. The same conclusion is correct, due to the periodicity of function D, if the minimum values are achieved at the interval ends $\alpha = 0$ and $\alpha = T$. Indeed, values of D along the interval $0 \leq \alpha \leq T$ exhaust all sets of values of this function. Thus, if another interval $\varepsilon \leq \alpha \leq T + \varepsilon$ is taken, where ε is a sufficiently small number, the said minimum value would remain the same, but would be located inside this new interval and would be a rigorous minimum of D.

With some modifications, this extends to the case of multiple variables α_s. Here also with T-periodicity of function D by α_s "as a rule" coarse minima of function D would exist; thus there will be a tendency toward synchronization.

A potential function would be called *m*-coarse, if it has only coarse minima. Then it can be said that, in systems with *m*-coarse periodic potential function, a tendency toward synchronization takes place.

Due to the periodicity of function D with one minimum, this function will also have an infinite number of minima at values of arguments α_s differing by $n_s T$ from one another (n_s is an integer). However, these minima would correspond to the same periodic (synchronous) motion, and there is no sense in distinguishing between them.

The conclusion of the existence of a coarse minimum for the potential function remains if the latter is not a truly periodic function but the sum of an *m*-coarse periodic function and a function which does not vary too much inside a cube

$$0 \leqslant \alpha_s \leqslant T \qquad (s = 1, \ldots, k) \tag{9.1}$$

relative to the amount of variance of its truly periodic component. Thus, if

$$D = D(\alpha) = D_1(\alpha) + B(\alpha), \tag{9.2}$$

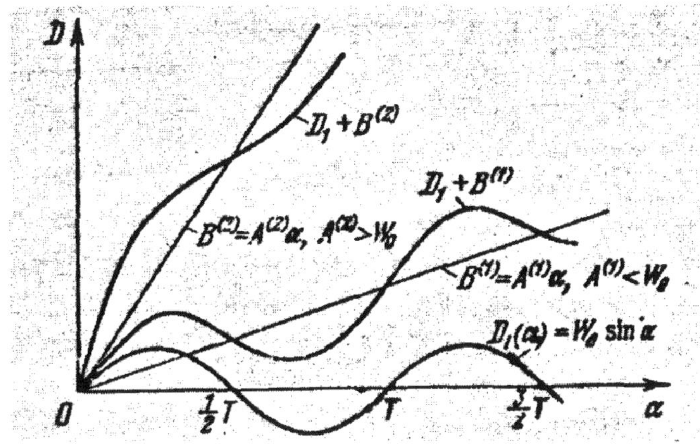

FIG. 52.

where $D_1(\alpha) = W_0 \sin \alpha$, $B = A\alpha$, then the conclusion of the existence of the coarse minima, correct for function $D_1(\alpha)$, would also remain correct for the function $D(\alpha)$ if (see Fig. 52)

$$|W_0| = \left(\frac{dD_1}{d\alpha}\right)_{\max} > |A| = \left|\frac{dB}{d\alpha}\right| \tag{9.3}$$

In such cases, it would be said that a *quasi-periodic m-coarse potential function exists.*

Moreover, let's assume that function D in a rigorous sense does not exist, but instead of (8.3), the following expressions hold

$$P_s(\alpha_1, \ldots, \alpha_k) = -\left[\frac{\partial D_0}{\partial \alpha_s} + A_s(\alpha_1, \ldots, \alpha_k)\right] \quad (s = 1, \ldots, k), \tag{9.4}$$

where functions $|A_s|$ are sufficiently small. Then, if function D_0 (which can be called a *quasi-potential function*) is periodic and m-coarse, the conclusion of the existence of stable synchronous solutions (i.e., solutions of equations $P_s = 0$), for which all the roots of equation (8.2) have negative real parts, remains correct. In other words, the existence of stable periodic or synchronous solutions (tendency toward synchronization) in the considered systems is, generally, a "coarse" property. It takes place both in the case of the existence of the m-coarse periodic potential function, and in cases of quasi-periodic potential and quasi-potential periodic functions.

From cases when the m-coarseness property of a periodic potential function D is not present, the most important practical ones are such, when all α_s (or, in an autonomous case, all differences $\alpha_s - \alpha_k$) exist determined from equations (8.1) or (8.4), i.e., from the stability conditions of function D. Equations (8.2) and (8.6) in such cases inevitably have some number (in an autonomous case more than one) zero roots, and the minimum of function D, if it exists, is nonrigorous in principle. In such cases, however, the stability conditions for function D and the conditions that all other roots of equation (8.6) are negative, give necessary conditions for the existence and stability of periodic or synchronous motions. To obtain sufficient conditions, the analysis of the next approximations is needed (see comment 8, § 7).

Chapter 11

SOME OTHER METHODS FOR SOLUTION OF PROBLEMS OF SYNCHRONIZATION OF WEAKLY COUPLED DYNAMIC SYSTEMS

§ 1) Introductory remarks.

This chapter describes some methods used for synchronization problems that are different from the Poincare/Lyapunov methods addressed in chap. 10. Most details given are for the methods used in this book, as well as for methods not yet described in available publications. For other methods, only their main concepts are described. Also a variational method by A. Lurie, which is illustrated in the example of the solution of one synchronization problem, is very briefly presented.

§ 2) Asymptotic methods and averaging principle.

Asymptotic methods, such as various versions of the so-called averaging principle, are an effective technique for solving problems of modern nonlinear vibration theory. These methods are described in (68–71, 157, 188, 195).

For analyses of the existing issues and for finding steady synchronous or periodic motions, asymptotic methods do not have any advantages over the Poincare method. However, these methods also allow us to study the processes for developing motions that are close to synchronous, and also some other steady motions, which is important for some cases.

The effectiveness of the asymptotic methods, as well as the Poincare/Lyapunov methods, is essentially correlated with the character of dependency of the right-hand parts of differential equations on the small parameter. The simplest case is one when differential equations, after appropriate transformations, can be presented in a so-called *standard* format,

$$\dot{x} = \mu X(x, t). \tag{2.1}$$

Here x and X are l-dimensional vectors, and $\mu \geq 0$. It is usually assumed that functions X are presented as

$$X(x, t) = \sum_{\nu} e^{i\nu t} X_{\nu}(x),$$

where ν are some real numbers. For equations like (2.1), N. Bogolubov has developed and substantiated an averaging scheme (68–71) consisting of substitution

$$x = \xi + \mu F_1(\xi, t) + \mu^2 F_2(\xi, t) + \ldots + \mu^m F_m(\xi, t), \tag{2.2}$$

by which equation (2.1) is reduced to the format

$$\dot{\xi} = \mu X_0(\xi) + \mu^2 P_2(\xi) + \ldots + \mu^m P_m(\xi) + \mu^{m+1} R(\xi, t). \tag{2.3}$$

Neglecting term $\mu^{m+1} R(\xi, t)$ in this equation, we arrive at an equation which is called an averaged equation of the m-th approximation, and corresponding to it, expression (2.2) is the m-th approximation. Functions $F_1(\xi,t)$, $F_2(\xi,t)$ are found as a result of substituting (2.2) into (2.1), and then expanding in a series by μ and equating terms at the same powers μ in both parts of the resulting expression. Functions

$X_0(\xi)$, $P_2(\xi)$, . . . are also determined by substitution of (2.2) into (2.1) and expansion by μ, but after averaging the right-hand side of this equation. Usually one is satisfied by determining one or two approximations. This is explained not only by the steep increase in the complexity of the correlations with the increasing number of approximation m, but also by the fact that usually the first or second approximation provides a solution of the required application problems. It should be remembered that the effectiveness of the asymptotic methods is determined not by the convergence properties of expression (2.2) at $m \longrightarrow \infty$, but by their asymptotic properties at $\mu \longrightarrow 0$ for this small preassigned m. The m-th approximation, according to the aforesaid, satisfies differential equation (2.1) with an accuracy up to terms of the order of μ^{m+1}.

The first approximation $x = \xi$ is determined from

$$\dot{\xi} = \mu X_0(\xi) \equiv \mu \underset{t}{M} \{X(\xi, t)\}, \tag{2.4}$$

where $\underset{t}{M}\{\ \}$ denotes an averaging operator by explicit time t. It is assumed that the average value exists. In the second approximation

$$x = \xi + \mu \sum_{\nu} \frac{e^{i\nu t}}{i\nu} X_s(\xi) \equiv \xi + \mu \widetilde{X}(\xi, t), \tag{2.5}$$

Here ~ denotes an integration by explicit time, with ξ being determined from the averaged equation of the second approximation

$$\dot{\xi} = \mu \underset{t}{M} \{X(\xi, t)\} + \mu^2 \underset{t}{M} \left\{\left(\widetilde{X} \frac{\partial}{\partial \xi}\right) X(\xi, t)\right\}. \tag{2.6}$$

Frequently, it is sufficient to find the so-called improved first approximation, in which x, which is the same as for the second approximation, is determined by (2.5), but ξ is found from the first approximation equation (2.4).

Thus, it can be said that equation (2.4) for the first approximation is obtained simply by averaging the right-hand side of the original equation (2.1) by the explicit time, and equation (2.6) for the second approximation is found by substitution in the right-hand side of expression (2.5) and by a similar averaging. It can be noted that equations (2.4) and (2.6) are, typically, easier to solve than the original equation (2.1).

Reduction of the equations for synchronization problems to the format of (2.1) is not always realizable. For example, synchronization problems for mechanical objects with rotational motions are such that the corresponding set of equations can be written in a more general format:

$$\dot{x} = \mu X(x, y, t, \mu), \quad \dot{y} = Y(x, y, t, \mu), \tag{2.7}$$

Here x and X are n-dimensional, and y and Y are m-dimensional vectors ($n + m = l$, $m \neq 0$). Equations like (2.2) are known as *systems with multidimensional fast and slow motions*. Variables x obviously describe "slow" and variables y, "fast" motions. For such systems a specific averaging routine (93, 94) was developed, generalizing the routine described before.

Asymptotic methods are described in (71, 94, 188). Using various modifications of these methods, some interesting, in a basic and application sense, problems were solved, such as synchronization of vacuum tube, transistor, and quantum generators (171, 181, 248), celestial bodies (23), mechanical vibro-exciters (148, 241, 242, 299, 300). A very general mathematical study of synchronization is presented in (109).

§ 3) Method of direct separation of motions.

Frequently, synchronous motions [see (1.6) in Chap. 1] and adjacent transient motions are of such a nature that they can be described as

$$x_j^{(s)} = \sigma_j^{(s)}\left[q_j^{(s)} n_j^{(s)} \omega(t)\, t + \alpha_j^{(s)}(t) + \tilde{y}_j^{(s)}\left(t, m_j^{(s)}\omega t\right)\right]$$
$$(j = 1, \ldots, r_s;\ s = 1, \ldots, k),$$
$$u_\rho = \sigma_\rho[q_\rho n_\rho \omega(t) t + \beta_\rho(t) + \tilde{v}_\rho(t, m_\rho \omega t)] \qquad (3.1)$$
$$(\rho = 1, \ldots, \nu),$$

Here ω, $\alpha_j^{(s)}$, β_ρ are "slow" (as a special case, stationary), and $\tilde{y}_j^{(s)}$, $\tilde{v}_\rho$ are"fast" oscillating components. Such a presentation is natural in many synchronization problems, when synchronous velocity or frequency ω is relatively high; thus independent variable t can be considered as "slow," and variable $\tau = \omega(t)t$, as "fast" time. In other words, in the studied cases it can be assumed that synchronous (or close to it) motion can be presented as a superposition of fast oscillations on the rotations of a slow motion.

In such an assumption, there is an effective method for synchronization problems which can be called a *method of direct separation of motions* (62, 65, 67). Conceptually, this method is close to the averaging principle and to the asymptotic methods. Since this method allows for a very convenient interpretation in mechanical problems, it will be described here in application to mechanical problems. The discussion will be made for a single degree-of-freedom (DOF) system. However, its generalization for multi-DOF systems would not create any difficulties.

It is assumed that equations of motion are as follows

$$m\ddot{x} = F(\dot{x}, x, t) + \Phi(\dot{x}, x, t, \omega t), \qquad (3.2)$$

Here x is a generalized coordinate; m is mass; t is "slow" and $\omega t = \tau$ is "fast" time ($\omega > 0$ is a "big" parameter); F is "slow" and Φ is "fast" force, having by $\tau = \omega t$ period 2π. The necessary assumptions in regard to smoothness of functions F and Φ will be clarified later.

It was noted that the basic precondition for the use of this method is the assumption that the analyzed "established by last time $\tau = \omega t$ " motion of the system has the format

$$x = X(t) + \psi(t, \omega t), \qquad (3.3)$$

where X is the slow, and ψ is the fast component. Function ψ is assumed here to be 2π-periodic by $\tau = \omega t$, and for certainty of presentation (3.3), it will be assumed to have the format of

$$\langle \psi(t, \tau)\rangle = 0 \qquad (3.4)$$

$\langle \ldots \rangle = \frac{1}{2\pi}\int_0^{2\pi} \ldots\, d\tau$ here and further denotes an averaging operation by the fast time $\tau = \omega t$. In other words, the average value of the fast component by the fast time will be assumed to be zero at the "frozen" slow time.

This method includes two steps. First, the original equation (2.2) (or the corresponding system of equations) is transformed into a set of integral differential equations of a "twice as high order" in relation to the slow and fast components X and ψ. Then, this system is approximately solved.

For the mentioned transformation of equation (3.2), expression (3.3) is substituted into the latter, and then the following expression is added to and substracted from its right-hand part:

$$F_1 - \langle \Phi(\dot{X}+\dot{\psi},\ X+\psi,\ t,\ \tau)\rangle, \tag{3.5}$$

Here

$$F_1 = F_1(\dot{X},\ X,\ \dot{\psi},\ \psi,\ t) = F(\dot{X}+\dot{\psi},\ X+\psi,\ t) - F(\dot{X},\ X,\ t) \tag{3.6}$$

denotes a fast-changing function which becomes zero at $\dot{\psi} = 0$, $\psi = 0$, and averaging $\Phi(\dot{X}+\dot{\psi}, X+\psi, t, \tau)$ is performed by the fast time τ which enters both explicitly and via function ψ.

Now fulfillment of the following equality between the terms of the resulting correlation will be required

$$\begin{aligned} m\ddot{\psi} = F_1(\dot{X},\ X,\ \dot{\psi},\ \psi,\ t) - \langle F_1(\dot{X},\ X,\ \dot{\psi},\ \psi,\ t)\rangle + \\ + \Phi(\dot{X}+\dot{\psi},\ X+\psi,\ t,\ \tau) - \langle \Phi(\dot{X}+\dot{\psi},\ X+\psi,\ t,\ \tau)\rangle. \end{aligned} \tag{3.7}$$

Then the following equation should be complied with

$$m\ddot{X} = F(\dot{X},\ X,\ t) + \langle F_1(\dot{X},\ X,\ \dot{\psi},\ \psi,\ t)\rangle + \langle \Phi(\dot{X}+\dot{\psi},\ X+\psi,\ t,\ \tau)\rangle. \tag{3.8}$$

The reason for such "splitting" of the original equation will be given below. Now it is only noted that the resulting set of integral-differential equations (3.7), (3.8) is equivalent to the original equation (3.2), independent of the assumption of the rate of change of functions X and ψ, at least in the sense that if some solution X and ψ of this system is found, then $x = X + \psi$ would be the solution of (3.2). It can also be noted that if a solution $\psi = \psi(\dot{X}, X, t, \tau)$ of equation (3.7) is known, then after its substitution into (3.8), we arrive at

$$m\ddot{X} = F(\dot{X},\ X,\ t) - W(\dot{X},\ X,\ t), \tag{3.9}$$

Here it is denoted

$$W(\dot{X},\ X,\ t) = -\langle \Phi(\dot{X}+\dot{\psi},\ X+\psi,\ t,\ \tau)\rangle - \langle F_1(\dot{X},\ X,\ \dot{\psi},\ \psi,\ t)\rangle. \tag{3.10}$$

Expression $W(\dot{X}, X, t)$, after P. Kapitsa (135), will be called *vibratory force.* Then equation (3.9) can be presented as an equation for the slow component X in which, together with the conventional slow force F, an additional slow force W (vibratory force) is present. Thus, it was shown that at quite general assumptions the statement is correct which is in some sense analogous to the known theorem of the dynamics of relative motion. That is: In the process of constructing equations of the relative motion, the addition of inertia forces is a "penalty" of sorts for using a noninertial coordinate frame, while in this case the addition of the vibratory force is a "penalty" for ignoring the fast force Φ and fast motion component ψ.

As was shown in subsection 6, § 8, chap. 3, and in subsection 2, § 6, chap. 12, use of the formulated statement is very convenient both in dealing with problems of the synchronization of objects with rotational motions, and for a physical interpretation of the results of their solution.

Vibratory force W is the result of averaging, by the fast time, the "proper fast force" Φ and that fast contribution, F_1, which is extracted from the slow force, F, on the trajectory of the motion system $x = X + \psi$. Accordingly, the *proper vibratory force* $W^{(s)} = -\langle\Phi\rangle$ and the *induced vibratory force* $W^{(i)} = -\langle F_1\rangle$ can be distinguished.

Coming now to the solution of the set of equations (3.7), (3.8), it can be noted that generally this is not more simple than the solution of the original equation (3.2). However, if the basic assumption about the rates of change of functions X and ψ is considered, then the following technique of approximate solution of this set seems to be a natural one. Initially, equation (3.7) is solved by parameters $\dot{X}$, X, and t, whose variation during the period of fast motion $2\pi/\omega$ is relatively small, being considered as constant ("frozen"). Let's assume that this equation indeed has, with "frozen" $\dot{X}$, X, and t, "fast developing" asymptotic stable 2π-periodic solution by ωt which complies with (3.4). It can be seen that the considered equation is created in just such a form during the "splitting" of the original equation (3.2) and that the necessary condition for the existence of said solution is satisfied. To see this, it is sufficient to integrate equation (3.7) term after term by ωt.

Substituting the resulting solution $\psi = \psi(\dot{X}, X, t, \omega t)$ into the right-hand side of equation (3.8) would lead to an equation like (3.9) for the slow component X. Now, however, this equation is only an approximate one.

The described approximate solution technique for the solution of the set (3.7), (3.8) allows justification in the spirit of the asymptotic methods. Namely, by introducing a small parameter $\mu = 1/\omega$ and a simple transformation of equations, they can be reduced to a set of equations like (2.7) with multidimensional fast and slow motions. Equation (3.9), thus obtained, corresponds to the first approximation of the averaging scheme which was discussed in § 2. This also solves the problem of computing the next approximations for functions ψ and X. However, this accuracy enhancement is not required, as a rule.

It is enough that function ψ enters expression for W under the integral sign. This allows us to limit the solution by determining the function ψ from equation (3.7), e.g., as a sum of a small number of harmonics or a small number of terms of a series by powers of the small parameter. Frequently, it can be considered that ψ is small in comparison with X ($\dot{X}$ is small in comparison with $\dot{\psi}$ per the original assumption). Also, in many cases we can consider only the linear terms in the expansion of function F_1 by powers $\dot{\psi}$ and ψ, by writing, according to (3.6),

$$F_1 = \left(\frac{\partial F}{\partial \dot{x}}\right)\dot{\psi} + \left(\frac{\partial F}{\partial x}\right)\psi, \tag{3.11}$$

where the derivatives are calculated at $\dot{\psi} = 0$ and $\psi = 0$. Then, according to (3.4), there is only proper vibratory force $W = W^{(s)} = -\langle\Phi\rangle$, and the induced vibratory force is absent. In other words, an induced component exists only if the slow force F is nonlinear by $\dot{x}$ and x. On the other hand, when the fast force Φ is absent in the original equation (2.2), the vibratory force can be nonzero because of its induced component. Such a situation is typical for the case of self-excited vibrations in systems with slow forces (see below).

This method of separation of motion can be considered a further development of the heuristic technique used by Kapitsa for the problem of a pendulum with a vibrating support (135). Some elements of this technique are present in (69), also for a pendulum problem, and in (292) for the concept of the enhancement of the stability of elastic systems by the application of vibrations. In these publications asymptotic methods are used, but terms corresponding to vibratory forces are present in the final equations. Later, the Kapitsa technique was used for the study of a particle drift in a standing sound wave (123), and for the study of the dynamics of the mechanisms on vibrating foundations (241, 242). The concept of vibratory forces (moments) was used in (36, 42, 57, 95, 133, 146) for interpreting the behavior of the various systems under vibratory excitations (see also chaps. 3 and 4). Some elements of this approach can be traced in older publications on turbulence, nonlinear acoustics, radioelectronics, and recent studies of vibrations in actively controlled nonlinear systems.

A full description of the method and some examples of its application are given in (62, 65, 67). Some general comments about this method follow:

1) An assumption has been made on the asymptotic stability and the fast rate of establishment of a periodic solution for (2.7) with frozen $\dot{X}$, X, and t. This seemingly strong assumption, however, is correct in many cases and, secondly, can be weakened. If this assumption is correct, some limitations related to the applicability of the averaging scheme of § 2 are automatically satisfied.

2) The separation of forces between the fast and slow ones is somewhat conditional, since an error would not occur if some or all slow forces are put in the fast group. Therefore, this should be done in doubtful cases.

3) Fast motion in a nonlinear system can develop even when the fast forces $\Phi(\dot{x}, x, t, \omega t)$ are absent. Specifics of such an "autonomous in relation to fast time" (or autonomous in a conventional sense) case lie in the fact that the period of fast motion $T = 2\pi/\omega$ is not known in advance and has to be determined in the solution process. In addition, as was already stated, in this case there is only the induced component of vibratory force $W = W^{(i)}$. Also, equation (3.7) would certainly allow for a trivial solution $\psi \equiv 0$ for the fast component, which corresponds to a zero vibratory force $W = W^{(i)}$. Thus, for an autonomous (at least, by the fast time ωt) system, such cases are especially important when equation (3.7) allows nontrivial, asymptotically stable solutions of a self-excited vibratory character.

4) After construction of equation (3.9), it is possible to perform an a posteriori control of the initial assumption of the separability of motions. This control is always necessary, since motions described by this equation might turn out to be fast, notwithstanding the fact that motions described by this equation at $W = 0$ were slow.

As follows from the aforesaid, the method of direct separation of motions in its part related to solution of the set of integro-differential equations (3.7), (3.8), is naturally combined with use of asymptotic methods. To find periodic solutions of equation (3.7) with "frozen" $\dot{X}$, X, and t can be used also perturbation method of Poincare/Lyapunov.

Application of the method of direct separation of motions for the analysis of solutions like (3.1), allows us to go from the original equations of motion for variables $x_j^{(s)}$ and u_ρ to simpler equations for the slow components of the process, ω, $\alpha_j^{(s)}(t)$, $\beta_\rho(t)$. Usually, just these slow components are of interest. In such cases if, as frequently happens, the phase coordinates of the connection system u_ρ are fast, or $\beta_\rho(t) = 0$, then the order of the set of equations (3.9) for determining $\alpha_j^{(s)}$ becomes lower (by ν) than the order of the original set (e.g., see subsection 6, § 8, chap. 3).

It should be noted that for steady motions ($\dot{\alpha}_j^{(s)} = 0$), the abovementioned equations for slow motion coincide with the equations for the synchronization problem [see equations (8.4) and (8.68), chap. 3] obtained by the Poincare method, from which parameters α_s in synchronous motions are determined. At sufficiently general assumptions, the appropriate stability conditions of synchronous motions are also obtained from equations for slow motion. At the same time, equations for slow motion (as well as the corresponding equations of the asymptotic methods) allow us to study not only steady synchronous regimes (as with the use of the Poincare method), but also the slow processes of the development of these regimes.

§ 4) Method of point imaging.

The concept of the method of point imaging in the format of the so-called *method of cutting surface (segment)* is due to Poincare (239, 325). It was widely used by D. Birkhof in the theory of dynamic systems. The use of the method of point imaging for problems of nonlinear mechanics is associated with A. Andronov, and its further development and application with Y. Neumark (218). Here the basic concepts of the method on the level of the method of cutting surface are described.

The motion of the system is considered as the motion of the imaging point in the phase space of this system. A smooth surface in this space, which is crossed in all its points by phase trajectories without contact, is called a *no-contact surface.* A *cutting surface* is a no-contact surface which is repeatedly crossed by phase trajectories both with increasing and decreasing time, in such a way that the time intervals between the subsequent crossings are limited.

Let the metrics of the phase space be n. Let's select a certain no-contact surface S in this space with the metrics $p \leq n - 1$ and attach to this surface a coordinate frame $y_1, \ldots, y_p$. Let's take a point M on the surface S, with coordinates $y_1, \ldots, y_p$, and consider a phase trajectory passing through this point (see Fig. 53). If, after some finite time interval, this trajectory again crosses the surface S at some point

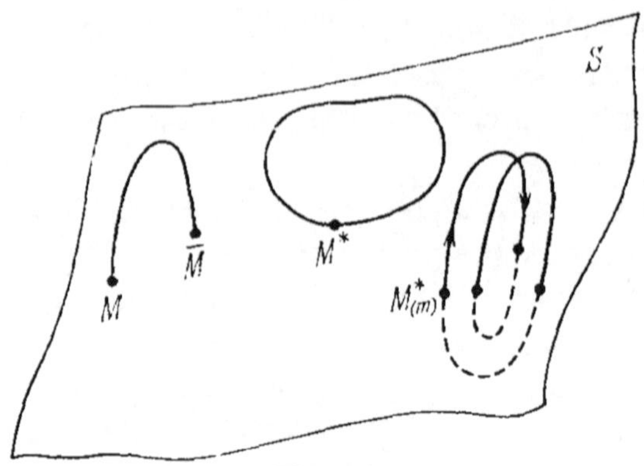

FIG. 53.

$\overline{M}$ with coordinates $\overline{y}_1, \ldots, \overline{y}_p$, the point is called the consequent point* for the point M. A transformation establishing a unique correspondence between all the points of surface S and their consequent points is called a *point image* of surface S, and is written as $\overline{M} = TM$, where T is the *operator of point imaging. A stationary*

* It may happen, of course, that the phase trajectory emerging from point M would never (in a finite time) cross the surface S. In such cases, it is said that point M does not have a consequent point.

point of the point image T is such a point M^* which reflects into itself, i.e., satisfies condition $M^* = TM^*$. Of course, there may exist so-called *m-times stationary points*, or points $M^*_{(m)}$, for which $M^*_{(m)} = T^m M^*_{(m)}$, but $T^k M^*_{(m)} \neq M^*_{(m)}$ if $k < m$. It is important that closed-phase trajectories, i.e., periodic motions of the system, correspond to the stationary points. Thus, the search for periodic regimes of motion of the system is reduced to finding the stationary points of the image if, of course, such an image is constructed. Moreover, knowledge of an accurate image allows us to reach a conclusion about the stability of periodic regimes.

Indeed, let m-times stationary point $M^*_{(m)}$ correspond to a periodic motion. A perturbed motion is considered, starting at some other point M_0 in the vicinity of point $M^*_{(m)}$. A consequent point for M_0 would be point $M_1 = T^m M_0$, and for point M_1, would be point $M_2 = T^m M_1 = T^{2m} M_0$, etc. For asymptotic stability of the considered periodic motion, it is necessary and sufficient that

$$\lim_{n\to\infty} M_n = \lim_{n\to\infty} T^{mn} M_0 = M^*_{(m)}.$$

The easiest application of this method is for a system with a two-dimensional phase space, when surface S is a planar curve and the position of the imaging point on this curve is given by one coordinate s (e.g., by the length of the arc counted from some reference point). In this case, the analysis is reduced to studying the point imaging of the segment into itself. Thus, the operator of the point imaging is a scalar function of one argument $\bar{s} = f(s)$, determining the correlation between the coordinate s of some point of the segment and coordinate $\bar{s}$ of its consequent point. This function is called the *function of following*. If function $\bar{s} = f(s)$ is presented as a plot (see Fig. 54), then its intersection points $s^*, s^{**}, \ldots$ with straight line $\bar{s} = s$ are stationary points of the point imaging. It is obvious that if $|d\bar{s}/(ds|_{s=s^*} = |df(s)/ds|_{s=s^*} < 1$, then stationary point s^* and the corresponding periodic motion are stable. In Fig. 54, stationary point s^* is stable and s^{**} is unstable. The graphic construction in this figure is called the *Koenigs-Lemerey diagram.* Point imaging was used for some application problems for the synchronization of the generators of electromagnetic oscillations (128, 129). These problems were studied as problems of entrainment. The point imaging method can also be used for more complicated multidimensional synchronization problems, and also as a means studying the general features of the synchronization in dynamic systems [e.g., see (77, 217, 218)].

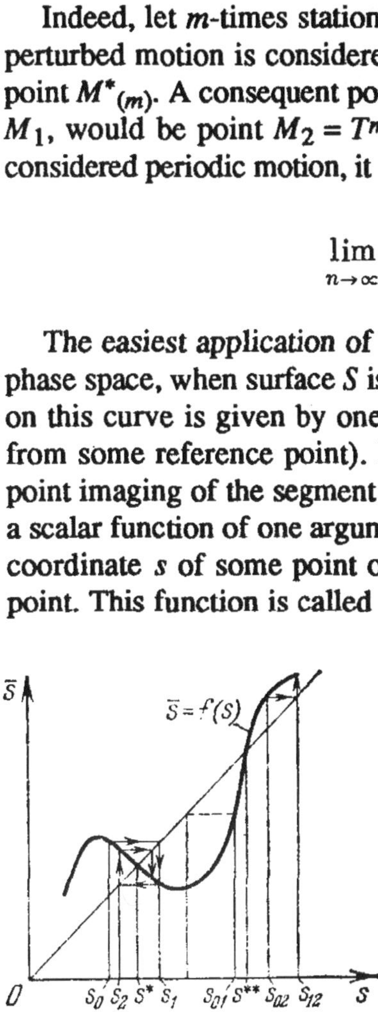

FIG. 54

§ 5) Stroboscopic method by N. Minorsky.

The stroboscopic method suggested by Minorsky (321) is based on ideas which are close both to the ideas of asymptotic methods and the methods of separation of motions and to the method of point imaging. These ideas are as follows. Let's assume an analysis of an oscillatory process that is near-periodic with some (may not be known in advance) period T. The process is observed in the phase space of the system as if it were in stroboscopic light, i.e., at discrete moments of time separated by time intervals T. If the process were strictly T-periodic, then imaging (phase) point M (called a stroboscopic point) would seem stationary. If the process is close to an asymptotically stable T-periodic one, then this point would be observed as "slowly" moving toward point M^*, corresponding to the strictly periodic process (see Fig. 55). It is natural to expect that if equations could be obtained which describe the trajectory not of the studied motion itself but of the stroboscopic point (*stroboscopic equations*), then these equations would be simpler than the original one, and also would allow us to study the motion characteristics which are usually of primary interest. Such stroboscopic equations are actually possible to construct, at least in cases when the original differential equations are close to the ones which can be fully integrated, for example, to linear equations with linear coefficients. It is important in this case that, instead of a nonautonomous system with T-periodic right-hand parts, an autonomous system be obtained.

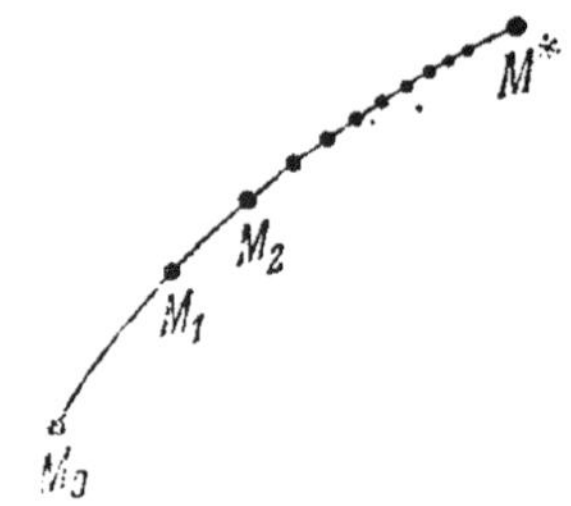

FIG. 55.

Let's consider the system

$$\dot{x} = X(x, y, t, \mu), \quad \dot{y} = Y(x, y, t, \mu), \tag{5.1}$$

This system is assumed to be close to a linear autonomous one, with the small parameter $\mu \geq 0$ characterizing the degree of closeness, and X and Y being T-periodic functions of t. By introducing new variables

$$\rho = x^2 + y^2, \quad \psi = \operatorname{arctg}(y/x) \tag{5.2}$$

it is frequently possible to transform this system into

$$\dot{\rho} = \mu f(\rho, \psi, t), \quad \dot{\psi} = -1 + \mu g(\rho, \psi, t), \tag{5.3}$$

Here f and g are periodic functions of t with period T. At $\mu = 0$, the imaging point $\rho = \rho_0$, $\psi = -t + \psi_0$, is moving with constant velocity, clockwise, around the circumference of radius ρ_0 (ρ_0 and ψ_0 are constants determined by the initial parameters).

At $\mu \neq 0$, the solution by integer powers of μ in series is

$$\rho(t) = \rho_0 + \mu\rho_1(t) + \ldots, \quad \psi(t) = \psi_0 - t + \mu\psi_1(t) + \ldots, \tag{5.4}$$

where

$$\begin{aligned} \rho_1(t) &= \int_0^t f(\rho_0, \psi_0 - t_1, t_1)\, dt_1, \\ \psi_1(t) &= \int_0^t g(\rho_0, \psi_0 - t_1, t_1)\, dt_1. \end{aligned} \tag{5.5}$$

From here

$$\rho(T) = \rho_0 + \mu T K(\rho_0, \psi_0), \quad \psi(T) = \psi_0 - T + \mu T L(\rho_0, \psi_0), \tag{5.6}$$

with

$$K(\rho_0, \psi_0) = \frac{1}{T}\int_0^T f(\rho_0, \psi_0 - t, t)\,dt,$$
$$L(\rho_0, \psi_0) = \frac{1}{T}\int_0^T g(\rho_0, \psi_0 - t, t)\,dt. \tag{5.7}$$

Thus, displacement of the stroboscopic point during period T will be

$$\Delta\rho = \rho(T) - \rho_0 = \mu K(\rho_0, \psi_0)T,$$
$$\Delta\psi = \psi(T) - (\psi_0 - T) = \mu L(\rho_0, \psi_0)T. \tag{5.8}$$

If an increment $\Delta\tau = T$ of "stroboscopic time" were introduced, then the latter equations, which are essentially equations in finite differences, would correspond to the following stroboscopic differential equations

$$\frac{d\rho}{d\tau} = \mu K(\rho, \psi), \quad \frac{d\psi}{d\tau} = \mu L(\rho, \psi), \tag{5.9}$$

These equations are already autonomous. These stroboscopic equations coincide with equations obtainable by the averaging method in an appropriate approximation. It can also be noted that consideration of the points separated by time intervals T is, essentially, equivalent to an analysis of the corresponding point image.

Using the stroboscopic method, Minorsky analyzed the problem of the mutual synchronization of two pendulum clocks, which were idealized as inductively coupled generators (van-der-Pol autogenerators) (194). It can be expected that the geometric images correlated to the concept of this method would also find application for more complex synchronization problems.

§ 6) Variational method by A. Lurie.

The basic idea of this method consists of using the Hamilton variational principle in conjunction with the perturbation method and the Galerkin method for problems of the synchronization of weakly coupled mechanical objects. This method was used by Lurie for problems of the synchronization of quasi-linear oscillators and unbalanced rotors (mechanical vibro-exciters) which are attached to a system of solid bodies elastically connected between themselves and the stationary supporting structure (175). The general part of this study is described in subsection 2 of § 6, chap. 12, and the results related to vibro-exciters are given in subsection 6 of § 8, chap. 3. Thus only the basic main points are described below.

The Hamilton variational principle is written as follows

$$\delta\Lambda + \frac{2\pi}{\omega}\int_0^{2\pi/\omega} \delta' A\,dt = 0, \tag{6.1}$$

where

$$\Lambda = \frac{2\pi}{\omega}\int_0^{2\pi/\omega} L\,dt = \frac{2\pi}{\omega}\int_0^{2\pi/\omega} (T - \Pi)\,dt \tag{6.2}$$

is average during the period value $T_0 = 2\pi/\omega$ of the Lagrange function (T is the kinetic and Π the potential energy), and $\delta' A$ is the incremental work of the nonconservative forces.

Then, for derivation of the principal variational correlation from (6.1), it is considered that the Lagrange function for a synchronization problem for weakly coupled objects can be presented as

$$L = \sum_{s=1}^{k} L_s + \mu\left(\tilde{L}^{(\mathrm{I})} + \tilde{L}^*\right), \tag{6.3}$$

Here L_s is the Lagrange function of the objects at the stationary supporting system of the bodies; $L^{(\mathrm{I})} = \mu\tilde{L}^{(\mathrm{I})}$ is the Lagrange function of the supporting system with objects attached to it in a motionless way; $L^* = \mu\tilde{L}^*$ is the Lagrange function that relates to interaction between the objects and the supporting system; and μ is, as before, a small positive parameter. In the equation, small terms are also separate for $\delta' A$.

In using the Galerkin algorithm, a solution is sought in the format of the fundamental solution of the problem of the synchronization of weakly coupled objects [see formula (1.3), chap. 2]. Parameters $\alpha_1, \ldots, \alpha_k$ are used as unknown (being varied) parameters to be determined by using the variational principle.

As Lurie pointed out, if parameters $\alpha_1, \ldots, \alpha_k$ are considered constant, then the equations for their determination follow from the basic variational correlation. These equations are exactly the same as the basic equations for the same synchronization problem—see equations (2.1), chap. 2, which were obtained by means of the perturbation method. If parameters $\alpha_1, \ldots, \alpha_k$ are considered as slowly varying time functions, then the same correlation leads to the approximate differential equations that describe a slow process of developing and establishing the synchronous regime. These equations, for example, lead to the same stability conditions for synchronous regimes as the ones which can be obtained for the same class of problems by the Poincare/Lyapunov methods.

Some components of this method were used before for other problems [e.g., see (153), (239)]. However, here a judicious combination of these components is important. A similar method was used in (265) for the problem of wave propagation in nonlinear media without energy dissipation. Lurie himself considers this method an heuristic technique that does not pretend to be rigorous. However, the coincidence of the solutions obtained by this method and by other "rigorous" methods (see subsection 2 of § 6, chap. 12), definitely speaks in its favor. It could be said that this method would find wide application for studies of periodic and synchronous motions in complex dynamic systems.

Chapter 12
SYNCHRONIZATION OF SOME BASIC CLASSES OF DYNAMIC OBJECTS

§ 1) Preliminary remarks.

This chapter considers synchronization of some basic weakly coupled dynamic objects.

For each class of object, expressions for the fundamental functions $P_s(\alpha_1, \ldots, \alpha_k)$ are given, as are (without derivation, as a rule) basic equations for the problem $P_s = 0$. Also, appropriate stability conditions and expressions for the potential function D, which is used in the formulation of the integral stability criterion (extremal property) for synchronous motions, are presented. Some of the considered classes are "mutually overlapping," or some objects could belong to two or more classes.

§ 2) General case of weakly coupled objects.*

1. External synchronization problem. Let the system whose structure is shown in Fig. 4 be described by general differential equations of the problem of synchronization of weakly coupled object [see equations (2.1), chap. 1]:

$$\begin{gathered}\dot{x}_j^{(s)} = X_j^{(s)}(x_1^{(s)}, \ldots, x_{r_s}^{(s)}) + \\ + \mu F_j^{(s)}(x_1^{(1)}, \ldots, x_{r_1}^{(1)}; \ldots; x_1^{(k)}, \ldots, x_{r_k}^{(k)}; u_1, \ldots, u_\nu; \omega t, \mu) \\ (j = 1, \ldots, r_s;\ s = 1, \ldots; k), \\ \dot{u}_\rho = U_\rho(x_1^{(1)}, \ldots, x_{r_1}^{(1)}; \ldots; x_1^{(k)}, \ldots, x_{r_k}^{(k)}; u_1, \ldots, u_\nu; t, \mu) \\ (\rho = 1, \ldots, \nu).\end{gathered} \tag{2.1}$$

These equations are written here in a scalar format, and the insignificant indexes ~ and * at coupling functions $F_j^{(s)}$, U_ρ are omitted. It is assumed that these functions explicitly depend on time and have period $T = 2\pi/\omega$ by this argument. Let, further, a fundamental system

$$\begin{gathered}\dot{x}_{j0} = X_j^{(s)}(x_{10}^{(s)}, \ldots, x_{r_s 0}^{(s)}) \qquad (j = 1, \ldots, r_s;\ s = 1, \ldots, k), \\ \dot{u}_{\rho 0} = U_\rho(x_{10}^{(1)}, \ldots, x_{r10}^{(1)}; \ldots; x_{10}^{(k)}, \ldots, x_{r_k 0}^{(k)}; u_{10}, \ldots, u_{\nu 0}; \omega t, 0) \\ (\rho = 1, \ldots, \nu)\end{gathered} \tag{2.2}$$

have a synchronous solution

$$x_{j0}^{(s)} = \varphi_{j0}^{(s)}(t + \alpha_s), \quad u_{\rho 0} = u_{\rho 0}(t, \alpha_1, \ldots, \alpha_k), \tag{2.3}$$

This is a solution with a format as characterized by expressions (1.3), chap. 2.

Equations in variations, corresponding to set (2.2) and to solution (2.3), are as follows:

$$\begin{gathered}\dot{z}_j^{(s)} = p_{j1}^{(s)}(t + \alpha_s) z_1^{(s)} + \ldots + p_{jr_s}^{(s)}(t + \alpha_s) z_{r_s}^{(s)} \\ (j = 1, \ldots, r_s;\ s = 1, \ldots, k),\end{gathered} \tag{2.4}$$

$$\dot{z}_\rho = q_{\rho 1} z_1 + \ldots + q_{\rho\nu} z_\nu + \sum_{m=1}^{k} \left[q_{\rho 1}^{(m)} z_1^{(m)} + \ldots + q_{\rho r_m}^{(m)} z_{r_m}^{(m)} \right],$$

*See (57, 207).

Here

$$p_{j\beta}^{(s)}(t) = \frac{\partial X_j^{(s)}}{\partial \varphi_\beta^{(s)}(t)},$$

$$q_{\rho h} = \left(\frac{\partial U_\rho}{\partial u_h}\right), \quad q_{\rho\delta}^{(m)} = \left(\frac{\partial U_\rho}{\partial x_\delta^{(m)}}\right) \tag{2.5}$$

$$(j, \beta, \delta = 1, \ldots, r_s; \ \rho, \ h = 1, \ldots, \nu; \ s = 1, \ldots, k),$$

As before, the parentheses around the derivatives of $X_j^{(s)}$ and U_ρ mean that these derivatives are calculated for the fundamental solution.

Let's assume that the equations in variations (2.4) allow for exactly k periodic solutions (δ_{sj} is the Kronecker symbol)

$$z_{j\gamma}^{(s)} = \dot{\varphi}_{j0}^{(s)}(t + \alpha_s)\,\delta_{\gamma s}, \quad z_{\rho\gamma} = \frac{\partial u_\rho}{\partial \alpha_\gamma}$$
$$(j = 1, \ldots, r_s; \ s, \gamma = 1, \ldots, k), \tag{2.6}$$

These solutions are obtained, according to the Poincare theorem (see § 1, chap. 10), by sequential differentiating of functions (2.3) by $\alpha_1, \ldots, \alpha_k$. Let all other solutions of the system in variations, independent of (2.6), be infinitely diminishing at $t \longrightarrow \infty$. Then theorem of § 4, chap. 10, considering comment 4, § 7, of the same chapter, can be applied to this system. For this purpose, a system conjugate in relation to the system in variations (2.4) is considered. The last ν equations of this conjugate system are as follows:

$$\dot{\overset{*}{z}}_\rho + q_{1\rho}\overset{*}{z}_1 + \ldots + q_{\nu\rho}\overset{*}{z}_\nu = 0 \qquad (\rho = 1, \ldots, \nu) \tag{2.7}$$

and thus contain only the variables corresponding to the coordinates of the system connecting the objects. This fact is not accidental but is typical for all synchronization problems for weakly coupled objects. Due to the assumptions made about the character of the solutions for system (2.4), system (2.7) allows only a trivial T-periodic solution $z_\rho^* \equiv 0$. Therefore, a T-periodic solution of the system which is conjugate relative to the system in variations, should be sought from the equations sets

$$\dot{\overset{*(s)}{z_j}} + p_{1j}^{(s)}(t + \alpha_s)\,\overset{*(s)}{z_1} + \ldots + p_{r_s j}^{(s)}(t + \alpha_s)\,\overset{*(s)}{z_{r_s}} = 0 \tag{2.8}$$
$$(j = 1, \ldots, r_s; \ s = 1, \ldots, k).$$

which correspond to coordinates of the objects and are independent for each object. Due to the same assumptions about the character of the solutions of the system in variations (2.4), each of the subsystems (2.8) has only unique T-periodic solution, denoted by $\eta_j^{(s)}(t + \alpha_s)$. Then system (2.8), as a whole, has k linearly independent T-periodic solutions

$$\overset{*(s)}{z_{j\beta}} = \eta_j^{(s)}(t + \alpha_s)\,\delta_{s\beta} \qquad (\beta = 1, \ldots, k). \tag{2.9}$$

Therefore, and also because $\overset{*}{z}_{\rho\beta} \equiv 0$ ($\rho = 1, \ldots, \nu$), expressions (4.9) of chap. 10 are reduced to conditions

$$\sum_{j=1}^{r_s} \dot{\varphi}_{j0}^{(s)}(t)\,\eta_j^{(s)}(t) = 1 \qquad (s = 1, \ldots, k), \tag{2.10}$$

to which the selection of periodic solutions $\eta_j^{(\beta)}(t)$ should be subordinated.

If now it is considered that summation by index s from 1 to l, in the notation of § 4, chap. 10, here corresponds with summation by j from 1 to r_s, and then by index s from 1 to k, and then, finally, by ρ from 1 to ν, then the application of the theorem of the said paragraph would lead to the following *basic equations* for determining parameters $\alpha_1, \ldots, \alpha_k$.:

$$P_s(\alpha_1, \ldots, \alpha_k) \equiv \sum_{j=1}^{r_s} \langle (F_j^{(s)}) \eta_j^{(s)} (t + \alpha_s) \rangle = 0 \qquad (s = 1, \ldots, k). \tag{2.11}$$

Stability conditions for the respective synchronous motion thus are reduced to the requirement that the real parts be negative for all the roots of the following algebraic equation*

$$\left| \frac{\partial P_s}{\partial \alpha_j} - \delta_{sj} \varkappa \right| = 0 \qquad (s, j = 1, \ldots, k). \tag{2.12}$$

2. Internal synchronization problem. Let's assume that for the same conditions as above, the system is autonomous, i.e., it is described by equations (2.1) whose right-hand parts, however, do not explicitly depend on t. If period $T = T_0$ of the fundamental solution is not known in advance and is determined (a frequent practice) by equating to zero the first correction $\delta(\mu)$ to the period of the solution being sought(see comment 5, § 7, chap. 10), then one phase α_s, e.g., α_k, can be assumed to be zero, and phases $\alpha_1, \ldots, \alpha_{k-1}$ and period T_0 are calculated from the same set of basic equations. However, the stability condition for this solution would now require a negative sign of only the $(k-1)$th root χ of equation (2.12). In this case, one root of this equation is certainly equal to zero; its presence does not influence the stability judgment. A case with the period of the fundamental solution known in the internal synchronization problem is considered in (57, 207).

§ 3) Quasi-linear single DOF objects (quasi-linear oscillators).

Equations of motion for the considered system in the nonautonomous case are

$$\begin{aligned} \ddot{x}^{(s)} + a_s \dot{x}^{(s)} + \omega_s^2 x^{(s)} &= f_s(\omega t) + \mu F_s(x, x, u, \omega t, \mu) \\ &\qquad (s = 1, \ldots, k), \\ \dot{u}_\rho = U_\rho(\dot{x}, x, u, \omega t, \mu) &\qquad (\rho = 1, \ldots, \nu), \end{aligned} \tag{3.1}$$

Here $x^{(s)}$ are scalar generalized coordinates of the objects; u_ρ are phase coordinates of the connection system; a_s and ω_s^2 are nonnegative constants; f_s, F_s, and U_ρ are assumed to be $2\pi/\omega$-periodic functions of t; and x and u represent the whole set of all variables $x^{(s)}$ and u_ρ, respectively. To such systems, the most simple analysis of many problems of the synchronization of weakly coupled systems are reduced in mechanics, electronics, radiophysics, and biology (see chaps. 5, 6, and 8).

Solution results are substantially dependent on the character of the fundamental system, as well as on the properties of the set of equations in variations

*For the notation, see the footnote on p. 196.

$$\ddot{z}^{(s)} + a_s\dot{z}^{(s)} + \omega_s^2 z^{(s)} = 0 \qquad (s = 1, \ldots, k),$$

$$\dot{z}_\rho = \sum_{j=1}^{k} [p_{\rho j}(t)\, z^{(j)} + q_{\rho j}(t)\, \dot{z}^{(j)}] + \sum_{f=1}^{\nu} r_{\rho f}(t)\, z_f,$$

$$\left[p_{\rho j} = \left(\frac{\partial U_\rho}{\partial x^{(j)}}\right), \quad q_{\rho j} = \left(\frac{\partial U_\rho}{\partial \dot{x}^{(j)}}\right), \quad r_{\rho f} = \left(\frac{\partial U_\rho}{\partial u_f}\right)\right], \tag{3.2}$$

which is derived for the fundamental system and fundamental synchronous solution. The system consists of k independent linear equations with constant coefficients, and of a set of ν equations, the behavior of whose solutions at $t \longrightarrow \infty$ coincides with the behavior of the solutions for the simpler system

$$\dot{z}'_\rho = \sum_{f=1}^{\nu} r_{\rho f}(t)\, z'_f \qquad (\rho = 1, \ldots, \nu). \tag{3.3}$$

If it is assumed that all the solutions of this system diminish unlimitedly at $t \longrightarrow \infty$, then the character of the synchronization problem solution would depend mainly on coefficients a_s and ω_s^2 in the equations for the objects.

The highest application interest give such cases when coefficient a_s is either small or equal to zero, and coefficient ω_s^2 is not small (quasi-conservative quasi-linear oscillator), or when ω_s^2 is zero (object with a quasi-uniform rotation), in which case the corresponding coefficient can be either small (quasi-conservative idealization) or not small (non-quasi-conservative idealization). Corresponding cases in a somewhat specified format are embraced by the cases considered in §§ 4 and 5, and also in chaps. 3–6.

The general analysis is given in (57). It can also be noted that the solution of many synchronization problems for quasi-linear objects is reduced to the application of theorems from §§ 2 and 3, chap. 10.

Many publications such as (6, 179, 222, 305, 319, 334, 335) treat entrainment and synchronization of quasi-linear quasi-conservative oscillators in application to radioelectronics problems. Some general results of the studies of the synchronization of the quasi-linear, quasi-conservative oscillators are given at the end of § 5.

§ 4) Systems with quasi-uniform rotational motions.*

In many synchronization problems for mechanical systems, the generalized coordinates can be chosen such that a part of them $\varphi_1, \ldots, \varphi_k$ are rotational and vary close to a uniform rotation in synchronous motions, with couplings between the respective degrees of freedom which could be considered weak. In other words, synchronous motions of such systems are as follows

$$\begin{aligned} \varphi_s &= \sigma_s[n_s\omega t + \alpha_s + \mu\psi_s(\omega t)] \qquad (s = 1, \ldots, k), \\ u_r &= u_r(\omega t) \qquad (r = 1, \ldots, \nu), \end{aligned} \tag{4.1}$$

Here, $\sigma_s = \pm 1$; n_s are mutually prime integers; α_s are constants; and ψ_s and u_r are periodic functions of t with period $2\pi/\omega$, and it can be assumed that $\langle\psi_s(\omega t)\rangle = 0$. This model represents systems such as many vibratory machines and installations; flexible shafts with unbalanced discs; devices for dynamic balancing of unbalanced

*See (41–43, 57).

rotors; electromechanic systems with synchronous machines operating in parallel; and some systems in celestial mechanics (see Part 2).

In accordance with the assumptions about the character of the considered synchronous motion, it can be assumed that, at least in the vicinity of this motion, Lagrange equations can be written as

$$I_s\ddot{\varphi}_s + k_s(\dot{\varphi}_s - \sigma_s n_s\omega) = \mu\Phi_s \qquad (s = 1, \ldots, k), \tag{4.2}$$

$$\mathscr{E}_{u_r}(L) = Q_r^{(0)} + \mu Q_r^{(1)} \quad (r = 1, \ldots, \nu), \tag{4.3}$$

Here $L = T - \Pi$ is a Lagrange function;

$$\mu\Phi_s = I_s\ddot{\varphi}_s + k_s(\dot{\varphi}_s - \sigma_s n_s\omega) - \mathscr{E}_{\varphi_s}(L) + Q_{\varphi_s}^{(1)},$$
$$\mu Q_r^{(1)} = Q_r - Q_r^{(0)}, \tag{4.4}$$

$\mathscr{E}_q = \frac{d}{dt}\frac{\partial}{\partial\dot{q}} - \frac{\partial}{\partial q}$ is a Euler operator corresponding to generalized coordinate q; I_s and k_s are positive constants; and $Q_{\varphi s}^{(1)}$ and $Q_r^{(0)} + \mu Q_r^{(1)}$ are nonconservative generalized forces. It is considered that functions L, $Q_{\varphi s}^{(1)}$, and Q_r may be $2\pi/\omega$-periodic functions of t, and that their dependence on φ_s and t is such that after substitution of expressions (4.1) they become (or remain) $2\pi/\omega$-periodic functions of t.

The fundamental system corresponding to (4.2) and (4.3) would have, for rotational coordinates, a set of solutions

$$\varphi_s^0 = \sigma_s(n_s\omega t + \alpha_s) \qquad (s = 1, \ldots, k), \tag{4.5}$$

which depends on k arbitrary parameters $\alpha_1, \ldots, \alpha_k$. Let

$$u_r^0 = u_r^0(\omega t, \alpha_1, \ldots, \alpha_k) \qquad (r = 1, \ldots, \nu) \tag{4.5'}$$

be the respective $2\pi/\omega$-periodic solution of the fundamental equations for (4.3), which is assumed to exist and be asymptotically stable for all the considered values of $\alpha_1, \ldots, \alpha_k$.

For such assumptions, using the theorem from § 2, chap. 10, would easily lead to the following basic equations of the problem [see also (57)]

$$P_s(\alpha_1, \ldots, \alpha_k) \equiv \frac{\sigma_s\mu}{k_s}\langle(\Phi_s)\rangle \equiv \frac{1}{k_s}\left(\frac{\partial\Lambda}{\partial\alpha_s} + A_s\right) \equiv$$
$$\equiv \frac{1}{k_s}\left(\frac{\partial\Lambda_0}{\partial\alpha_s} + A_s\right) = 0 \qquad (s = 1, \ldots, k), \tag{4.6}$$

Here it is denoted

$$\Lambda = \Lambda(\alpha_1, \ldots, \alpha_k) = \langle(L)\rangle, \quad \Lambda_0 = \langle(L_0)\rangle,$$
$$L_0 = L - \sum_{s=1}^{k} L_s, \quad A_s = \sigma_s\langle(Q_{\varphi_s}^{(1)})\rangle + \sum_{r=1}^{\nu}\left\langle(Q_r^{(0)})\frac{\partial u_r^0}{\partial\alpha_s}\right\rangle. \tag{4.7}$$

Parameters $L_s = L_s(\dot{\varphi}_s, \varphi_s)$ are "proper" Lagrange functions, and $\langle(L_s)\rangle = 0$ is due to (4.5). Equations (4.6) can be treated as equations of equilibrium of the average moments, or as equations of budget of energy supplied to the s-th object and used by the object. Derivatives $\partial\Lambda/\partial\alpha_s = \partial\Lambda_0/\partial\alpha_s$ have a physical meaning of average moments which frequently are not correlated with the supply and loss of energy in the system, but only characterize redistribution of energy between the objects, which is necessary for synchronization. This happens, for example, if Lagrange function L of the system does not depend explicitly on t; thus replacement of ωt by $\omega t + \alpha_0$ in solutions (4.5) should not change values Λ, i.e.,

$$\Lambda(\alpha_1 + n_1\alpha_0, \ldots, \alpha_k + n_k\alpha_0) \equiv \Lambda(\alpha_1, \ldots, \alpha_k).$$

Differentiating this identity by α_0, we arrive at

$$\sum_{s=1}^{k} n_s \frac{\partial\Lambda}{\partial\alpha_s} = 0; \tag{4.8}$$

and then, adding up equations (4.6) and multipling by $n_s k_s$, we arrive at

$$\sum_{s=1}^{k} n_s A_s = 0, \tag{4.9}$$

This is an equation of energy budget in the system (in the fundamental approximation), which serves in the external synchronization problem to determine the initial approximation of synchronous frequency ω. Function $\Lambda(\alpha_1, \ldots, \alpha_k)$, because of the natural assumptions about the character of the dependence of the Lagrange function L on variables φ_s, is 2π-periodic by each of the phases $\alpha_1, \ldots, \alpha_k$. This fact, along with functions A_s being sufficiently small and compliance with quite general requirements, assures the tendency toward synchronization in the considered systems (see § 7).

Parameter A_s, which can be called the *excess moment (torque) of the s-th object,* is an average moment of nonconservative forces reduced to coordinates φ_s. It is a difference between the supply and loss of energy during the synchronous motion of the s-th object. Only this difference is compensated by moment $\partial\Lambda_0/\partial\alpha_s$.

In the internal synchronization problem for self-exciting objects at

$$\sum_{r=1}^{\nu} \left\langle (Q_r^{(0)}) \frac{\partial u_r^0}{\partial\alpha_s} \right\rangle = 0, \quad \text{[w]}$$

when the presence of nonconservative forces along vibratory coordinates does not cause energy dissipation, equations

$$A_s \big|_{u_1^0 = \ldots = u_\nu^0 = 0} \equiv \sigma_s \langle (Q_{\varphi_s}^{(1)}) \rangle = 0 \qquad (s = 1, \ldots, k) \tag{4.10}$$

are equilibrium equations for the averaged nonconservative forces, or equations of energy budget inside uncoupled objects. Usually these equations are used to determine the *partial frequencies* ω_s *of the objects,* i.e., the frequencies (average angular velocities) which the objects would have in an isolated condition (but, of course, considering "proper" nonconservative forces $\mu Q_{\varphi s}^{(1)}$).

One of the most important features of the mutual synchronization phenomenon is the fact that it can occur even at a significant difference between frequencies $n_s\omega$ and partial frequencies ω_s (see § 3 of the introduction, and §§ 5, 6, and 14 of chap. 3).

If a function $B = B\ (\alpha_1, \ldots, \alpha_k)$ exists such that

$$\partial B/\partial \alpha_s = A_s \tag{4.11}$$

which will be called *a potential of excess averaged nonconservative forces,** then the potential function can be expressed as

$$D = -(\Lambda + B). \tag{4.12}$$

If the Lagrange function can be presented as

$$L = L^* + L^{(I)} + L^{(II)},$$

$$L^* = \frac{1}{2}\sum_{s=1}^{k}\sum_{j=1}^{k} d_{sj}\dot{\varphi}_s\dot{\varphi}_j + \sum_{r=1}^{\nu} f_r(\dot{\varphi}_1, \ldots, \dot{\varphi}_k; \varphi_1, \ldots, \varphi_k)\,\dot{u}_r + \sum_{s=1}^{k} F_s(\varphi_s),$$

$$L^{(I)} = \frac{1}{2}\sum_{r=1}^{\nu}\sum_{j=1}^{\nu} a_{rj}\dot{u}_r\dot{u}_j - \frac{1}{2}\sum_{r=1}^{\nu}\sum_{j=1}^{\nu} b_{rj}u_r u_j, \tag{4.13}$$

$$L^{(II)} = \Psi(\dot{\varphi}_1, \ldots, \dot{\varphi}_k; \varphi_1, \ldots, \varphi_k),$$

where a_{rj}, b_{rj}, d_{rj} are constants, and f_r, F_r and Ψ are functions of the listed arguments, with f_r and F_r being periodic by φ_s with period 2π, and also $Q_r^{(0)} = 0$ ($r = 1, \ldots, \nu$), then the following correlation exists

$$\frac{\partial \Lambda}{\partial \alpha_s} = \frac{\partial(\Lambda^{(II)} - \Lambda^{(I)})}{\partial \alpha_s} \quad \left(\Lambda^{(I)} = \langle(L^{(I)})\rangle, \;\; \Lambda^{(II)} = \langle(L^{(II)})\rangle\right) \tag{4.14}$$

In this case, the potential function can be presented in one of two formats:

$$D = \Lambda^{(I)} - \Lambda^{(II)} - B, \quad D = -\left(\Lambda^{(II)} + \frac{1}{2}\Lambda^* + B\right)$$
$$(\Lambda^* = \langle(L^*)\rangle). \tag{4.15}$$

Terms $L^{(I)}$ and $L^{(II)}$ in (4.13) represent the Lagrange functions of, respectively, the first and second kind of the connections, see § 5.

For external synchronization problems, the stability conditions of the synchronous motion are reduced, as before, to the requirement that the real parts of all the roots of equations like (2.12) be negative. For internal synchronization problems, the statement in subsection 2 of § 2 is correct.

§ 5) Quasi-conservative objects.

The results of § 4 allow generalization for the case of weakly coupled conservative objects.† Let's consider a system whose block diagram is given in Fig. 56. The condition of the s-th object is determined by the vector column of generalized coordinates $x^{(s)} = [x_1^{(s)}, \ldots, x_{r_s}^{(s)}]$, where r_s is the number of DOFs of the s-th object.

*Of course, function B can exist notwithstanding the absence of the potential for forces Q_r and Q_s.
†It is a brief but modified and broader version of (210); see also (57).

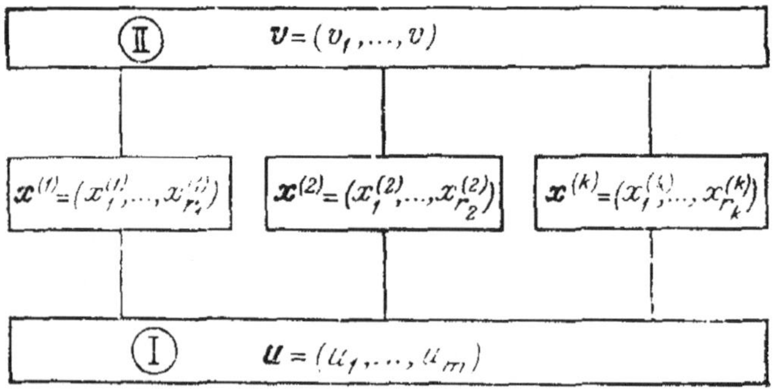

FIG. 56.

In many applications, especially in mechanics, it is useful to distinguish two types of connections (interactions, constraints) between the objects. *Connections (constraints) of the first kind,* whose state is determined by a column vector of generalized coodinates $u = [u_1, \dots, u_m]$, can be treated as caused by the presence of some *"carrying (supporting) body"* or *"system of carriers"* with m significant DOFs. Being connected with these bodies, the objects acquire some additional mobility. Thus their total kinetic and potential energies become

$$T^* = \sum_{s=1}^{k} T_s + \Delta T^*, \quad \Pi^* = \sum_{s=1}^{k} \Pi_s + \Delta\Pi^*, \tag{5.1}$$

Here $T_s = T_s\ (\dot{x}^{(s)}, x^{(s)}), \Pi_s = \Pi_s(x^{(s)})$ are proper kinetic and potential energies of the objects which can be treated, respectively, as kinetic and potential energies of the objects when the supporting system is stationary (175).

Additional kinetic and potential energies associated with the mobility of the supporting systems are denoted as

$$\Delta T^* = \Delta T^* (\dot{x}, x; \dot{u}, u) \text{ and } \Delta\Pi^* = \Delta\Pi^* (x, u)$$

where x represents a whole set of vectors $x^{(s)}$. Proper kinetic and potential energies of the connections of the first kind are denoted, respectively, as $T^{(I)} = T^{(I)}\ (\dot{u}\ , u)$ and $\Pi^{(I)} = \Pi^{(I)}\ (u)$.

Connections (constraints) of the second kind can be conditionally called *supported.* They do not lead to increasing the mobility of the objects but their presence can also be (although, not necessarily) associated with increasing the number of DOFs of the whole system. Thus, for determining the condition of the system it is also necessary, besides x and u, to specify a generalized coordinate vector $v = [v_1, \dots, v_n]$. The kinetic and potential energies of the connections of the second kind would be denoted as $T^{(II)} = T^{(II)}\ (\dot{x}, x, \dot{u}\ , u, \dot{v}\ , v)$ and $\Pi^{(II)} = \Pi^{(II)}\ (x, u, v)$.

The physical meaning of distinguishing between the connections of the first and second kind can be illustrated by examples of the synchronization problems for orbital systems (see § 6).

In agreement with the assumption of the weakness of the coupling between the objects, it should be assumed that after the application of the connections (constraints), the overall kinetic and potential energies of the system change only insignificantly, i.e.,

$$T = T^* + T^{(I)} + T^{(II)} = \sum_{s=1}^{k} T_s + O(\mu),$$
$$\Pi = \Pi^* + \Pi^{(I)} + \Pi^{(II)} = \sum_{s=1}^{k} \Pi_s + O(\mu), \tag{5.2}$$

Here $O(\mu)$ denotes terms on the order of the small parameter μ. The generalized non-conservative forces corresponding to the objects' coordinates are assumed to have an order of magnitude not less than μ. Components $Q^{(s)}$ of these forces have an order of μ, and are considered to be dependent only on the coordinates and velocities of the s-th objects and also, perhaps (in the case of an external synchronization problem), on time t with period $2\pi/\omega$ of the external excitation, or $Q^{(s)} = Q^{(s)}$ $(\dot{x}^{(s)}, x^{(s)}, \omega t)$. Generalized nonconservative forces Q_u and Q_v, associated with the coordinates of the constraint system, can generally not be small and depend on all the coordinates of the system.

The fundamental system in this case consists of k separate autonomous systems that describe motions of isolated objects, and of the fundamental equations of the constraint system. It is assumed that each of the k systems has in some domain G_s of the space $(\dot{x}^{(s)}, x^{(s)})$ solutions of the format

$$x_{j0}^{(s)} = x_{j0}^{(s)}(\psi_s, c_s) = \sigma_j^{(s)}\left[q_j^{(s)}\psi_s + y_{j0}^{(s)}(\psi_s, c_s)\right],$$
$$p_{j0}^{(s)} = p_{j0}^{(s)}(\psi_s, c_s), \quad \psi_s = \tilde{\omega}_s(c_s)\,t + \alpha_s \quad (j = 1, \ldots, r_s,\ s = 1, \ldots, k), \tag{5.3}$$

Here α_s and c_s are arbitrary constants; $y_{j0}^{(s)}$ and $p_{j0}^{(s)}$ are periodic functions ψ_s with period 2π; $\sigma_j^{(s)} = \pm 1$; and $q_j^{(s)}$ is zero for vibratory and one for rotational coordinates. Constants α_s, as before, are the initial phases of the object motions. Constants c_s on the trajectories corresponding to solution (5.3) are mutually uniquely correlated with energy constant,

$$h_s(c_s) = T_s\left(\dot{x}_0^{(s)}, x_0^{(s)}\right) + \Pi_s\left(x_0^{(s)}\right). \tag{5.4}$$

Constant c_s determines frequency $\tilde{\omega}_s$ of solution (5.3), as does h_s. This frequency changes for solutions (5.3) located in the domain G_s in some range*

$$\tilde{\omega}_s^{(1)} < \omega_s < \tilde{\omega}_s^{(2)}. \tag{5.5}$$

The synchronous motions of the objects with frequencies $\tilde{\omega}_s = n_s\omega$, which are multiples of excitation frequency ω, are possible if frequencies $\tilde{\omega}_s = n_s\omega$ are located inside these ranges.

Let's assume first that for each solution (5.3) at all considered α_s and c_s there is at least one $2\pi/\omega$-periodic solution of the fundamental equations of the constraint (connections) system

$$u_0 = u_0(\omega t;\ c_1, \ldots, c_k;\ \alpha_1, \ldots, \alpha_k),$$
$$v_0 = v_0(\omega t;\ c_1, \ldots, c_k;\ \alpha_1, \ldots, \alpha_k), \tag{5.6}$$

---

*Parameters $\tilde{\omega}_s$ should not be confused with the partial frequencies ω_s of the objects introduced earlier.

This depends on the same $2k$ constants α_s and c_s of solution (5.3) and is asymptotically stable. Then, let's assume that the objects are essentially nonisochronous inside domains G_s, i.e., widths of the frequency ranges $\tilde{\omega}_s(2) - \tilde{\omega}_s(1)$ as well as the derivatives $d\tilde{\omega}_s/dc_s$ are not small. The case of the isochronous objects in the fundamental approximation requires a special consideration (see below).

With the formulated assumptions, the use, for example, of the theorem of § 6 of chap. 10, leads to the following results.

Parameters $c_1, \ldots, c_k$ in the synchronous motion are determined from the expressions

$$\tilde{\omega}_s(c_s) = n_s\omega \quad (s = 1, \ldots, k), \tag{5.7}$$

The equations for the initial phases α_s (basic equations) can be written as

$$P_s(\alpha_1, \ldots, \alpha_k) \equiv \frac{1}{e_s(n_s\omega)}\left(\frac{\partial \Lambda}{\partial \alpha_s} + A_s\right) \equiv \frac{1}{e_s(n_s\omega)}\left(\frac{\partial \Lambda_0}{\partial \alpha_s} + A_s\right) = 0 \quad (s = 1, \ldots, k), \tag{5.8}$$

where

$$\Lambda_0 = \langle (L_0) \rangle, \quad L_0 = \Delta L^* + L^{(I)} + L^{(II)}, \tag{Y}$$

with $\Delta L^* = \Delta T^* - \Delta\Pi^*$, $L^{(I)} = T^{(I)} - \Pi^{(I)}$ and $L(II) = T^{(II)} - \Pi^{(II)}$ being Lagrange functions of the respective connections (constraints):

$$e_s(\tilde{\omega}_s) = \frac{1}{\tilde{\omega}_s}\frac{dh_s\left[c_s(\tilde{\omega}_s)\right]}{d\tilde{\omega}_s} \tag{5.9}$$

is the so-called *slope of the frequency characteristic of the object* (in the case of $e_s > 0$, it is a *hardening anisochronous* object, in the case of $e_s < 0$ it is *a softening anisochronous*, and if $e_s = 0$, it is *isochronous*).

Functions Λ, Λ_0, $\Delta\Lambda^* = \langle(\Delta L^*)\rangle$, $\Lambda^{(I)} = \langle(L^{(I)})\rangle$, and $\Lambda^{(II)} = \langle(L^{(II)})\rangle$, as in § 4, are 2π-periodic by each of the phases $\alpha_1, \ldots, \alpha_k$. This fact is very important since it determines whether function A_s is sufficiently small and whether some additional conditions are complied with, which is a tendency toward synchronization in such systems (see § 7).

Parameters

$$A_s = \sum_{j=1}^{r_s}\left\langle \left(Q_j^{(s)}\right)\frac{\partial x_{j0}^{(s)}}{\partial \psi_s}\right\rangle + \sum_{p=1}^{m}\left\langle \left(Q_{u,p}^{(0)}\right)\frac{\partial u_{p0}}{\partial \psi_s}\right\rangle + \sum_{r=1}^{n}\left\langle \left(Q_{v,r}^{(0)}\right)\frac{\partial v_{r0}}{\partial \psi_s}\right\rangle, \tag{5.10}$$

as in § 4, are average for the period of nonconservative generalized forces reduced to the s-th object, with $Q_u^{(0)}$ and $Q_v^{(0)}$ being components of the generalized forces corresponding to the coordinates u and v, and do not contain μ.

Everything said in § 4 about equations (4.6) can be applied in relation to the physical meaning of equations (5.8) and their individual terms. For example, at $\partial L/\partial t = 0$, the equation of energy budget (4.9) is satisfied, while if the two latter in (5.10) are zeros, then expressions (4.10) for determining the partial frequencies ω_s of the objects are satisfied. If there is a potential of the averaged nonconservative forces $B = B(\alpha_1, \ldots, \alpha_k)$, that is, a function satisfied by expressions (4.11), and if anisochronism of all the objects is the same,

$$\operatorname{sgn} e_1 = \ldots = \operatorname{sgn} e_h = \sigma, \tag{5.11}$$

then the potential function can be expressed as

$$D = -(\Lambda_0 + B)\sigma, \tag{5.12}$$

This is a generalization of (4.12), since objects with quasi-uniform rotations are hardening anisochronous ($\sigma = 1$).

If the equations of the carrying constraints in the fundamental approximation are a set of linear differential equations with constant coefficients, and the nonconservative forces corresponding to the system coordinates are small ($Q_u^{(0)} = 0, Q_v^{(0)} = 0$), then

$$\Lambda_0 = \Lambda^{(II)} - \Lambda^{(I)}, \tag{5.13}$$

corresponding to (4.14), and then expression (5.12) becomes

$$D = (\Lambda^{(I)} - \Lambda^{(II)} - B)\sigma. \tag{5.14}$$

According to (5.12) and (5.14), the character of the extremum of function $\Lambda_0 + B$ or $\Lambda^{(I)} - \Lambda^{(II)}$ corresponding to stable motions changes depending on the type of anisochronism of the objects, and the action of the connections (constraints) of the first and second kinds is to a certain degree opposite to each other. At B = const and in the presence of only supporting connections, $D = \sigma\Lambda^{(I)}$, while in the presence of only supported connections, $D = -\sigma\Lambda^{(II)}$. This results in Table 6, which shows the influence of the connections and the anisochronism of the objects on the character of the extrema of functions $\Lambda^{(I)}$ or $\Lambda^{(II)}$, corresponding to stable synchronous motions (211).

Contrary to functions $\Lambda^{(I)}$, $\Lambda^{(II)}$, and Λ_0, potential function D is introduced (see § 2, chap. 2, and § 8, chap. 10) in such a manner that the stable motions always correlate with the minimum of this function.

TABLE 6. Extremal character for function $\Lambda^{(I)}$ or $\Lambda^{(II)}$ for stable synchronous motions (at B = const).

Kind of constraints	Type of anisochronism of objects: Hardening anisochronism $e(\omega) > 0,\ \sigma = 1$	Type of anisochronism of objects: Softening anisochronism $e(\omega) < 0,\ \sigma = -1$
Supporting connections ($\Lambda^{(II)} = 0,\ D = \sigma\Lambda^{(I)}$)	Minimum $\Lambda^{(I)}$ ($D = \Lambda^{(I)}$)	Maximum $\Lambda^{(I)}$ ($D = -\Lambda^{(I)}$)
Supported connections ($\Lambda^{(I)} = 0,\ D = -\sigma\Lambda^{(II)}$)	Maximum $\Lambda^{(II)}$ ($D = -\Lambda^{(II)}$)	Minimum $\Lambda^{(II)}$ ($D = \Lambda^{(II)}$)

Contrary to systems with quasi-uniform rotations, the stability conditions expressed by equations (2.12) or the conditions for the minimum of potential function D in this case are only necessary. In addition, the stability roots of equation(2.12) should now be real and negative. Additional correlations leading to a set of necessary and sufficient conditions follow from inequalities (6.16), chap. 10. In the specific case of quasi conservative objects with a single DOF, and with constraints (connections) not adding new DOFs to the system, the said additional stability conditions become

$$\sum_{s=1}^{k} e_s(n_s\omega)\left\langle\left(\frac{\partial Q^{(s)}}{\partial p_s}\right)\right\rangle M_s^2 \Big/ \sum_{s=1}^{k} M_s^2 e_s(n_s\omega) < 0, \tag{5.15}$$

These conditions should be satisfied for each set of constants M_s determined from a linear homogeneous set of equations

$$\sum_{j=1}^{k} M_j\left[\frac{\partial P_s}{\partial \alpha_j} - \delta_{sj}\varkappa\right] = 0 \qquad (s = 1, \ldots, k) \tag{5.16}$$

at each one from k values of roots χ of equation (2.12). The generalized impulse of the object is denoted p_s. Since in this case parameters $A_s = \left\langle (Q^{(s)}) \frac{\partial x_0^{(s)}}{\partial \psi_s} \right\rangle$ do not depend on $\alpha_1, \ldots, \alpha_k$, then (5.16) can be also written as

$$\sum_{j=1}^{k} M_j\left[\frac{1}{e_s(n_s\omega)}\frac{\partial^2 \Lambda_0}{\partial \alpha_s \partial \alpha_j} - \delta_{sj}\varkappa\right] = 0 \qquad (s = 1, \ldots, k). \tag{5.17}$$

In the important case when $\partial Q_s/\partial p_s < 0$ and the type of anisochronism of all objects is identical [see (5.11)], additional conditions (5.15) are certainly satisfied, and the previously stated necessary conditions are also sufficient ones. For quasi-conservative objects with linear supporting connections, additional stability conditions are obtained in (213).

In relation to the internal synchronization problem, the statement at the end of § 2 applies.

In the more general case of nonisochronous conservative objects, when the fundamental solution depends not only on phases $\alpha_1, \ldots, \alpha_k$, but also on other arbitrary parameters, the theorem in § 6, chap. 10, can be used. It follows from this theorem that the stability conditions given above remain necessary for a comparatively broad group of cases. The solution of several specific synchronization problems shows that these conditions are critical in the selection of the stable phase sets of objects [see Part 2, and (57)].

Some results are available for the special case of objects which are *isochronous in fundamental approximation*, i.e.,when a fundamental system can be chosen in such a way that the frequencies do not depend on the energy, and frequency ranges (5.5) of the objects are reduced to a point corresponding (in the case of the external synchronization problem) to the excitation frequency ω. One of important classes of such objects are quasi-linear oscillations (see § 3), if they are quasi-conservative.

Such oscillators model various actual objects in mechanics, radio engineering, radiophysics, and biology (see chaps. 5, 6, 8).

The fundamental synchronous solution depends in this case not only on phases $\alpha_1, \ldots, \alpha_k$, but also on constants $c_1, \ldots, c_k$, thus it contains $2k$ arbitrary constants. Their values, which at sufficiently small μ may correspond to the synchronous motions of the system, are determined from

$$A_s + \frac{\partial \Lambda_0}{\partial \alpha_s} = 0, \quad B_s + \frac{\partial \Lambda_0}{\partial c_s} = 0 \qquad (s = 1, \ldots, k). \tag{5.18}$$

Here, parameters A_s and Λ_0 have the same meaning as above and can be calculated from the same formulas, and

$$B_s = \sum_{j=1}^{r_s} \left\langle \left(Q_j^{(s)}\right) \frac{\partial x_{j0}^{(s)}}{\partial c_s} \right\rangle + \sum_{p=1}^{m} \left\langle \left(Q_{u,p}^{(0)}\right) \frac{\partial u_{p0}}{\partial c_s} \right\rangle + $$
$$+ \sum_{r=1}^{n} \left\langle \left(Q_{v,r}^{(0)}\right) \frac{\partial v_{r0}}{\partial c_s} \right\rangle. \tag{5.19}$$

In the case of a purely conservative system (and also under the assumption that nonconservative forces have a higher order of smallness), the integral stability criterion for synchronous motions still applies. It states that stable motions can correspond to the coarse maxima of function $\Lambda_0 = -D$.

Similar problems were considered in (87). It was shown in (87) that for the considered quasi-linear, as well as for quasi-canonic purely conservative systems, stable synchronous motions are not only associated with the maxima, but also with the minima of function Λ_0 by $2k$ initial parameters or by other, related parameters of the fundamental solution (see also § 8, chap. 10).

The most general case—when the fundamental approximation objects are both isochronous and nonisochronous— is covered by the theorem from § 6, chap. 10.

§ 6) Orbital systems.

1. Definition of orbital systems; free and constrained (structured) orbital systems; their synchronization. All systems considered in this book, as well as many other mechanical and electromechanical systems with synchronizing objects, can be included in one broad class of *orbital systems*.

In the general case, an orbital system is one consisting of $(k + 1)$ interacting solid or deformable bodies, in which centers of mass or other characteristic points $C_1, \ldots, C_k$ of bodies $B_1, \ldots, B_k$ (*supported* or *carried bodies*), can perform motions by close trajectories in relation to some body B_0 which is called a *supporting* or *central body* (see Fig. 44). Any constitutive body of the system can be assumed to be central. However, a very specific body differing from other bodies by some characteristic feature, such as substantially larger mass, is usually assumed to be central. Besides interaction forces, each body of an orbital system can be acted upon by the prescribed forces of both conservative and nonconservative character.

Two types of orbital systems can be distinguished: *free* and *constrained (structured)*. In free orbital systems, the motion of the characteristic points $C_1, \ldots, C_k$

*See (63, 66).

of the supported bodies is not subordinated to any kinematic constraints. Systems such as these are considered, for example, in celestial mechanics (see chap. 7). In constrained orbital systems, the supporting body is usually idealized as one or several solid bodies, elastically connected with each other's body and with the stationary foundation. In such cases, the characteristic points can move along fixed closed trajectories inside solid bodies constituting a supporting body. Such systems play a significant role in vibratory technology. They correspond, for example, to a dynamic model of a vibratory machine with several mechanical vibro-exciters (see Figs. 5 and 6). Other examples are the gases of antifriction bearings (see Fig. 39) and pendulum clocks attached to a moving foundation (see Fig. 40).

The generalized coordinates determining the positions of supported bodies B_s relative to the supporting (central) body B_0 may include both vibratory (libration) and rotational coordinates φ_s (t) $(s = 1, \ldots, p)$. The most important role in the latter is played by coordinates φ_s^* (t) $(s = 1, \ldots, k)$, not necessarily angular, which characterize the positions of points C_s on their trajectories (orbits). These coordinates are called *orbital*. For a constrained system whose supported bodies have only one DOF in relation to body B_0 (being, for example, massive points or pendulums), orbital coordinates φ_s^* (t) fully determine the positions of supported bodies $B_1, \ldots, B_k$ relative to supporting body B_0.

The principal problem of the synchronization of orbital systems is to study the motions at which the characteristic points of supported bodies $B_1, \ldots, B_k$ move along close trajectories in relation to supporting body B_0 with identical or multiple periods, at the same time also performing synchronous motions along all or some other generalized coordinates.*

An overwhelming majority of synchronization problems for orbital systems can be considered as synchronization problems for weakly coupled quasi-conservative objects or objects with quasi-uniform rotations.† In this case, synchronizing objects are bodies $B_1, \ldots, B_k$, the supporting constraint (connection) is body B_0 with which the supported bodies interact, and the supported constraints are determined by the interactions of the supported bodies between themselves. For example, in problems of celestial mechanic, the interaction of bodies $B_0, \ldots, B_k$ is characterized by the law of gravitation. It can be stressed that the quasi-conservative idealization for orbital systems is not the only possible idealization.

For noninteracting rotational coordinates, such as for independently moving supported bodies B_s (not interacting with each other and with the supporting body B_0), or in the fundamental approximation, parameters φ_s are determined with accuracy up to arbitrary initial phases α_s:

$$\varphi_s(t) = \varphi_s^0(t + \alpha_s) \qquad (s = 1, \ldots, p), \tag{6.1}$$

Here p is the number of rotational coordinates.

All of what was said in §§ 4 and 5 can be applied to the synchronization of the orbital systems being idealized as objects with quasi-uniform rotations or as

*In other words, orbital coordinates φ_s^* (t) are periodic functions with the same or multiple periods.

†In the latter case, coordinates φ_s^* (t) are chosen, if at all possible, in such a manner that in the considered motions they would vary in a close resemblance to uniform rotation [see expressions (4.1)]. An example of such a selection of coordinates is the transition from true anomaly to average anomaly in the theory of Kepler orbits in celestial mechanics.

quasi-conservative objects. For example, "selection" of phases α_s in stable synchronous motions is performed by means of the solution of the appropriate basic equations (4.6), (5.8), and stability analysis is done using equation (2.12). This selection can also be made by using the integral stability criterion of the synchronous motions (their extremal property). It is important in this process that, due to the 2π-periodicity of the Lagrange functions by φ_s for the whole system and for its individual parts, said functions, after averaging during one period (on fundamental solutions), become 2π-periodic for all phases α_s. This fact, with some additional conditions, assures a tendency toward synchronization in the considered systems (see § 7).

In case of a free orbital system, and also when supported bodies $B_1, \ldots, B_k$ have several DOFs in relation to the supporting body B_0, the stability conditions for the synchronous motions which are expressed via equations (2.12) or via the integral stability criterion, are only necessary while playing the principal role in most cases. In case of a constrained system, and when the supported bodies have a single DOF in relation to the supporting body (see Figs. 5 and 45), these conditions are also sufficient.

2. Constrained orbital system with weakly interacting primitive bodies. Principal variation correlation and example of its use. Let's consider a constrained orbital system (see Fig. 5) in which the *"supporting body"* B_0 is a system of solid bodies $B_{01}, \ldots, B_{0n}$ which are connected to each other and to a stationary foundation by a system of elastic and damping elements, and supported bodies $B_1, \ldots, B_k$ have only one degree of freedom in relation to the bodies of the supporting systems. Such supported bodies, for example, can be unbalanced rotors driven by some motors (vibro-exciters); pendulums; or spring-suspended weights. The supported connections (constraints) will be assumed absent, and the system as a whole, not necessarily quasi-conservative.

The synchronization problem in this system is considered in chap. 3, and the synchronization problem for pendulum clocks (in somewhat less general formulation), in chap. 5. Here is derived, following A. Lurie (175), the so-called *basic variational correlation* for the general case. An example of its application will be given for a synchronization problem with quasi-uniform rotations. In the already mentioned chap. 3 (subsection 6, § 8), this correlation was used as an illustration of the Lurie method, whose basic ideas were described in § 5, chap. 11, for the solution of a corresponding specific problem.

In the following, $\boldsymbol{u}$ denotes the vector of the generalized coordinates of the supporting system of bodies, and φ_s, the generalized coordinate characterizing the positions of body B_s in relation to a body from the supporting system.

Kinetic energy of the system contains three terms

$$T = \sum_{s=1}^{k} T_s + \mu \widetilde{T}^{(\mathrm{I})} + \mu \Delta \widetilde{T}^{*}; \qquad T_s = \frac{1}{2} a_s \dot{\varphi}_s^2,$$
$$T^{(\mathrm{I})} = \mu \widetilde{T}^{(\mathrm{I})} = \frac{1}{2} \dot{\boldsymbol{u}}' \cdot \boldsymbol{M} \cdot \dot{\boldsymbol{u}}, \qquad \Delta T^{*} = \mu \Delta \widetilde{T}^{*} = \frac{1}{2} \sum_{s=1}^{k} \dot{\varphi}_s \boldsymbol{b}_s' (\varphi_s) \cdot \dot{\boldsymbol{u}}, \tag{6.2}$$

The first term is the kinetic energy of the supported bodies $B_1, \ldots, B_k$ when the supporting bodies are stationary; the second one is the kinetic energy of the supporting bodies when the supported bodies $B_1, \ldots, B_k$ are motionless on them; and third one is the additional kinetic energy associated with the mobility of the supporting body. Here M is the constant symmetric matrix of the inertia coefficients of the supporting body; a_s are inertia coefficients of bodies B_s, and $b_s(\varphi_s)$ is the vector which is a periodic function of φ_s. The "prime"(′) indicates the transposed matrix and the small parameter μ before terms $\tilde{T}^{(I)}$ and $\Delta\tilde{T}^*$ in (6.2) corresponds, as in § 5, to the assumption about weak coupling of the supported bodies $B_1, \ldots, B_k$ by means of the bodies of the supporting system; the oscillations of the supporting bodies are also considered to be small.

The potential energy and virtual work of the nonconservative forces are, respectively, as follows

$$\Pi = \sum_{s=1}^{k} \Pi_s(\varphi_s) + \mu\tilde{\Pi}^{(I)}(u) \qquad \left(\Pi^{(I)}(u) = \mu\tilde{\Pi}^{(I)}(u) = \frac{1}{2} u' \cdot C \cdot u\right);$$

$$\delta' A = -\delta u' \cdot R\dot{u} + \sum_{s=1}^{k} \left(Q_{\varphi_s}^{(0)} + \mu\tilde{Q}_{\varphi_s}^{(1)}\right) \delta\varphi_s, \tag{6.3}$$

Here C and R are the constant symmetric matrices of the stiffness and dissipation coefficients; thus $\dot{u}' \cdot R \cdot \dot{u}$ is the doubled dissipative function of the supporting body; $Q_s^{(0)}$ and $\mu\tilde{Q}_s^{(1)} = Q_s^{(1)}$ are, respectively, the finite and the small part of the nonconservative generalized forces corresponding to the objects' coordinates φ_s.

Thus, the Lagrange function of the system is

$$L = \sum_{s=1}^{k} L_s + \mu\left(\tilde{L}^{(I)} + \Delta\tilde{L}^*\right), \quad L_s = \frac{1}{2} a_s\dot{\varphi}_s^2 - \Pi_s(\varphi_s),$$

$$L^{(I)} = \mu\tilde{L}^{(I)} = \frac{1}{2}(\dot{u}' \cdot M \cdot \dot{u} + u'Cu), \quad \Delta\tilde{L}^* = \mu\Delta\tilde{L}^* = \sum_{s=1}^{k} \dot{\varphi}_s b_s'(\varphi_s) \cdot \dot{u}, \tag{6.4}$$

Here L_s are the "proper" Lagrange functions of bodies B_s. The respective generalized impulses $p_s = \partial L/\partial\dot{\varphi}_s$ and $z = \partial L/\partial\dot{u}$ will be

$$p_s = a_s\dot{\varphi}_s + b_s' \cdot \dot{u}, \quad z = M \cdot \dot{u} + \sum_{s=1}^{k} b_s\dot{\varphi}_s. \tag{6.5}$$

Up to terms of higher order of smallness than μ, the Lagrange function (6.4) can be rewritten as

$$L = \sum_{s=1}^{k} p_s\dot{\varphi}_s - H_0 + \mu(\tilde{L}^{(I)} + \Delta\tilde{L}^*), \tag{6.6}$$

where

$$H_0 = \sum_{s=1}^{k}\left[\frac{1}{2}\frac{p_s^2}{a_s} + \Pi_s(\varphi_s)\right] \tag{6.7}$$

is the Hamilton function of the supported bodies (the nonsmall term in the expression for the Hamilton function of the system). The average of the Lagrange function during period $T = 2\pi/\omega$, that is, the average for the period value of the Hamilton action will thus be

$$\Lambda = \langle L \rangle = \left\langle \sum_{s=1}^{k} (p_s \dot{\varphi}_s - H_0) + \mu (\widetilde{\Lambda}^{(\mathrm{I})} + \Delta \widetilde{\Lambda}^*) \right\rangle, \tag{6.8}$$

where

$$\mu \widetilde{\Lambda}^{(\mathrm{I})} = \Lambda^{(\mathrm{I})} = \langle L^{(\mathrm{I})} \rangle \quad и \quad \mu \Delta \widetilde{\Lambda}^* = \Delta \Lambda^* = \langle \Delta L^* \rangle. \tag{6.9}$$

In accordance with the Hamilton principle, this should be

$$\delta \Lambda + \langle \delta' A \rangle = 0. \tag{6.10}$$

From this, and using (6.3), (6.8), and (6.9), we arrive at

$$\sum_{s=1}^{k} \left[\delta (p_s \dot{\varphi}_s - H_0) + Q_s^{(0)} \delta \varphi_s \right] + \delta (\Lambda^{(\mathrm{I})} + \Delta \Lambda^*) +$$

$$+ \left\langle \sum_{s=1}^{k} Q_s^{(1)} \delta \varphi_s - \delta \dot{u}' \cdot R \cdot \dot{u} \right\rangle - \frac{1}{T} \left(\sum_{s=1}^{k} p_s \delta \varphi_s + z' \cdot \delta u \right) \Bigg|_0^T = 0$$

$$\left(Q_s^{(1)} = \mu \widetilde{Q}_s^{(1)} \right) \tag{6.11}$$

After the known transformation using the "integration by parts" formula, we arrive at the principal variational correlation

$$\left\langle \sum_{s=1}^{k} \left(\dot{\varphi}_s - \frac{p_s}{a_s} \right) \delta p_s - \left(\dot{p}_s + \frac{\partial \Pi}{\partial \varphi_s} - Q_s^{(0)} \right) \delta \varphi_s \right\rangle +$$

$$+ \delta (\Lambda^{(\mathrm{I})} + \Delta \Lambda^*) + \left\langle \sum_{s=1}^{k} Q_{\varphi_s}^{(1)} \delta \varphi_s - \delta u' \cdot R \cdot \dot{u} \right\rangle = \frac{1}{T} z' \cdot \delta u \Bigg|_0^T. \tag{6.12}$$

The equations of motion for the supported bodies and the supporting system, corresponding to (6.4) and the assumption discussed above, are

$$a_s \ddot{\varphi}_s + \frac{\partial \Pi_s}{\partial \varphi_s} + \mu \mathscr{E}_{\varphi_s} (\Delta \widetilde{L}^*) = Q_s^{(0)} + \mu \widetilde{Q}_s^{(1)} \qquad (s = 1, \ldots, k), \tag{6.13}$$

$$\mathscr{E}_u (L^{(\mathrm{I})} + \Delta L^*) + R \cdot \dot{u} = 0. \tag{6.14}$$

Here $\mathscr{E}_q = \frac{d}{dt} \frac{\partial}{\partial \dot{q}} - \frac{\partial}{\partial q}$ — is the Euler operator corresponding to the generalized coordinate q; thus

$$\mathscr{E}_{\varphi_s} (L^*) = b_s' \ddot{u}, \qquad \mathscr{E}_u (L^{(\mathrm{I})} + \Delta L^*) = M \ddot{u} + C u + \sum_{s=1}^{k} b_s \dot{\varphi}_s. \tag{6.15}$$

As an example of its application, the principal variational correlation is used for the synchronization problem of a constrained orbital system with quasi-uniform rotational motions. As in § 4, it is assumed that the variation of the generalized coordinates φ_s in the first approximation are determined by (4.5), and that the corresponding time history of $2\pi/\omega$ steady periodic vibrations is determined by (4.5′),

$$\varphi_s = \varphi_s^0 = \sigma_s(n_s\omega t + \alpha_s) \qquad (s = 1, \ldots, k),$$
$$u = u^0(\omega t, \alpha_1, \ldots, \alpha_k). \tag{6.16}$$

Comparing equations (6.13) and (6.14) with (4.2) and (4.3) and considering that, according to (6.4), $\mathscr{E}_u(L) = \mathscr{E}_u(L^{(I)} + \Delta L^*)$, in this case

$$a_s = I_s, \quad \Pi_s(\varphi_s) = 0, \quad Q_{\varphi_s}^{(0)} = -k_s(\dot{\varphi}_s - \sigma_s n_s \omega), \quad Q_u^{(0)} = -R \cdot \dot{u}. \tag{6.17}$$

Expressions (6.16) are considered to be a form of the approximate solution of the synchronization problem, after using (6.12) for determining $\alpha_1, \ldots, \alpha_k$. Having in mind that according to (6.5), $p_s^0 = I_s\sigma_s n_s\omega$ or $\delta p_s^0 = 0$, and also considering the $T = 2\pi/\omega$-periodicity of functions $u^0(t)$ and $z(t)$, and also the arbitrary character of variations $\delta\alpha_s$, the following equations result:

$$\frac{\partial(\Lambda^{(I)} + \Delta\Lambda^*)}{\partial\alpha_s} + \sigma_s\langle(Q_{\varphi_s}^{(1)})\rangle - \left\langle\frac{\partial u_0'}{\partial\alpha_s}\cdot R\cdot\dot{u}_0\right\rangle = 0$$
$$(s = 1, \ldots, k). \tag{6.18}$$

Since, according to (4.7) and (6.4),

$$\frac{\partial\Lambda_0}{\partial\alpha_s} = \frac{\partial\Lambda}{\partial\alpha_s} - \sum_{s=0}^{k}\frac{\partial\Lambda_s}{\partial\alpha_s} = \frac{\partial(\Lambda^{(I)} + \Delta\Lambda^*)}{\partial\alpha_s} \tag{6.19}$$

and since in our case $Q^{(0)} = -R\cdot\dot{u}$, equations (6.18) would coincide identically with equations (4.6), derived in § 4 by the perturbation method.

It is also easy to derive from the variational correlation equation which describes the slow process of developing and establishing the synchronous regime and its stability conditions. To do this, parameters α_s in expressions (6.16) are assumed to be not constant, but slow varying functions of time $\alpha_s = \alpha_s(t)$. Then,

$$\dot{\varphi}_s^0 = \sigma_s(n_s\omega + \dot{\alpha}_s), \quad Q_{\varphi_s}^{(0)} = -k_s(\dot{\varphi}_s^0 - \sigma_s n_s\omega) = -\sigma_s k_s\dot{\alpha}_s,$$
$$p_s^0 = I_s\sigma_s(n_s\omega + \dot{\alpha}_s), \quad \dot{p}_s^0 = \sigma_s I_s\ddot{\alpha}_s. \tag{6.20}$$

Next, these expressions will be substituted into (6.12), and the variation of α_s, $\dot{\alpha}_s$, $\ddot{\alpha}_s$, during one period will be neglected, which is an accepted practice in the averaging principle. In addition, it is assumed that vector u is determined by the current values of α_s as described earlier by expressions (6.16). This corresponds to the assumption about the slowness of the change of α_s and the fast rate of the process of establishing oscillations by variables u (see also subsection 6, § 8, chap. 3). This all results in the following equations:

$$I_s\ddot{\alpha}_s + k_s\dot{\alpha}_s = \frac{\partial(\Lambda^{(I)} + \Delta\Lambda^*)}{\partial\alpha_s} + \sigma_s\langle(Q_{\varphi_s}^{(1)})\rangle - \left\langle\frac{\partial u_0'}{\partial\alpha_s}\cdot R\cdot\dot{u}_0\right\rangle \tag{6.21}$$
$$(s = 1, \ldots, k).$$

If the potential of excess averaged nonconservative forces exists, i.e., a function $B(\alpha_1, \ldots, \alpha_k)$ that (see § 4)

$$\frac{\partial B}{\partial \alpha_s} = \sigma_s \langle (Q_{\varphi_s}^{(1)}) \rangle - \left\langle \frac{\partial u_0'}{\partial \alpha_s} \cdot R \cdot \dot{u}_0 \right\rangle, \tag{6.22}$$

then the following function can be introduced.

$$D = -(\Lambda^{(1)} + \Delta\Lambda^* + B) \tag{6.23}$$

After this, equation (6.21) can be written as

$$I_s \ddot{\alpha}_s + k_s \dot{\alpha}_s = -\frac{\partial D}{\partial \alpha_s} \qquad (s = 1, \ldots, k). \tag{6.24}$$

From well-known stability theorems (183, 294), it follows that the stable equilibrium positions of this system that correspond to the stable synchronous motions (6.16), are associated with the minima of function D, which plays the role of potential energy here. However, this function coincides in the considered case with the potential function introduced in § 4. This explains the complete identity of the results of the analyses performed by both methods.

If a function complying with (6.22) does not exist, but terms $I_s\ddot{\alpha}_s$, in equations (6.21) could be omitted due to their smallness,* then the steady values of constants α_s determined from these equations, as well as the corresponding stability criteria, would also coincide with the ones determined in accordance with equations (4.6) and (2.12). In other words, in this case the appropriate results also coincide.

§ 7) Tendency toward synchronization in the considered dynamic systems.

The results obtained in this chapter suggest the important conclusion that the tendency toward synchronization (see § 1, chap. 1) is characteristic for quite a broad class of dynamic systems, regardless of their specific nature. To prove it, one has to agree that, both for objects with quasi-uniform rotations and for essentially nonisochronous quasi-conservative objects, as well as for the supported bodies in the orbital systems, the equations for phases $\alpha_1, \ldots, \alpha_k$ in the synchronous motions (*basic equations*) are written as

$$\frac{\partial \Lambda}{\partial \alpha_s} + A_s = 0 \qquad (s = 1, \ldots, k), \tag{7.1}$$

Here $\Lambda = \Lambda(\alpha_1, \ldots, \alpha_k)$ is the averaged Lagrange function of the system, or a simple linear combination (with "+" or "–" signs) of the averaged Lagrange functions of the individual parts of the system. The functions $A_s = A_s(\alpha_1, \ldots, \alpha_k)$ are the averaged magnitudes of the nonconservative generalized forces. Function here is cer-

* It can be noted that just such assumptions were made for derivation of equations like (6.21) in (175). The appropriate section of this publication is described in this book with some modifications.

tainly a $2\pi/\omega$-periodic function of each phase $\alpha_1, \ldots, \alpha_k$. But then, according to § 9, chap. 10, (locally) stable synchronous motions exist at quite general assumptions, if only the averaged nonconservative generalized forces A_s are sufficiently small. The latter condition is satisfied, for example, for objects like those considered in §§ 4 and 5 on the condition of sufficient closeness between their partial frequencies (angular velocities) ω_s.

Bibliography

1. Avotinya, K., Tomashuns, I., "On synchronization in ball bearings," Trudy Latv. S.-Kh. Akad. Elgava, 1977, No. 109.

2. Akalovsky, I.V., "Investigation of mutual synchronization of two auto-generators using method of generalized frequency-phaseous diagram," Radiotekhnika, 1950, Vol. 17, No. 3.

3. Akselrod, Z.M., "Clock mechanisms: theory, calculation and design," M.-L.: Mashgiz, 1947.

4. Akselrod, Z.M., "Electro-mechanical clocks," M.-L.: Mashgiz, 1952.

5. Alabuzhev, P.M., Zuev, A.K., Kirnarsky, M.Sh., "Self-synchronization of a rotational pair," In: Trudy po teorii i primeneniyu Yavlenia Samosinkhronizatzii v mashinakh i ustroystvakh, Vilnius, Mintis, 1966.

6. Andronov, A.A., Witt, A.A., "Mathematical theory of entrainment," Zhurnal Prikladnoy Fisiki, 1934, Vol. 7, No. 4.

7. Andronov, A.A., Witt, A.A., "Stability analysis by Lyapunov," Zhurnal eksperimentalnoy i teoreticheskoy Fisiki, 1933, Vol. 3, No. 5.

8. Andronov, A.A., Witt, A.A., "To mathematical theory of auto-oscillation systems with two degrees of freedom," Zhurnal Technisheskoy Fisiki, 1934, Vol. 4, No. 1.

9. Andronov, A.A., "Collected works," M: A.A. SSSR Publish. House, 1956.

10. Andronov, A.A., Witt, A.A. Khaykin, C.E., "Vibration theory," M: Fizmatgiz, 1959.

11. Arkadiev, V., "On electro-magnetic hypothesis of telepathy," Zhurnal Prikladnoy Fisiki, 1924, Vol. 1, No. 1-4.

12. "Astronomical year-book of the USSR, 1979," Astronomichesky Ezhegodnik SSSR, 1979, Vol. 58, M: Nauka, 1976.

13. Babsky, E.V., Ulianovsky, L.S., "Electrical stimulation of the heart," M: AN SSSR Publish. House, 1961.

14. Barzukov, O.I., Denisov, G.A., Timofeev, I.G., "Dynamics of vibro-elevator with elliptical trajectories of oscillations," In: Trudy in-ta "Mekhanobr," Isledovanie i raschety obogatitelnogo oborudovania, 1971.

15. Barzukov, O.P., "Multiple synchronization in the system with weakly connected objects and one degree of freedom," PMM, 1972, Vol. 36, No. 2.

16. Barzukov, O.P., "Multiple double synchronization of mechanical vibrators, coupled with linear oscillating system," Izv. AN SSSR, MTT, 1973, No. 6.

17. Barzukov, O.P., Vaysberg, L.H., "On stability of self-synchronous rotation of two vibro-exciters in the devices with a 3-D dynamic system," In: Vibratsionaya tekhnika, M: MDNTI im. F.E. Dzerzhinskogo, 1977.

18. Bautin, N.N., "On the synchronization theory," Zhurnal Tekhnicheskoy Fisiki, 1939, Vol. 9, No. 6.

19. Bautin, N.N., "Behavior of dynamic systems near boundaries of stability region," M.: Gostekhizdat, 1949.

20. Bautin, N.N., "Dynamic models of free clock escapements," In: Pamyati A.A. Andronova, M.: A.N. SSSR Publish. House, 1955.

21. Belenov, E.M., Oraevsky, A.N., "Molecular oscillator with two resonators in series," Radiotekhnika i electronika, 1963, Vol. 8, No. 1.

22. Beletsky, V.V., Shlakhtin, A.N., "Extremal properties of resonance motions," DAN SSSR, 1976, Vol. 231, No. 4.

23. Beletsky, V.V., "Essays on motions of celestial bodies," 2nd edition, M: Nauka, 1977.

24. Beletsky, V.V., "Extremal properties of resonance motions," In: Tesisy dokladov 3 Vsesousnoy Chetaevskoy Konferentsii po ustoychivosti dvizhenia, analiticheskoy mekhanike i upravleniu dvizheniem, Irkutsk; AN SSSR, Sib. otd., 1977.

25. Beletsky, V.V., Kasatkin, G.V., "On extremal properties of resonance motions," DAN SSSR, 1980, Vol. 251, No. 1.

26. Bellustin, S.V., "To a theory of two weakly coupled contours," Uch. zap. Gorkovskogo Gos. un-ta, 1939, No. 12.

27. Bernshtein, I.L., "Entrainment frequency band of a laser resonator," DAN SSSR, 1965, Vol. 163, No. 1.

28. "Biological Clocks," M: Mir, 1964.

29. "Bionics," M: Nauka, 1965.

30. Bitseno, K.B., Grammel, R., "Technikal dynamics," Vol. II, M: Gostekhizdat, 1952 (from German).

31. Blekhman, I.I., "Self-synchronization of vibrators in certain vibro-machines," Ingenerny Sbornik, 1953, Vol. 16.

32. Blekhman, I.I., "Rotation of unbalanced rotor due to harmonic oscillations of its axis," Isv. AN SSSR, OTN, 1954, No. 8.

33. Blekhman, I.I., "To the question of stability of periodic solutions of quasi-linear non-autonomous systems with many degrees of freedom," DAN SSSR, 1955, Vol. 104, No. 6.

34. Blekhman, I.I., "On stability of periodic solutions of quasi-linear autonomous systems with many degrees of freedom," DAN SSSR, 1957, Vol. 112, No. 2.

35. Blekhman, I.I., "Inertia screen," USSR Certif. of Invention 112,448; 1958.

36. Blekhman, I.I., "On self-synchronization of mechanical vibrators," Isv. AN SSSR, OTN, 1958, No. 6.

37. Blekhman, I.I., "Vibratory ball mill," USSR Certif. of Invention 125,117, 1959.

38. Blekhman, I.I., "Vibratory machines with mechanical vibro-exciters," In: Primenenie vibrotekhniki v gornom dele, M.: Gosgortekhizdat, 1960.

39. Blekhman, I.I., "Dynamics of the drive for vibratory machines with many mechanical vibrators," Isv. AN SSSR, OTN, Mekhanika i mashinostroenie, 1960, No. 1.

40. Blekhman, I.I., Lavrov, B.P., "On one integral indication of motion stability," PMM, 1960, vol. 24, No. 5.

41. Blekhman, I.I., "Proof for the integral criterion of motion stability in problem of vibrator synchronization," PMM, 1960, Vol. 24, No. 6.

42. Blekhman, I.I., "Synchronization of dynamic systems," Auto-referat dissertatsii na soiskanie uchenoy stepeni doctora fiz.-mat. nauk., L.: Leningradsky Politekhnichesky institut im, M.I. Kalinina, 1962.

43. Blekhman, I.I., "Integral stability criterion for periodic motions of certain non-linear systems and its application," Trudy Mezhdunarodnogo Simposiuma po nelineynym kolebaniam, Vol. 2, Kiev: Isd-vo AN USSR, 1963. (Available in English as Proc. of International Symposium on Nonlinear Vibrations.)

44. Blekhman, I.I., Dzhanelidze, G. Yu., "Vibration conveyance," M: Mauka, 1964.

45. Blekhman, I.I., Lavrov, B.P., "Vibro-conveyor," USSR Certif. of Invention 163,524; 1964.

46. Blekhman, I.I., "Synchronization problems in dynamic systems," PMM, 1964, Vol. 28, No. 2.

47. Blekhman, I.I., "Synchronization problems for oscillating and rotational motions," In: Trudy no teorii i primeneniyu Yavlenia Sinkhronizatsii v mashinakh i ustroystvakh, Vilnius, Mintis, 1966.

48. Blekhman, I.I., Denisov, G.A., Lavrov, B.P., "Vibro-conveyor of 'Mekhanobr' type with self-synchronizing mechanical vibrators," Obogachshenie rud, 1966, No. 5.

49. Blekhman, I.I., Rudin, A.D., "Stability of the working regime of vibratory machines," In: Vibratsionaya tekhnika. Materialy nauchno-tekhnicheskoy konferentsii, M.: NIIINFSTROY-DORKOMMUNMASH, 1966.

50. Blekhman, I.I., "Mathematical problems of synchronization theory of dynamic systems," In: Mezhdunarodny kongres matematikov, Tesisy kratkikh nauchnykh xoobchshenii, sektsia 12, M., 1966. (Available in English as International Mathematical Congress, paper abstracts, Section 12.)

51. Blekhman, I.I., "Dynamics of common foundations under unbalanced machines," Ingenerny zhuznal. Mekhanika tverdogo tela, 1967, No. 1.

52. Blekhman, I.I., Nagaev, R.F., "Optimal stabilization of synchronous motions for mechanical vibrators," In: Tesisy dokladov i soobchshenii, na Vsesousnom mezhvusouskom simposiume po prikladnoy matematike i kibernetike, Gorky, 1967.

53. Blekhman, I.I., "Synthesis of machines and devices with self-synchronizing vibrators," Obogachshenie rud, 1968, No. 4.

54. Blekhman, I.I., "Method of a small parameter," In: Mekhanika V SSSR za 50 let, Vol. 1, Nouka, 1968.

55. Blekhman, I.I., Lavrov, B.P., "Mechanical vibratory stand," USSR Certif. of Invention 243,923; 1969.

56. Blekhman, I.I., Marchenko, Yu. I., "Influence of rheological properties of the working medium on self-synchronization of mechanical vibrators," Isv. AN SSSR, MTT, 1969, No. 5.

57. Blekhman, I.I., "Synchronization of dynamic systems," M.: Nauka, 1971.

58. Blekhman, I.I., Frolov, K.V., "Synchronization of parametrically coupled vibrators," In: Tesisy dokladov konferentsii po kolebaniam mekhanich. sistem., Kiev: Naukova dumka, 1971.

59. Blekhman, I.I., "Inertial vibrator," USSR Certif. of Invention 388,974; 1973.

60. Blekhman, I.I., Marchenko, Yu. I., "Synchronization of quasi-linear oscillators connected by a complex oscillating system," Isv. AN SSSR, MTT, 1973, No. 6.

61. Blekhman, I.I., Myshkis, A.D., Panovko, Ya. G., "Applied Mathematics: subject, logic, specifics of approaches," Kiev: Naukova dumka, 1976.

62. Blekhman, I.I., "Method of the direct separation of motions in the problems of vibratory excitation of non-linear mechanical systems," Isv. AN SSSR, MTT, 1976, No. 6.

63. Blekhman, I.I., "Stability of orbital systems," In: Tesisy dokladov 3 Vsesousnoy Chetaevskey Konf. po ustoychivosti dvizhenia, analitischeskoy mekhanike i upravleniu dvizheniem, Isd-vo AN SSSR Sib. otd. 1977.

64. Blekhman, I.I., Lugovaya, I.N., "Synchronization phenomena and the problems of design of common foundations for unbalanced machines," In: Dinamika osnovanii, fundameutov i nadzemnykh sooruzhenii, Materialy IV vsesousnoy konferentsii, Tashkent: FAN Usbekskoy SSR, 1977.

65. Blekhman, I.I., "Development of the concept of direct motion separation in non-linear mechanics," in: Sovremennye problemy teoreticheskoy i pricladnoy mekhaniki., Kiev: Naukova dumka, 1978.

66. Blekhman, I.I., "Synchronization and entrainment," In: Vibratsii v tekhnike (spravochnik) T.2. Kolebaniya nelineynikh mekhanicheskikh sistem. M.: Mashinostroenie, 1979.

67. Blekhman, I.I., "Actions of vibration on non-linear mechanical systems (mechanics of slow motions, vibro-transportation, vibro-rheology)," In: Vibratsia v tekhnike (spravochnik) T.2. Kolebania nelineynykh mekhanicheskikh sistem. M.: Mashinostroenic, 1979.

68. Bogolubov, N.N., "On certain statistic methods in mathematical physics," Kiev: Isd-vo AN USSR, 1945.

69. Bogolubov, N.N., "Perturbation theory in non-linear mechanics," Sbornik trudov in-ta stroitelnoy mekhaniki, AN USSR, 1950, No. 14.

70. Bogolubov, N.N., Mitropolsky, Yu. A., "Asymptotic methods in theory of non-linear oscillations," M.: Fizmatgiz, 1958.

71. Bogolubov, N.N., "Selected works in 3 volumes," Vol. 1, Kiev: Naukova dumka, 1969.

72. Bocharov, V.V., Zubarev, I.G., "Entrainment of the solid-state laser with non-homogeneous spectral line of amplification in the regime of free generation by external force," Isv. vusov, Radiofizika, 1974, Vol. 17, No. 7.

73. Bremsen, A.S., Feinberg, I.S., "Analysis of operation for two coupled relaxation oscillators," Zh, T.F., 1941, Vol. 11, No. 10.

74. Bulgakov, B.V., "Vibrations," M.: Gostekhisdat, 1954.

75. Bulgakov, N.G., "Oscillations of quasi-linear autonomous systems with many degrees of freedom and non-analytical characteristics of non-linearity," PMM, 1965, Vol. 19, No. 3.

76. Butenin, N.V., "On the theory of forced synchronization," In: Pamyati A.A. Andronova, M.: Isd-vo AN SSSR, 1955.

77. Butenin, N.V., Neymark, Yu, I., Fufaev, N.A., "Introduction to the theory of non-linear oscillations," M.: Nauka, 1976.

78. Bykhovksy, I.I., "Progress in vibratory technology and the research problems," In: Vibratsionaya tekhnika, Materialy nauchno-tekhnicheskoy konferentsii, M.: NIIINFSTROYDOR-KOMMUNMASH, 1966.

79. Bykhovsky, I.I., "Fundamentals of vibratory technology," M.: Mashinostroenie, 1969.

80. Bykhovsky, I.I., "Vibratory maintenance of rotation for a double pendulum," In: Vibratsionye mashiny proisvodstvenogo nasnachenia, Chap.1., M.: MDNTII im. F.E. Dzherzhinski, 1971.

81. Valeev, K.G., Ganiev, R.F., "Study of vibrations in non-linear systems," PMM, 1969, Vol. 33, No. 3.

82. Van der Pol, B., "Non-linear theory of electrical oscillations," M.: Svyasisdat, 1935.

83. Vasiliev, L.L., "Enigmatic phenomena of human psychology," M.: Gospolitisdat, 1963.

84. Agadjanian, N.A., Gorshkov, M.M., et al., "Your efficiency of today," M.: Sovetskaya Rossia, 1978.

85. Veksler, A.M., et al., In: "Abstracts of the International Congress of Biophysics," M., 1972.

86. Ventsel, E.C., "Theory of probability," M.: Fismatgis, 1958.

87. Vetukov, M.M., Khodjaev, K. Sh., "Equations of slow non-steady regimes of synchronous oscillators," Electrichestvo, 1978, No. 11.

88. Wiener, N., "Non-linear problems in the theory of random processes," M.: IL. 1961.

89. Wiener, N., "New chapters in cybernetics," M.: Sovetskoe Radio, 1963.

90. Wiener, N., "Dynamic systems in physics and biology," Vestnik AN SSSR, 1964, No. 7.

91. Barzukov, O.P., et al., "Influence of technological load on self-synchronization of vibro-exciters," Obogachshenie rud, 1978, No. 2.

92. Zeiger, S.G., et al., "Wave fluctuation processes in lasers," M.: Nauka, 1974.

93. Volosov, V.M., Morgunov, B.I., "Calculation methods for stable resonance oscillatory and rotational motions of certain non-linear systems," ZhVM i MF, 1968, Vol. 2, No. 2.

94. Volosov, V.M., Morgunov, B.I., "Averaging method in the theory of non-linear oscillatory systems," M.: MGU, 1971.

95. Ganiev, R.F., Ukrainsky, L.E., "Dynamics of particles under vibratory action," Kiev: Naukova duka, 1975.

96. Ganiev, R.F., Fasullin, F.F., "On the non-linear synchronous oscillation and stability of turbine blades," Trudy Ufimsk aviats. in-ta, 1975, No. 98.

97. Gantmakher, F.F., Krein, M.G., "Oscillation matrices and kernels and small oscillations of mechanical systems," M.: Gostekhisdat, 1950.

98. Gaponov, V.I., "Two coupled oscillators with soft self-excitements," Zhurn. tekhn. fisiki, 1936, Vol. 6, No. 5.

99. Getsiavichus, Yu.Yu., Ragulskis, K.M., "Self-synchronization of pneumo-vibrators in a three directional vibro-stand," In: Trudy po teorii i primeneniu Yavlenia sinkhronizatsii V Mashinakh i ustroystvakh, Vilnius: Mintis, 1966.

99a. Godovannik, B.V., "Synchronization of vibrators during longitudinal-rotational oscillations of the shell being submerged in soil," Mashinovedenie, 1973, No. 5.

100. Goldreich, P., "Explanation of a frequent encounterment of commensurable average motions in solar system," In: Tides and resonances in solar systems, M.: Mir, 1975.

101. Gorev, A.A., "Selected works on stability of electrical systems," M.-L.: Gosenergoisdat, 1960.

102. Gorelik, G.S., "Life and works of A.A. Andronov," In: Pamyati A.A. Andronova. M.: Isd-vo AN SSSR, 1955.

103. Gorozhankin, B.N., "On the mutual entrainment in the distance," Isv. AN SSSR, ser. fix. 1947, Vol. 11, No. 2.

104. Gorshkov, S.N., Lavrov, B.P., "Vibrating screen," USSR Certif. of Invention 155,391; 1963.

105. Gorshkov, S.N., Lavrov, B.P., Timofeev, N.G., USSR Certif. of Invention 212,046; 1968.

106. Grebennikov, E.A., Ryabov, Yu.A., "Resonances and small denominators in celestial mechanics," M.: Nauka, 1978.

107. Gudvin, B., "Temporal organization of a cell," M.: Mir, 1966.

108. Guzev, V.V., "Synchronization of oscillators for harmonic forces interacting by extensive carrying body of a beam type," Abstract of Candidate of Science Thesis L.: Leningrad, politekhn. in-t, im. M.I. Kalinina, 1977.

109. Gurtovnik, A.S., Neumark, Yu.I., "On synchronization of dynamic systems," PMM, 1974, Vol. 38, No. 5.

110. Huygens, Ch., "Three memoirs on mechanics," M.: Isd-vo AN SSSR, 1951.

111. Demin, V.G., "The fate of the solar system," M.: Nauka, 1969.

112. Demyanchenko, A.G., "Synchronization of harmonic oscillators," M.: Energia, 1976.

113. Den-Gartog, J.P., "Mechanical vibrations," M.: Fismatgis, 1960.

114. Denisov, G.A., Lauzov, B.P., "Vertical helical vibrating conveyor," USSR Cert. of Invention 152,408; 1962.

115. Denisov, G.A., Lavrov, B.P., Nagaev, R.F., "Dynamics of multi-vibrator vibro-elevator with a synchronizer," In: *Tesisy dokl. Vsesousn. konf. po kolebaniyam mekhanicheskikh sistem*, Kiev: 1968.

116. Denisov, G.A., Lavrov, B.P., "Application of multi-vibrator drive for a helical vibro-elevator," Obogashchenie rud, 1968, No. 2.

117. Denisov, G.A., Lavrov, B.P., Nagaev, R.F., "Drive for horizontal vibro-conveyor," USSR Certif. of Invention 233,523; 1969.

118. Denisov, G.A., "On stability of a working regime of vertical helical vibrating conveyor with self-synchronizing vibrators," Mashinovedenie, 1969, No. 1.

119. Detinko, F.M., "On stability of operation of auto-balancer for dynamic balancing," Isv. AN SSSR, OTN, Mekhanika i mashinostroenie, 1959, No. 4.

120. Dimentberg, F.M., "Bending oscillations of rotating shafts," M.: Isd-vo AN SSSR, 1959.

121. Duboshin, G.N., "Celestial mechanics, basic problems and methods," M.: Nauka, 1968.

122. Dubrovin, B.N., Blekhman, I.I., "On a critical slot width in the inertial crushers," Obogashchenie rud, 1960, No. 6.

123. Dukhin, S.S., "Drift theory of aerosol particle in a stationary acoustic wave," Kolloidny Zhurnal, 1960, Vol. 22, No. 1.

124. Esafov, N.I., "Experimental investigation of mutual synchronization of two coupled Thomson's oscillators," ZhTF, 1947, Vol. 17, No. 7.

125. Zhabotinsky, A.M., "Concentration of auto-oscillations," M.: Nauka, 1974.

126. Zhadanov, P.S., "Stability of electrical systems," M.: Gostekhenergoisdat, 1948.

127. Zuravlev, V.G., Lapin, A.A., "Self-synchronization phenomena in high school giroscopic supports," Isv. AN SSSR, MTT, 1979, No. 4.

128. Zanadvorov, P.N., "Analysis of the synchronization process for generators of discontinuous oscillations," Vestn. Leningrad un-ta, 1972, Vol. 2, No. 10.

129. Zanadvorov, P.N., "Method of point imaging in synchronization problems," Vestn. Leningrad un-ta., 1972, Vol. 2, No. 10.

130. Zaretsky, L.B., "On self-synchronization of centrifugal vibro-exciters of vibro-impact mechanism," Mashinovedenie, 1967, No. 1.

131. Zaretsky, L.B., "Synchronization of centrifugal vibro-exciters in systems with discontinuous characteristics," Inzh. Zhurn. MTT, 1968, No. 1.

132. Zuev, A.K., Ivanov, E.N., Kirnarsky, M.Sh., "Giroscopic oscillator of poly-harmonic oscillations," In: Mashiny i mekhanismy vibroudarnogo deystvia, Trudy Novosibirskogo electro tekhnich. In-ta, Novosibirsk, 1959.

133. Iorish, Yu.I., "One-sided drift and rotation of an instrument's hands due to vibration," Priborostroenie, 1956, No. 4.

134. Kazhinsky, B.B., "Biological radio communication," Kiev: 1962.

135. Kapitsa, P.L., "Dynamic stability of pendulum with oscillating suspension point," ZhETF, 1951, Vol. 25.

136. Kartvelishvili, N.A., "Influence of interaction between hydraulic, mechanic and electrical processes on the operation stability of power stations," Isv. AN SSSR, OTB, 1958, No. 2.

137. Kartvelishvili, N.A., Galaktionov, Yu.I., "Idealization of complex dynamic systems," M.: Nauka, 1976.

138. Kimbark, E., "Synchronous machines and stability of electrical systems," M.-L.: Gosenergoisdat, 1960.

139. Kirichenko, A.P., Rundkvist, K.A., "Inertial cone crusher," USSR Certif. of Invention 88,805; 1951.

140. Kirnarsky, M.Sh., "Synchronization of giroscopic generators," In: Mashiny i mekhanismy vibroudaznogo deystvia. Trudy Novosibirskoyo electro-tekhnicheskogo in-ta, Novosibirsk, 1969.

141. Kitaygorodsky, A.I., "Order and disorder in the atom's world," M.: Nauka, 1977.

142. Klibanova, I.M., "Mutual synchronization of oscillators with integer-ratios of frequencies," Isv. vusov, Radiofisika, 1968, Vol. 11, No. 11.

143. Klibanova, I.M., Malakhov, A.N., Maltsev, A.A., "Fluctuations in multi-frequency oscillators (review)," Isv. vusov, Radiofisika, 1971, Vol. 14, No. 2.

144. Klimentovich, Yu.L., "Quantum light oscillators and non-linear optics," M.: Prosvechshenie, 1966.

145. Klimentovich, Yu.L., Kuryatov, V.N., Landa, P. S., "On wave synchronization in a gaseous laser with circular resonator," ZhETF, 1966, Vol. 51, No. 1(7).

146. Kobrinsky, A.E., "Mechanisms with elastic connections," M.: Nauka, 1964.

147. Coddington, E.A., Levinson, N., "Theory of ordinary differential equations," M.-L.: IL, 1958.

148. Ragulskis, K.M., et al., "Oscillations of complex mechanical systems," Vilnius, Mintis, 1969.

149. "Oscillation processes in biological and chemical systems," M.: Nauka, 1967.

150. "Oscillation processes in biological and chemical systems," T.2. Puchshino-na-Oke; in-ta biofisiky AN SSSR, 1971.

151. Kolovsky, M.Z., Pervozvansky, A.A., "On linearization by the distribution function in theory of non-linear oscillations," Machinostroenie, 1962, No. 2.

152. Kolovsky, M.A., "On application of a small parameter method for determination of discontinuous periodical solutions," In: Trudy Mezhdunarodnogo simposiuma po nelineynym kolebaniyam, Vol. 2, Kiev: Isd-vo AN USSR, 1963 [see (43)].

153. Kolovsky, M.Z., "Non-linear theory of vibration control systems," M.: Nauka, 1966.

154. Conti, R. Graphy, D., Sansone, T., "Contribution of Italian mathematicians to the theory of non-linear ordinary differential equations during 1951-61," In: Trudy Mezhdunarodnogo simposiuma po nelineynym kolebaniyam, Vol. 1, Kiev: Isd.-vo AN USSR, 1963 [see (43)].

155. Kopnin, Yu.M., "Periodical oscillations of non-linear non-automatic systems with multi-degrees of freedom," Inzhenerny Zhurnal, MTT, 1965, Vol. 5, No. 2.

156. Kostin, I.K., Romanovsky, Yu.F, "Fluctuations in the systems of multi-coupled autooscillators," Vestnik MGU, 1972, No. 6.

157. Krylov, N.M., Bogolubov, N.N., "1. On the oscillations of synchronous machines. 2. On the stability of *n* synchronous machines in parallel operation," Kharkiv—Kiev: ONTVU-ENERGOVIDAV, 1932.

158. Kumpikas, A.L., Ragulskis, K.M., "Self-synchronization and dynamic vibration absorption for machine foundations," In: Trudy po teorii i primeneniyu yavlenia sinkhronizatsii v mashinakh i ustroystvakh, Vilnius: Mintis, 1966.

159. Kumpikas, A.L., Ragulsky, K.M., "On the entrainment regions synchronous rotations of pendulums with the vibrating suspension point," Vibrotekhnika, 2(2), Vilnius: Mintis, 1968.

160. Kupriyanovich, L.I., "Biological rhythms and sleep," M.: Nauka, 1976.

161. Kuryatov, V.N., Landa, P.S., Lariontsev, E.G., "Frequency characteristics of the ring laser on a vibrating support," Isv. vusov, Radiofisika, 1968, Vol. II.

162. Lavrov, B.P., "Vibrating conveyor," USSR Cert. of Invention 135,401; 1961.

163. Lavrov, B.P., "Three-dimensional problem of synchronization of mechanical vibrators," Isv. AN SSSR, OTN, Mekhanika i mashinostroenie, 1961, No. 5.

164. Lavrov, B.P., "Application of synchronization phenomenon in designing of new types of vibratory transport machines," Leningradsky gorny institut im G.V. Plekhanova, L.: 1966.

165. Lavrov, B.P., "New formulation of the integral stability criterion of synchronous motions of mechanical vibrators and its applications," In: Vibratsionnaya tekhnika (materialy nauchnotekhnicheskoy konferentsii), M.: NIIINFSTROIDORKOMMUNMASH, 1966.

166. Lavrov, B.P., "Vibratory machines with self-synchronizing vibrators (design schematics and specific features of design)," In: Trudy po teorii i prilozheniyu yavlenia sinkhronizatsii v mashinakh i ustroystvakh. Vilnius: Mintis, 1966.

167. Lavrov, B.P., Nagaev, R.F., Timofeev, N.G., "Dynamics of the shaking slit conveyor with self-synchronizing vibrators," Inzhenerny Zhurnal, MTT, 1970, No. 6.

168. Lavrov, B.P., Goldin, L.A., "Investigation of synchronization of mechanical vibrators in typical designs of two-mass screens," Obogachshenie rud, 1977, No. 3.

169. Landa, P.S., Lariontsev, E.G., "Beating regimes and synchronization of meeting waves in rotating ring gaseous lasers," Radiotekhnika i electronika, 1970, Vol. 15, No. 6.

170. Landa, P.S., Vygodin, V.A., "On the modal self-synchronization in lasers," Kvantovaya elektronika, 1977, Vol. 4, No. 4.

171. Landa, P.S., "Self-excited vibrations in systems with a finite number of degrees of freedom," M.: Nauka, 1980.

172. Lesin, A.D., "Experimental investigation of self-synchronization of mechanical vibrators in vibratory mills," In: Trudy po teorii i primeneniyu yavlenia sinkhronisatsii v mashinakh i ustroystvakh, Vilnius: Mintis, 1966.

173. Lugovaya, I.N., "To the problem of oscillation damping in common foundations for unbalanced machines," In: Sbornik Trudov Leningrad. Inzhenerno-stroitelnogo in-ta, L.: 1974, No. 89.

174. Lurie, A.I., "Analytical mechanics," M.: Fismatgis, 1961.

175. Lurie, A.I., "Certain problems of self-synchronization," In: Trudy V Mezhdunarod. Konf. po Helineynym kolebaniam, Vol. 3, Kiev: In-T, matematiki AN USSR, 1970 [see (43)].

176. Lurie, A.I., "Synchronization of dynamic systems (review of the book by I.I. Blekhman, M.: Nauka, 1971)," ISV. AN SSR, MTT, 1975, No. 4.

177. Lubimov, G.M., Khokhlov, R.V., "On polarization of a molecular ray by an alternating field with changing amplitude and phase," Zhetf, 1957, t. 33, vyp. 6/12.

178. Lyapunov, A.M., "General problem of motion stability," M.-L.: Gostekhisdat, 1950.

179. Mayer, A.G., "To the theory of forced oscillations in the complicated oscillator," Uchenye zapisky Gorkovskogo gos. un-ta, 1937, No. 6.

180. Malakhov, A.N., Kaksimov, A.S., Nefedov, Yu.Ya., "On electro-magnetic hypothesis of biological communication," In: Bionika, M.: Nauka, 1965.

181. Malakhov, A.N., "Fluctuations in auto-oscillating systems," M.: Nauka, 1968.

182. Malakhov, A.N., Maltsev, A.A., "On the spectral line width of a system of n mutually synchronized oscillators," DAN SSSR, 1971, Vol. 196, No. 5.

183. Malkin, I.G. "Theory of motion stability," M.-L.: Gostekhisdat, 1952.

184. Malkin, I.G., "Certain problems of theory of non-linear oscillations," M.: Gostekhisdat, 1956.

185. Maneshin, P.K., Khokhlov, R.V., "Mutual synchronization of two molecular oscillators at weak coupling," Nauchnye doklady vysshey shkoly, Radiotekhnika i elektronika, 1958, No. 3.

186. Marchenko, Yu.N., Rubanik, V.P., "On the mutual synchronization of molecular oscillators," Isv. vusov, Radiofisika, 1965, Vol. 4, No. 8.

187. Marchenko, Yu.N., Rubanik, V.P., "On the interaction of oscillating systems at the presence among them of wave connections," Inzhenerny Zhurnal, mekhanika tverdogo tela, 1966, No. 2.

188. Mitropolsky, Yu.A., et al., "Methods of mathematical analysis of non-linear oscillating systems," In: Vibratsii v tekhnike (spravochnik), Vol. 2 "Volebania nelineynykh mekhanicheskkh sistem," M.: Mashinostroenie, 1979.

189. Matukhin, V.M., "On the stability of complex energosystems," Isv. AN SSSR OTN, 1958, No. 11.

190. Machabeli, L.I., "On the motion of a disc with two pendulum," Isv. AN SSSR, Mekhanika, 1965, No. 2.

191. Merman, G.A., "New class of periodic solutions in the limited problem of three bodies and in Hill's problems," Trudy in-ta teoretich, astronomii, 1952, No. 1.

192. Ishlinsky, A.Yu., et al., "Balancing method of rotating bodies on the string drive," Isv. AN SSSR, MTT, 1979, No. 5.

193. Milovsky, N.D., Yastrebova, G.V., "On the operation of a ring oscillator synchronized by external force," Kvantovaya electronika, 1974, Vol. 1, No. 11.

194. Minorsky, N., "On synchronization," In: Trudy mezhdunarodnogo simposiuma po nelineynym kolebaniyam, Vol. 1, Kiev: Isd-vo., AN SSSR, 1961.

195. Mitropolsky, Yu.A., "Problems of asymptotic theory of non-stationary oscillations," M.: Nauka, 1964.

196. Mitulis, A.A., "Character of stationary motion of the mathematical pendulum with vibrating suspension point depending on selection of initial conditions," In: Trudy po teorii primeneniyu yavlenia sinkhronizatskii v mashinakh i ustroystvakh, Vilnius: Mintis, 1966.

197. Matyshin, V.M., et al., "On synchronous change of cell nuclea," In: Kolebatelnye portsesy v. viologicheskikh i khimicheskikh sistemakh, M.: Nauka, 1967.

198. Mikhnevich, G.V., Kozlovsky, G.F., "Stability and quality of transient processes in control system of excitation of a multi-aggregate power station," M.: Isv. AN SSSR, 1960.

199. Moiseev, N.N., "Asymptotic methods of non-linear mechanics," M.: Nauka, 1969.

200. Molchanov, A.M., "A possible role of oscillatory processes in evolution," In: Kolebatelnye protsessy v biologicheskikh i khimicheskikh sistemakh, M.: Nauka, 1967.

201. Molchanov, A.M., "Thermodynamics and evolution," in: Kolebatelnye protsessy v biologicheskikh a i khimicheskikh sistemakh, M.: Nauka, 1967.

202. Molchanov, A.M., "On the resonance structure of the solar system," In: Sovremennye problemy nebesnoy mekhaniki i astrodinamiki, M.: Nauka, 1973.

203. Monin, A.S., "On the nature of turbulence," Uspekhi fisicheskikh nauk, 1978, Vol. 125, No. 1.

204. Morosovsky, V.T., "On the operational stability of synchronous oscillators of the same type," Isv. AN SSSR, Energetika i automatika, 1959, No. 2.

205. Muyzhniek, A.I., "Stability investigation of the automatic dynamic balancer," Uchenye zapiski Rizhskogo Politekhnicheskogo instituta, Vol. 1, 1959.

206. Nagaev, R.F., "Dynamics of a vibro-impact crusher with a pair of synchronizing vibrators," Isv. AN SSSR OTN, Mekhanika i mashinostroenie, 1963, No. 5.

207. Nagaev, R.F., "On the internal synchronization of almost similar dynamic objects under action of weak linear connections," PMM, 1964, Vol. 28, No. 2.

208. Nagaev, R.F., "On synchronization of almost similar dynamic systems which are close to Lyapunov's systems," MMM, 1964, Vol. 28, No. 3.

209. Nagaev, R.F., "Synchronization in a system of essentially non-linear one degree of freedom objects," MMM, 1965, Vol. 29, No. 2.

210. Nagaev, R.F., "General problem of synchronization in almost conservative systems," MMM, 1965, t. 29. vyp 5.

211. Nagaev, R.F., "Connection types among the objects and criteria of stability of synchronous regimes," in: Trudy po teorii i primeneniyu yavlenia sinkhronisatsii v mashinakh i ustroystvakh. Vilnius: Mintis, 1966.

212. Nagaev, R.F., Popova, I.A., "Synchronization of several mechanical vibrators mounted on the common operating organ of beam type," Inzhenerny Zhurnal, MTT, 1967, No. 1.

213. Nagaev, R.F., Khodzhaev, K.Sh., "Synchronous motions in the system with carrying (supporting) connections," MMM, 1967, Vol. 31, No. 2.

214. Nagaev, R.F., "Synchronization of oscillators of finite-dimensional "forces," MMM, 1968, Vol. 32, No. 5.

215. Nagaev, R.F., "Quasi-conservative synchronizing systems," Kiev: in-t mekhaniki AN USSR, 1973.

216. Nagaev, R.F., "A case of fundamental family of quasi-periodic solutions in the theory of small parameters," MMM, 1973, Vol. 37, No. 6.

217. Neumark, Yu.I., Shilnikov, L.P., "On application of a small parameter method to the system of differential equations with disconnected right-hand sides," Isv AN SSSR, OTN, Mekhanika i mashinostroenie, 1959, No. 6.

218. Neumark, Yu.I., "The point imaging method in theory of non-linear oscillations," M.: Nauka, 1972.

219. Neumark, Yu.I., "On development of stochastic behavior in dynamic systems," Isv. vusov, Radiofisika, 1974, Vol. 17, No. 4.

220. Nemytsky, V.V., Stepanov, V.V., "Qualitative theory of differential equations," M.-L.: Gostekhisdat, 1949.

221. Novoselsky, P.I., et al, "Synchronous-synphaseous drive of debalanced vibro-stands," Stroitelnue i dorozhnoe mashinostroenie, 1956, No. 5.

222. Mandelstam, L.I., et al., "Complete works," Isd.-vo AN SSSR, 1950.

223. Olekhnovich, K.A., "Use of synchronization effect in large vibratory vibro-impact

units with elastic cushions," In: Trudy po teorii i primeneniyu yavlenia sinkhronizatsii v mashinakh i ustroystvakh, Vilnius: Mintis, 1966.

224. Alabuzhev, P.M., et al., "On synchronization of giroscopic vibrators," In: Mashiny i mekhanismy vibro-udarnogo deystvia, Trudy novosibirskogo electrotekhnichestogo in-ta, No. 1, Novosibirsk, 1969.

225. Panovko, Ya.G., "Fundamentals of applied theory of elastic oscillations," M.: Mashinostroenie, 1967.

226. Parygin, V.N., "Mutual synchronization of three connected auto-oscillating oscillators in case of weak connection," Radiotekhnika i electronika, 1956, Vol. 1, No. 2.

227. Parygin, V.N., "Revision of the mutual synchronization theory for two auto-oscillators," Vestnik MGU, ser mat., mekh., astr., fis., khim., 1957, No. 2.

228. Pervosvansky, A.A., "Random processes in non-linear automatic systems," M.: Fismatgis, 1962.

229. Petrova, I.M., "Hydro-bionics in ship building," L.: Isd. Tsentraln. NII informatsii i tekhniko-ekonomich isledovanii (TSNIITEIS), 1970.

230. Pirtskhalashvili, O.G., "Self-synchronization of mechanical vibrators in the system of two elastically connected solid bodies," Trudy Grusinskogo selskokhosyastvennogo in-ta, 1976, Vol. XCVI.

231. Pirtskhalashvili, O.G., "Self-synchronization of mechanical vibro-exciters on the pendulum-suspended solid body," Isv. AN SSSR, Mashinostroenie, 1978, No. 3.

232. Pliss, X.A., Abramovich, I.M., "Development of a vibratory method of dry classification of minerals," Report VNII "Mekhanobr," L.: 1948.

233. Polyakova, M.S., "On mutual synchronization of generators through conductivity," Vestnik MGU, fisika i astronomia, 1971, No. 3.

234. Popov, E.P., Paltov, I.P., "Approximate methods of investigation of non-linear automatic systems," M.: Fismatgis, 1960.

235. Poturaev, V.N., et al., "On the realization of synchronous and synphaseous motion regimes of inertial exciters located on the ends of the load-bearing organ of a vertical vibro-conveyor and synthesis of the system's basic dynamic parameters," In: Problemy vibratsionnoy tekhniki, Kiev: Naukova dumka, 1970.

236. Pressman, A.S., "Vernadsky's ideas in modern biology (planet-cosmic fundamentals of life organization)," M.: Znanie, 1976.

237. Proskuryakov, A.N., "Poincare's method in the theory of non-linear oscillations," M. Nauka, 1977.

238. Prokhorov, A.M., et al., "Laser thermonuclear synthesis," Uspekhi fisich. nauk., 1976, Vol. 119, No. 401.

239. Poincare, A., "Selected works in three volumes," M.: Nauka, Vol. 1, 1971; Vol. 2, 1972; Vol. 3, 1974.

240. Rabinovich, M.I., "Stochastic oscillations and turbulence," Uspekhi fisich. nauk., 1978, Vol. 125, No. 1.

241. Ragulskis, K.M., "Mechanisms on vibrating foundations (problem of dynamics and stability)," Kaunas: In-t energetiki i electrotekhniki an Litovskoy SSR, 1963.

242. Ragulskis, K.M., et al., "Self-synchronization of mechanical systems (1. Self-synchronous and vibro-impact systems)," Vilnius: Mintis, 1965.

243. Rapoport, G.N., "To the problem of mutual synchronization of auto-oscillators with comparable power," Radiotekhnika, 1951, Vol. 6, No. 4.

244. Rodygin, L.V., "To the theory of coupled relaxation oscillators," In: Pamyati A.A. Andronova, M.: Isd-vo AN SSSR, 1955.

245. Rosanov, M.N., "Analysis of re-synchronization and self-synchronization regimes of hydro-generators," Isv. vusov, energetika, 1956, No. 4.

246. Romanovsky, V.E., "Analysis of nominal power and mechanical characteristics of a vibratory pile-driver," Stalingrad: Stalingradsky selskokhosyast, in-t, 1958.

247. Romanovsky, V.E., et al., "Mathematical modeling in biophysics," M.: Nauka, 1975.

248. Rubanik, V.P., "Oscillations of quasi-linear systems with retardation," M.: Nauka, 1969.

249. Rudin, A.D., "Dynamic properties and development prospects for vibratory machines of a three mass system," In: Vibratsionnaya tekhnika (materialy nauchno-tekhnicheskoy konferentsii), M.: NIIINFSTROYDORKOMMUNMASH, 1966.

250. Rundkvist, K.A., Kirichenko, A.I., "Gyration crusher," USSR Cert. of Invention 76,141, 1949.

251. Rundkvist, A.K., "Mechanics of inertial crusher," In: Mekhanika i raschet mashin vibratsionnogo tipa, M.: Isd-vo AN SSSR, 1957.

252. Rundkvist, A.K., et al., "To the theory of a critical slot width in inertia crushers," Obogachshenie rud, 1961, No. 2.

253. Rudenberg, R., "Transient processes in electro-energetic systems," M.: IL, 1955 (from English).
254. Ryabov, Yu.A., "On the evaluation of application area of small parameter method in the problems of non-linear oscillations," In: Trudy mezhdunarodnogo simposiuma no nelineynym kolebaniam, Vol. II, Kiev: Isd-vo, AN USSR, 1963 [see (43)].
255. Savinov, O.A., "Modern design of foundations for machines and their calculations," L: Stroyisdat, 1979.
256. Ragulskis, K.M., et al., "Self-synchronization of mechanical systems," Vol. 2, Vilnius: Mintis, 1967.
257. Sidorova, G.A., "Mutual synchronization of two relaxation oscillators," Vest. MGU, 1971, Vol. 12, ser. III, No. 6.
258. Sidorova, G.A., "Synchronization of relaxation oscillators by external delta-connected voltage," Vest. MGU, 1972, Vol. 13, ser. III, No. 4.
259. Abalakin, V.K., et al., "Reference book on celestial mechanics and astrodynamics," M.: Nauka, 1971.
260. Starzhinsky, V.M., "Applied methods in non-linear oscillations," M.: Nauka, 1977.
261. Stratonovich, R.L., "Selected problems of fluctuation theory in radio engineering," Sov. radio, 1961.
262. Strett, G. (Lord Rayleigh), "Theory of sound," Vol. 2, M.-L: Gostekhisdat, 1944.
263. Syromyatnikov, I.A., "Operation regimes of synchronous generators," M.: Gosenergoisdat, 1952.
264. Teodorchik, K.F., "Thomson's oscillating systems with two degrees of freedom," Radiotekhnika, 1946, Vol. 1, Nos. 3–4.
265. Whisem, G., "Linear and non-linear waves," M.: Mir, 1977.
266. Ulam, S., "Not solved mathematical problems," Book series of "Modern problems of mathematics," M.: Nauka, 1964.
267. Bukaty, G.B., et al., "Control device for granular content of crushed ore," USSR Cert. of Invention 189,218; 1967.
268. Utkin, G.M., "Mutual synchronization of auto-oscillators on the multiple frequencies," Radiotekhnika i electronika, 1957, Vol II, No. 1.
269. Utkin, G.M., "Investigation of a two-contour oscillator at frequency ratio 1:2," Nauchnye doklady vyssh. shkoly, Radiotekhnika i elektronika, 1958, No. 1.
270. Utkin, G.M., "Synchronization of auto-oscillators on the combined frequencies for band frequency stabilization," Radiotekhnika i electronika, 1959, Vol. 4, No. 2.
271. Collected articles, "Phaseous synchronization," M.: Svyas Publish. House, 1975.
272. Fazullin, F.F., "To synchronization of mechanical systems with circular symmetry," Prikladnaya mekhanika, 1976, Vol. 12, No. 1.
273. Fayn, V.M., "On the equations of molecular oscillation generator," ZhETF, 1957, Vol. 33, No. 4(10).
274. Fedorov, B.F., et al., "Optic quantum gyroscope," M.: Mashinostroenie, 1973.
275. Denisov, G.A., et al., "Flotation machine," USSR Cert. of Invention 306,104; 1972.
276. Frank-Kamenetsky, D.A., "Mechanism of two-stage combustion," Zhurn. fisich. khimii, 1940, Vol. 14, No. 1.
277. Frolov, K.V., Khodzhaev, K.Sh., "Interaction of source of excitation with the oscillatory system," In: Vibratsia v tekhnike (spravochnik), Vol. 2, Volebania nelineynykh mekhanicheskikh sistem. M.: Mashinostroenie, 1979.
278. Hagikhara, Yu., "Stability of the solar system," In: Planety i sputniki, M.: IL., 1963.
279. Krindach, A.A., et al., "Radiation characteristics of a continuous laser on a coloring agent with synchronous pumping," ZhETF, 1979, Vol. 5, No. 23.
280. Heil, G., "Oscillations in non-linear systems," M.: Mir, 1966.
281. Henton, A.A., "Satellite synchronization," In: Dinamika sistem, Mezhvysovsky sbornik, Gorsky, 1974, No. 4.
282. Khodzhaev, K.Sh., "Oscillations in the system with many electro-magnetic exciters," Inzhenerny zhurnal, Mekhanika tverdogo tela, 1966, No. 2.
283. Khodzhaev, K.Sh., "Synchronization of mechanical vibrators coupled with a linear oscillating system," Inzhenerny zhurnal, Mekhanika tverdogo tela, 1967, No. 4.
284. Khodzhaev, K.Sh., "On excitation of vibrations," Inzhenerny zhurnal, Mekhanika tverdogo tela, 1968, No. 2.
285. Khodzhaev, K.Sh., "Integral stability criterion for systems with quasi-cyclic coordinates and energy correlations for oscillations of conductors carrying currents," PMM, 1969, Vol. 33, No. 1.
286. Khodzhaev, K.Sh., "Oscillations of non-linear electro-mechanical systems," In: Vibratria v tekhnike (spravochnik), Vol. 2, Kolebania nelineynykh mekhanicheskikh sistem, M.: Mashinostroenie, 1979.

287. Kholodenko, L.P., "To the synchronization theory of non-continuous auto-oscillations," Zhurnal tekhnicheskoy fisiki, 1940, Vol. 10, No. 2.

288. Khokhlov, R.V., "To the entrainment theory at a small amplitude of the external force," DAN SSSR, 1954, Vol. 97, No. 3.

289. Khokhlov, R.V., "On one case of mutual synchronization of reflecting klystrons," Radiotekhnika i elektronika, 1956, Vol. 1, No. 1.

290. Tseytln, A.I., Guseva, N.I., "Determination of loads on the foundations for group mounting of non-balanced machines with synchronous motors," Osnovania, fundamenty i mekhanika gruntov, 1972, No. 3.

291. Chesari, L., "Asymptotic behavior and solution stability of ordinary differential equations," M.: Mir, 1964.

292. Chelomey, V.I., "On the possibility of stability enhancement in the elastic systems by vibrations," DAN SSSR, 1956, Vol. 110, No. 3.

293. Chervonenko, A.G., et al., "On stability of synchronous regimes of non-conservative oscillators," In: Problemy vibratsionnoy tekhniki, Kiev: Naukova dumka, 1970.

294. Chetaev, I.G., "Stability of motion," M.: Gostekhisdat, 1955.

295. Chetaev, I.G., "Stability of motion. Works of analytical mechanics," M.: Isd-vo AN SSSR, 1962.

296. Chizhevsky, A.L., Shashina, Yu.G., "In the sun's rhythm," M.: Nauka, 1969.

297. Chizhevsky, A.L., "Earth echo of solar storms," M.: Mysl, 1973.

298. Shakhgildian, V.V., Lyakhovsky, A.A., "Phaseous automatic frequency tuning," M.: Svyas, 1966.

299. Shekhter, O. Ya., "On one example of sub-harmonic oscillation," Trudy sovechshania po primeneniyu vibratsy pri ustroystve sooruzheny i burenii skvazhin v stroitelnykh tseliakh, L.: NTO stroitelnoy industrii SSSR, 1959.

300. Shekhter, O. Ya., "On submersion of heavy concrete shells," In: Dynamics of grounds, No. 44, M.: Gosstroyisdat, 1961.

301. Shimanov, S.I., "To the oscillation theory of quasi-linear systems with retardation," PMM, 1959, Vol. 23, No. 5.

302. Shimanov, S.I., "On almost periodic oscillations of quasi-linear systems with retardation in case of degeneration," DAN SSSR, 1969, Vol. 133, No. 1.

303. Shnol, S.E., "Conforming oscillations of molecules," In: Kolebatelnye protsessy v biologicheskikh i khimicheskikh sistemakh, M.: Nauka, 1967.

304. Aizawa, Y., "Synergetic Approach to the Phenomena of Mode-Locking on Nonlinear Systems," Progr. Theor. Phys., 1976, Vol. 56, No. 3.

305. Appleton, E.V., "The automatic synchronization of triode oscillator," Proceedings of the Cambridge Philosophical Society, Math. and Phys. Sciences, 1922, Vol. 21.

306. Blekhman, I., "Die selbstsynchronisation mechanischen Vibratoren und das Problem der Gruppenfundamente fur unbalancierte Maschinen, (Maschinen mit Unwuchten)," International Tagung "Verfahren und Gerate der mechanischen Schwingungstechnika," Teil II, Technische Hochschule Otto von Guerike Magdeburg, September 1965.

307. Bogusz W., Engel, Z. Badania doswiadczalne urzadzen samosynchronizujacych, Przeglad mechaniczny, 1965, No. 8.

308. Coughey, T.K., "Hula-hoop: one example of heteroparametric excitation," American Journal Physics, 1960, Vol. 28, No. 2.

309. Diliberto, S.P., Hufford, G. "Perturbation theorems for non-linear ordinary differential equations. Contributions to the theory of non-linear oscillations," V. III, Princeton, 1956.

310. Faure, R., "Sur les systemes d'equations differentielles non lineaires a coefficents periodiques. Etude d'un cas particulier." C. r. Acad. Sci., 1957, Vol. 245, No. 19.

311. Faure, R., "Sur la synchronisation des systemes oscillants. Solutions voisines de points singuliers," C. r. Acad. Sci., 1958, Vol. 247, No. 15.

312. Friedrichs, K.G., "Fundamentals of Poincare's theory," Proc. Symp. Non-linear Circuit Analysis, 1953, Vol. 2.

313. Goldacre, R., "The control of rhythm and homeostasis in biology and medicine," Cibernetica, 1960, No. 2.

314. Haus, H., "Theory of forced-mode locking," Proc. IEEE, Q-e, 1975, Vol. 11.

315. Hills, J.G., "Dynamic relaxation of planetary systems and Bod's law," Nature, 1970, Vol. 225.

316. Jordan, H., "Uber den stationaren Betrieb und die Auslegung von elektrischen Ausgleichswellen," Elektrotechnische Z., 1952, Vol. 73, No. 2.

317. Lotka, A.J., "Elements of physical biology," Baltimore, 1925.

318. MacMillan, W.D., "An existence theorem for periodic solutions," Transactions of the American Mathematical Society, 1912, Vol. 13.

319. Mayer, A., "On the theory of coupled vibrations on two self-excited generators," Technical physics of the USSR, 1935, Vol. 11, No. 5.

320. Miklaszewski, R. "On the possibility of self-synchronization of rotating eccentric vibrators," Nonlin. Vibr. Probl., 1962, No. 4.

321. Minorsky, N., "Non-linear oscillations," New York: D. Van Nostrand Company, Inc., 1962.

322. Ollendorf, F., Peters, W. "Schwingungstabilität parallelarbeitender synchronmaschinen. Wissenschaftliche Veroffentlichungen aus dem Sciemens, Konzern," Vol. VI, No. 7-26, Springer, 1925–1926.

323. Yoshiaki, A., Junkichi, I., Yukimasa, O., et al., "On the self-synchronization of mechanical vibrators," Trans. Hapan. Soc. Mech. Eng., Part 1, 1966, Vol. 32, No. 234; Part 2, 1967, Vol. 33, No. 246; Part 3, 1969, Vol. 35; Part 4, 1975, Vol. 41, No. 350.

324. Ovenden, M.W., Feagin, T., Graff, O. "On the principal of least interaction action and the Laplacian satellites of Jupiter and Uranus," Celestial Mechanics, 1973, Vol. 8, No. 3.

325. Poincare, H., "Les methodes nouvelles de la mecanique celeste," Paris: 1892, Vol. I: 1893, Vol. II; 1899, Vol. III.

326. Roy A. E., Ovenden M.W, "On the occurrence of commensurable mean motions in the solar system. The mirror theorem," Mon. Noth., 1955, Vol. 115.

327. Signeul, R.A.O., "Apparat for behandling av ett objeki modelst riktade vibrationer," Sverige, Patent No. 163270, Kl. 80a: 49. Patenttid fran den 6 Augusti 1946; pablicerat den 13 maj 1958.

328. Signeul, R.A.O., "Vibrating device for a directed vibratory effect by means of rotatable vibratory members," United States Patent Office, No. 2, 531,706. Application May 25, 1948, in Sweden August 6, 1946, Patented November 28, 1950.

329. Sperling, L., "Selbstsynchronisation unwuchtbehafteter Rotoren an elastischen Ketten," Internationale Tagung "Verfahren und Gerate der mechanischen Schwingungstechnik." Magdeburg: Technischen Hochschule Otto von Guericke Magdeburg, 1965, Teil II.

330. Sperling, L., "Beitrag zur allgemeinen Theorie der Selbstsynchronisation umlaufender Unwuchtmassen im Nichtresonanzfall," Wissenschaftliche Zeitschrift der Technischen Hochschule Otto von Guericke Magdeburg, 1967, Vol. 1, No. 11.

331. Suezaci, T., Shiusaku, M., "Mutual synchronization of two oscillators," Summaries of Papers Int. Conf. Microwaves, Circuit Theory and Inform. Theory, Toyko, 1964, Part 2. Tokyo, Inst. Electr. Common Eng. Japan, s.a. 111-112.

332. Thearle, E.L., "A new type of dynamic balancing machine," Trans. ASME, 1932, Vol. 54, No. 12.

333. Thearle, E.L., "Automatic dynamic balancers," Machine Design, 1950, Vol. 22, Nos. 9–11.

334. Van der Pol, B., "On relaxation oscillation," Phil. Mag., 1926, Vol. 2, No. 7.

335. Van der Pol, B., "Ober Relaxationschwingungen," Zeitschr. fur Hochfrequenztechnik, 1926, Vol. 28; 1927, Vol. 29.

336. Van der Pol, B., Van der Mark, M., "Le battement du coeur considere comme oscillation de relaxation et un modele electrique de coeur," L'Onde electrique, 1928, No. 7.

337. Volterra V., "Lecons sur la Theorie Mathematique de la Lutte pour la Vie," Paris: Gauthier, Villars, 1931.